THE NEW SOLAR ELECTRIC HOME

Joel Davidson
Fran Orner

THE NEW SOLAR ELECTRIC HOME

THE COMPLETE GUIDE to PHOTOVOLTAICS for YOUR HOME

3RD EDITION

aatec publications
Ann Arbor, Michigan

First printing 2008

aatec publications
PO Box 7119   Ann Arbor, Michigan 48107
734.995.1470 PHONE / 734.418.2226 FAX
aatecpub@SolarElectricBooks.com
www.SolarElectricBooks.com

ISBN 978-0-937948-17-0

Library of Congress Control Number 2008931676

Cover design by Hesseltine & DeMason Design, Ann Arbor, Michigan
Illustrations by Fran Orner
Manufactured in the United States of America
Printed and bound by McNaughton & Gunn, Saline, Michigan

The authors and **aatec publications** assume no responsibility for any
personal injury, property damage, or other loss suffered in activities related
to the information presented in this book. Please practice caution and
follow proper safety procedures.

CONTENTS

LIST OF FIGURES, TABLES, WORKSHEETS & CHECKLISTS

FIGURES
Charts, graphs, diagrams, photographs

TABLES

WORKSHEETS & CHECKLISTS

PREFACE

PV has come out of the backwoods and into the mainstream. Your solar electric home can be anywhere—in the city or suburbs connected to the utility grid or far from the nearest power pole.

In 1983, when the first edition of *The Solar Electric Home* was published, there were only a few hundred, mostly rural, photovoltaic-powered homes. Now PV homes can be found everywhere. The growing acceptance of urban PV is just one of the major changes in photovoltaics since the first edition of *The Solar Electric Home*. Improved equipment and better installation methods make it easier to have a reliable and safe PV system.

In 1984, *The National Electrical Code* (NEC) added PV installation rules and Underwriters Laboratories (UL) started listing PV modules, both important steps toward PV acceptance by building inspectors. In 1987, *The New Solar Electric Home* was the first PV book to include the Code. This latest edition includes information about the new generation of DC-to-AC inverters that can connect to the utility grid or provide utility quality electricity anywhere.

Today, the vast majority of people want solar power. Almost everyone wants to protect and preserve non-renewable resources and reduce waste and pollution; both are goals that PV serves very well. Governments and utilities are still slow to adopt PV, but you do not have wait. The third edition of *The New Solar Electric Home* explains your options and tells you how to go solar now.

+ **Chapters 1 through 8** explain what PV is, explain the policies and regulations that affect your installation, and help you make the necessary decisions.
+ **Chapters 9 through 15** describe PV system components—how they work and how to select them.
+ **Chapters 16 through 19** explain how to install a code-compliant PV system.
+ **Chapter 20** is a glimpse into the future of PV.

The New Solar Electric Home: The Complete Guide to Photovoltaics for Your Home is your how-to handbook for clean, safe, reliable solar electricity.

Joel Davidson
Fran Orner

A WORD OF CAUTION

The purpose of *The New Solar Electric Home* is to encourage the wise use of photovoltaics. To this end, information regarding equipment and methods is presented in a clear, straightforward way. The reader is expected to use safe practices and good judgment in the selection of equipment, methods, and workers.

The application of any technology, especially photovoltaics, continually changes. This book will continue to be updated as new and better equipment and methods evolve. However, it is up to you to verify the safety and validity of any photovoltaics work to be done. Be sure to consult with a PV professional and the *National Electrical Code*.

The authors and **aatec publications** disclaim responsibility for any injury, damage, or other loss suffered related to the information presented in this book.

THINK SAFETY!

Chapter 1
INTRODUCTION TO PHOTOVOLTAICS
How & Why

The Age of Photovoltaics Is Here

Just look around you. PV is everywhere.

- Tiny solar modules have replaced throwaway batteries in your watch and calculator.
- PV panels quietly power caution signs at highway construction sites along your morning commute, eliminating noisy, fume-spewing generators.
- Rooftop panels on homes tied into the utility grid provide non-polluting electricity for you and your neighbors.
- Commercial buildings have PV panels on their roofs and transparent PV windows.
- PV powers the International Space Station and the satellites that transmit data and countless phone calls daily.

For years, photovoltaic cells have provided power for the space industry, for communications, and for remote homes and villages that have no access to utility power. Now we all benefit from PV daily.

PV is a fact of modern life.

Solar power is our most abundant natural resource. The solar power shining on 135 square miles, an area about the size of Philadelphia, is greater than the peak capacity of all the electric power plants on earth. Solar panels that are only 10% efficient on less than 4% of the world's deserts could produce all of the primary energy consumed on earth.[1]

We don't need giant PV power plants and thousands of miles of wire to deliver solar electricity to us. PV on only 7% of the 140 million acres covered by U.S. cities and residences—on roofs, parking lots, along highway walls and the sides of buildings—could supply the nation's total electrical energy requirement.[2] A clean, quiet, long-lasting PV array covering less than half the roof of a modern all-electric home can power every appliance inside. Far fewer solar cells are needed to power an energy-efficient home with the same conveniences.

A Brief History of Photovoltaics

In 1839 French physicist Edmond Becquerel discovered that copper oxide electrodes inserted into a liquid would produce an electric current when exposed to light. In the 1880s, the American inventor Charles Fritts made the first selenium solar cell; he predicted that roofs covered with solar cells would someday provide electric power for homes and businesses. In 1904, Albert Einstein formulated the mathematical model for the photovoltaic principle. Despite the efforts of these luminaries, interest in PV waned because even the best solar cells could convert only about 1% of light into electricity, and they cost more than fossil fuel generators.

PV remained a scientific curiosity until 1954 when Bell Laboratories researchers Gerald Pearson, Calvin Fuller, and Darryl Chapin made a crystal silicon cell with 6% light-to-electricity conversion efficiency. Their invention sparked the interest of Dr. Hans Ziegler, lead scientist for the U.S. Army Signal Corps, who saw its potential for use in "Operation Lunch Box," the artificial satellite whose 1958 launch entered the United States into the space race. The success of the PV-powered Vanguard I opened people's minds to the potentials of solar electricity, and the doors to further PV research and development.[3]

In 1955, Bell Telephone Company installed the first solar-powered telephone repeater near Americus, Georgia. By 1958, communications satellites powered by a few watts of solar cells were circling the earth. Dur-

Figure 1.1.
There was a lot of media excitement in 1954 when Bell Laboratories announced the development of silicon solar cells. Fifty years later, after many improvements and price reductions, clean PV is finally challenging fossil fuel dominance. Note the batteries (at right) for storing electricity at night. (AT&T)

ing the 1970s, PV was still too expensive for most terrestrial applications, but it was the lowest cost source of reliable electricity where no power lines existed. Then the United States experienced two crippling oil embargoes that prompted intense interest in alternative sources of energy. Government and private research, and a 40% tax credit for PV systems, resulted in substantial investment and growth in the PV industry. By 1986, even though oil was cheap again, research funding had been slashed, and the tax credits eliminated, over a megawatt of solar electric power systems was being installed every month in the U.S. By 1990, the PV industry was still small, but it was firmly established and growing rapidly. In 2007 over nine megawatts were installed every day worldwide.

During the last two decades of the twentieth century, awareness of the environmental impact of energy production was on the rise. The Chernobyl nuclear power plant explosion made Europeans acutely aware of the need for cleaner and safer energy. The Japanese, who already had a strong environmental ethic, were deeply concerned by their dependency on nuclear and foreign energy sources. Other countries, such as China and India, had a massive and growing demand for electrical distribution, coupled with a need to abate pollution. Almost every nation was researching its solar options.

Today, worldwide use of PV more than doubles every four years. This growth promises to continue. Multi-national corporations have launched campaigns to promote their commitment to solar energy. In

2007, investors put over $1 billion in venture capital into PV companies. In 2005, three of the largest technology public stock offerings were by solar companies. Worldwide silicon production is ramping up to meet the growing demand for solar cells, and the race toward a terawatt (one thousand billion—1,000,000,000,000—watts) annual PV production has begun.

In 2007, worldwide PV module shipments exceeded 3.7 billion watts. Although incentives for privately owned grid-connected PV are a mere fraction of the billions of dollars in subsidies given to polluting, non-renewable energy industries, PV incentive programs are helping to dramatically increase the use of PV and drive down prices.

Since 1994 incentives in Japan have resulted in over 1,500 MW of installed PV. In 2007 another 350 MW of PV were installed in Japan alone. In 2005, 2006, and again in 2007 Germany led the world with over 1,000 MW installed each year, thanks to their rate-based PV feed-in tariff incentive. Other countries are using their own versions of U.S., Japanese, and German-style PV incentives.

In 2005, the U.S. launced the Solar America Initiative, a government and industry research funding partnership to make PV electricity cost-competitive without government support by 2015. Californians voted for PV subsidies in 1998, before they were struck by electric energy crises, blackouts, and utility rate increases that drove people to seek alternative energy solutions. By 2007, over 360 MW of PV had been installed in California. In 2005, Californians made a 10-year, $3.2 billion commitment to encourage businesses and homeowners to install 3,000 megawatts of PV before 2017. Other states are challenging California's lead. New Jersey installed over 10 MW during the first two years of its solar initiative.

The reality of solar electric power is that it is practical—and available—right now.

How PV Cells Work: Light to Electricity
photo = light + *voltaic* = electricity

The photovoltaic process is completely solid-state and self-contained. There are no moving parts and no materials are consumed or emitted. Radiated light energy—either direct from the sun, diffused through the atmosphere, or from an artificial source, like a light bulb—consists of a stream of energy units called photons. When these photons strike a solar cell, the electrons in the cell circuit move, resulting in electrical current.

The development of silicon solar cells evolved from research in solid-state diodes and transistor semiconductors in the 1950s. Crystal cells are made from silicon refined at high temperature to over 99.9999% purity. Silicon has four electrons in its outer valence (the outer "shell" of the atom); when pure silicon crystallizes, it shares an electron with each of its four neighbors. This makes pure silicon a good insulator, like glass, with no "free" electrons.

The addition of minute amounts of other substances, a process called doping, adds impurities to the crystal to make silicon more conductive. When silicon is doped with an atom that has five electrons in its valence (phosphorous, arsenic, or antimony), it then has one more loosely held electron than it needs and so can share it with neighbors. This is an n-type semiconductor.

When silicon is doped with an atom with three electrons in its valence (boron, aluminum, gallium, or indium), the atom fits into the crystal matrix but has one too few electrons to share with neighbors, which means there is an electron "hole." This is a p-type semiconductor.

When the two are layered, the zone between the p-type material and the n-type surface layer creates a barrier called a p-n junction. There are free electrons on the "n" side of the junction and electron holes on the "p" side. When photons (light) strike the wafer, electrons move across the p-n junction. This electron movement is electrical current. The flow of current induces a voltage potential at external terminals.

Cell voltage potential under full-sun conditions is about 0.5 volt. Current varies with cell size and the sunlight-to-energy conversion efficiency of the cell material. A 6-inch (152.4-mm) round crystalline cell produces about 6 amperes in full sun. A 6-inch single or polycrystalline square cell

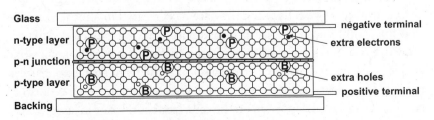

Figure 1.2. A solar cell.

of the same efficiency produces over 7 amps in full sun, again at 0.5 volt, simply because it is bigger. Break a cell in half and each half produces half the current, but the voltage for each half remains 0.5 volt.

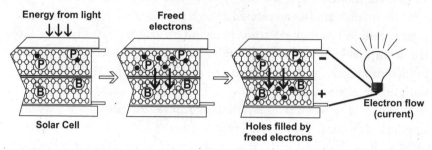

Figure 1.3. How a solar cell works.

The amount of current produced is proportional to the amount of light (photons) absorbed by the surface. Shade, shadows, fog, haze, snow, dirt, reflectivity, circuit resistance, and the angle of the sun's rays relative to the panel all affect power output.

Single-Crystal PV Production

Solar cell production is an exacting, energy-intensive process; it accounts for about 80% of a single-crystal PV module's cost. Single-crystal silicon solar cells start as a large crystal of pure silicon doped with a tiny amount of boron grown under very controlled conditions in a vacuum furnace at more than 2,500°F/1,400°C for several hours. The finished single crystal can be a solid cylinder over 6 inches (155 mm) in diameter, 5 feet (1.5 m) long, and weighing over 130 pounds (59 kg).

The large crystal is cooled to room temperature and sliced into 0.02-inch (2-mil) or thinner wafers. The wafer surface is acid-etched to eliminate the reflective planar surface created by sawing. The wafers become photoactive cells when they are next treated in another furnace to diffuse a thin layer of phosphorous onto the wafer surface. An anti-reflective coating is diffused onto the cell surface. Then a thin, wire-like grid is silk-screened or vapor-deposited onto the front surface of the cell to provide an electrical path. The grid may also be made of wires or other conductive material deposited on the cell's front surface or in laser-etched grooves. Solder pads are screened or deposited onto the back surface of the cell and the cells are then connected with wires that are soldered to the grids. Over 35% (1.3 GW) of the 3.73 gigawatts (GW)[4] of solar cells and modules produced in 2007 were single-crystal silicon.

Polycrystalline PV Production

Over 55% of solar cells and modules produced in 2007 were polycrystalline silicon. Polycrystalline modules are assembled in the same way as single-crystal modules, except that the cells are produced differently and have a very different appearance.

Polycrystalline silicon production begins with the same silicon raw material used to make single-crystal cells. Both processes require the same high temperature to melt the silicon. However, instead of pulling a crystal from the melt, as done with single-crystal silicon cells, the molten silicon is cast into an ingot cube. As the cube cools, a multi-crystal matrix forms, giving the cells their distinctive "jack frost" appearance.

The cooled cube is sawn into four ingots that are then sliced into wafers, similar to single-crystal wafers. The rest of the cell production process is also similar. Because polycrystalline cells also have reflectance losses due to the flat (planar) surfaces of the cell, an anti-reflective coating is applied. The anti-reflective coating, used on both single- and polycrystalline cells, may be thin and bright blue or thick and almost black.

Polycrystalline cells lose some power when first exposed to sunlight due to minute electrical shorts that occur along the crystal grain boundaries. Poly-cells with large crystallites have fewer grain or edge boundaries and act more like single-crystal cells. Polycrystalline cell output drops around 3% during the first few weeks in the sun. Manufacturers allow for this "field conditioning" when rating modules for power output.

The production of polycrystalline ingots requires less energy than single crystal, so polycrystalline cells cost less to produce. Polycrystalline modules do cost a little less per watt, but single-crystal module prices have kept pace through increased production volume and economy of scale.

Ribbon and sheet silicon cell production are variations on the crystalline growth process. A wide ribbon of silicon is pulled from a crucible of molten silicon, similar to the way a single-crystal ingot is pulled. The cooled ribbon is then cut into cells and assembled into standard modules. Some polycrystalline cells are also formed by a continuous-sheet process similar to the manufacture of plate glass or sheet metal. Ribbon cells account for less than 3% of world cell production.

Amorphous Silicon PV Production

Almost 8% of solar cells made in 2007 were amorphous silicon (a-Si), with about half of that output dedicated to indoor uses, such as solar calculators and small novelties. As the name implies, amorphous solar cells have no crystalline atomic structure. They may be divided into two types: amorphous silicon and thin film. Both types are thin films of semiconductor material heated to a gaseous state and then deposited on glass, metal, or plastic to form a thin photoactive film. Thin film generally refers to non-silicon semiconductor combinations such as cadmium telluride (CdTe) and copper indium gallium diselenide (CIS and CIGS).

For over two decades, PV researchers have been working on a-Si PV because it has the potential for very low production costs. Thin-film PV production consumes hundreds of times less silicon and requires fewer production steps than crystalline PV. However, amorphous silicon and other thin-film technologies can experience greater power degradation than crystal PV and have low sunlight-to-energy conversion efficiency. Scientists are beginning to understand what causes this degradation. Some a-Si modules now have 20-year power output warranties, which are comparable to crystal PV. Higher than 10% sun conversion efficiency has been achieved in the lab; however, standard a-Si production modules of about 7% efficiency are still half as efficient as crystalline modules.

a-Si cells produce about 0.8 volt before initial degradation. Although voltage and current continues to degrade, voltage remains slightly higher than in crystal cells. Most charge controllers are designed for a variety of charge voltage ranges, but some utility-interconnect inverters may require custom power tracking electronics when used with a-Si modules.

Some a-Si manufacturers claim their modules put more amp-hours into batteries and produce more energy in grid-connected applications than single and poly modules with the same rated power. Indeed, temperature does not affect a-Si cells as much as it affects crystalline cells. With each micron-thin layer designed to collect a different wavelength of light, triple-junction or 3-layered a-Si cells collect more infrared and ultraviolet energy than crystal cells tuned mostly to the visible wavelength of light.

Nevertheless, crystalline technology is likely to dominate through year 2015. Still, many people will buy lower efficiency a-Si and thin-film PV because of its uniform appearance, flexibility, and uniqueness—as long as prices and warranties are comparable to crystalline.

Researchers continue to hunt for the right combination of chemicals and processes that will result in PV cells that produce electricity cheaper than fossil fuel and nuclear power. One company makes crystalline cells that are over 20% efficient. Another company makes a combination crystal and amorphous silicon cell that is rated over 17% efficient. High-efficiency gallium arsenide cells are available for space satellite arrays and specialty applications, but they are three to five times too costly for residential systems. Cadmium telluride and copper indium gallium diselenide thin-film modules are also available. Perhaps other technologies will one day replace silicon-based PV, but for now crystalline silicon cells dominate the market.

For a more detailed description of solar cells , see *Practical Photovoltaics* by Dr. Richard Komp (**aatec publications**).

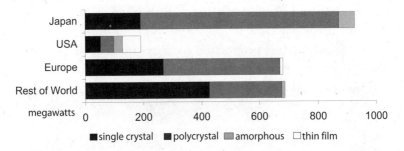

Figure 1.4. 2007 worldwide cell/module production: 3.73 total gigawatts. (Source: Prometheus Institute)

PV Production Cost & Energy Payback

Producing solar cells remains expensive, but costs are getting lower, making PV affordable and cost-effective for more and more applications. People in developing countries now purchase PV-powered lights with monthly payments equal to the amount they used to spend on fueling their smoky, unhealthy, inefficient kerosene lights. In industrialized nations PV provides cost-effective power for emergency telephones along highways, radio and TV repeater stations, lighted bus shelters, and security lights for backyards and businesses. Every price reduction expands the market for PV and furthers its replacing the polluting technologies.

Because PV modules consume no energy after they are made, they can quickly pay back the energy invested in their manufacture. In 1980, the energy used to refine silicon, the major raw material in solar module manufacture, was less than 2 kilowatt-hours (kWh) per watt of PV produced. Fabricating cells and assembling modules added another 1 kWh per PV watt. The energy investment in those 10% efficient solar cells was returned at the rate of about 1.5 kWh per watt per year, which constitutes a three-year payback. Thus, cells made in 1980 have paid for themselves over eight times in the first 25 years of their existence—and they continue to return the investment.

Today's cells can be over 16% efficient and the energy consumed in their manufacture has been reduced. During their first 30 years today's PV modules will produce well over 20 times more energy than was consumed in their production. Several life-cycle cost studies have confirmed one- to four-year payback times for installed systems of various PV technologies. By contrast, coal and gas power plants consume irreplaceable resources and have no energy payback.

PV cell and module production costs are influenced by three factors: energy conversion efficiency, material utilization, and economy of

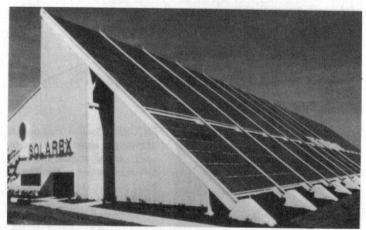

Figure 1.5. The proposed PV Breeder plant at Frederick, Maryland, was dedicated in 1982. Power from the 200-kW solar array was going to be used to make more solar modules. (Solarex Corporation)

scale. Solar modules are sold by their price per watt (150-watt module at $4 per watt = $600 per module). As their light-to-electricity conversion efficiency increases, module price per watt decreases. Similarly, costs go down when less photoactive material is used to produce the same power. Mass production methods further reduce some production costs.

In theory, single-crystal silicon cells can convert about 28% of sunlight into electricity, with the rest of the light energy that strikes a cell being reflected or converted into heat. In 1960, cells were 6% efficient and worldwide production was only 7 kilowatts. At that time PV cost over $1,000 per watt, but it was ideal for use in outer space because of its light weight and extreme reliability—and because cost was no object. By the 1970s, the price for terrestrial solar cells had dropped below $50 per watt and PV became the preferred power source for radio repeaters located on remote mountaintops and offshore oil derricks.

In 1976, the U.S. Department of Energy and the PV industry made overly optimistic cost projections based on the prospect of large utility PV purchases that were never made. PV prices did drop as a result of public and private research investment. In 1980 research money became scarce and inflation outpaced technical development cost reductions. By 1985, module prices stabilized at $8 to $10 per watt. Since then price reductions have continued, but at a slower rate. Even as cell prices drop, the relative costs of glass, mounting hardware, electronics, and other system parts increase.

Over the years, cell efficiency and production methods have improved, production capacity has increased to over 6 GW, and prices have steadily dropped. By 2007, terrestrial solar modules that convert over 15% of sunlight into electricity were available for approximately $4 per watt.

While higher production volume does reduce cost, significant savings remain elusive even though some companies are producing well over 100 MW annually. For one thing, PV production equipment is as expensive as computer chip production equipment, but PV profit margins are much less. In addition, the computer industry competes with the PV industry for silicon. Significant price reductions, first available to megawatt buyers, contribute to driving prices down for everyone, but PV prices will never drop the way computer prices have. Computers use tiny bits of expensive semiconductor material to perform high value calculations. Solar modules use large amounts of similar material to perform relatively low value electric generation.

Any real price breakthrough will require radical departures in the production process of solar cells. Past research has resulted in new ways to produce cells from cubes, ingots, and sheets, and by using lower-grade material. Research in polycrystalline solar cells has resulted in efficiencies and prices that match single-crystal cells. Manufacturers have developed combination crystalline and thin-film cells that can convert a wider spectrum of light into electricity. Others are developing lower cost thin-film amorphous and crystalline cells.

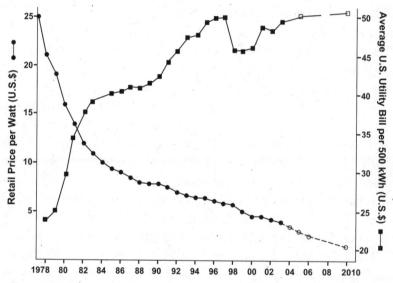

Figure 1.6. Retail cost of photovoltaics per watt, compared to the average U.S. urban utility bill.

Average utility bills are based on nationwide averages published by the U.S. Bureau of Labor and Statistics in the Consumer Price Index. The cost of electricity per 500 kWh is used because that was how the data were collected and presented. The actual national average residential monthly use in 2001 was 877 kWh, and the bill was $75.57. According to the Energy Information Agency (EIA), utility rates range from a low (in first quarter 2005) of 5.88¢ per kWh in Idaho to a high of 18.69¢ per kWh in Hawaii. These are averages and might not reflect the reality of your electric bill. For example, for the same time period EIA shows the average retail cost of residential electricity in California as 11.45¢ per kWh. Looking at our own bill we see rates ranging from 13.009¢ to 19.704¢ per kWh, with a tiered billing structure ranging up to 25.959¢ per kWh. Projected household expenditures are based on figures from EIA's "2003 Annual Energy Outlook."

If prices are dropping, should you wait for cheaper solar cells? News stories regularly announce revolutionary breakthroughs that will significantly lower solar cell prices. Few if any ever reach the marketplace. Press releases about these so-called "breakthroughs" are not news—they are investment solicitations. PV technology is evolutionary, building upon ongoing research by thousands of scientists and engineers. Even the prognosticators no longer consider a breakthrough necessary for further commercialization of PV.

PV prices will come down slowly. The 1976 prediction of 50¢ per watt by 1986 was off by a factor of ten. On the other hand, the price for utility power has increased by more than 300% during that same period.

The best time to buy PV is now so you can stop paying utility rate increases, start amortizing your costs, and begin to reap the environmental benefits of clean, renewable solar power.

Electric Power Production Options

The electric utility grid is a technological marvel. Electricity produced by thousands of generators is transmitted through millions of miles of wire to power an electric light so you can read this book. One hundred years ago most people could only dream about having electricity in their homes. Today people in industrialized countries take electricity for granted; meanwhile 1.6 billion people in developing countries may never have electricity in their homes or even in their villages.

The United States has vast natural resources. Still, Americans import over half the oil they consume. Some oil-exporting countries are openly anti-American and support terrorists, yet each year the U.S. spends more than $50 billion to protect oil fields and shipping lanes in the Persian Gulf region. Over $50 billion was spent in the 1990 Kuwait war. The Iraq war has been even more costly. These costs are not itemized on your electric bill, but they do add to the net cost of electricity. Endless war and terrorist attacks are enormous prices to pay for so-called cheap foreign oil.

Most of this oil is burned for transportation, but it is also a price benchmark for other non-renewable fossil fuels. Oil and coal are about equal as sources of carbon dioxide emissions. The industrialized world's large-scale electric production from the burning of coal and gas produces massive amounts of pollution.

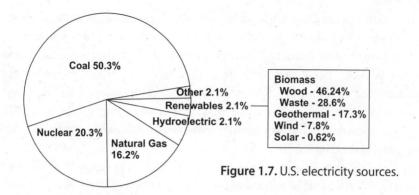

Figure 1.7. U.S. electricity sources.

Table 1.1. Emissions from Energy Consumption for Electricity Generation

Year	Million Metric Tons Carbon Dioxide	Thousand Short Tons Sulfur Dioxide	Thousand Short Tons Nitrogen Oxides
1989	1,944	17,079	8,969
1990	1,929	17,043	8,776
1991	1,933	16,912	8,723
1992	1,951	16,568	8,519
1993	2,034	16,497	8,815
1994	2,064	15,953	8,599
1995	2,080	13,113	8,692
1996	2,155	14,226	6,924
1997	2,223	14,904	6,971
1998	2,313	13,789	6,875
1999	2,327	13,717	6,319
2000	2,429	12,453	5,930
2001	2,380	12,088	5,561
2002	2,395	11,591	5,293
2003	2,416	11,732	4,768
2004	2,444	11,362	4,355

Data series 1989–2000: U.S. Department of Energy, Energy Information Agency
Data series 2000–2004: U.S. Environmental Protection Agency

Compare the fuel mix, emissions, and environmental impact of the electricity provided by your local utility by using the "Power Profiler" on the U.S. Environmental Protection Agency Web site.[5]

Mountains of coal and millions of cubic feet of gas are being consumed each year. Mineral resources that took hundreds of millions of years to develop will be depleted in just a few hundred years. Every year the U.S. burns more than 900 million tons of coal, which releases over 51 tons of mercury into the air, along with tons of other chemicals that damage the environment and cause health problems. These atmospheric pollutants include carbon dioxide (CO_2), which contributes to global warming, changes in precipitation patterns, and rising sea levels, and sulfur dioxide (SO_2) and nitrogen oxides (NO_x), which cause respiratory illness, smog, and "acid rain." There are other sources for these pollutants, but the U.S. EPA reports that electric generation and other human activities are adding greenhouse gases, pollutants that trap the earth's heat in the atmosphere, at a faster rate than at any time in the past several thousand years. Compare the emissions in 2001 alone—over 14 million tons of sulfur dioxide, nearly 7 million tons of nitrogen oxides, and over 2.5 billion tons of CO_2—to the mere 1.7 million tons of ruble left by the fallen World Trade Center towers in New York City, an area 10 stories high on a 15-block site.[6]

Newer natural gas and coal power plants are more efficient and pollute less than did the old power plants, yet the U.S. Department of Energy "Annual Energy Outlook for 2003" estimates that carbon dioxide emissions will increase by 1.5% each year through 2050 in spite of clean air and emissions control standards that are already proposed for implementation over the next half century.

The federal government and the utility industry plan that 357 gigawatts of new natural gas-fired generating capacity will go online over the next 20 years. They expect that the retail price of electricity will remain the same or decrease slightly. This does not factor in seasonal spikes in gas demand that regularly push prices higher—where they remain.

Nuclear energy is used to produce electricity, though controversy and concern about safety, accidents, nuclear weapon proliferation, and terrorism have put its practicality into question. The transportation and disposal of radioactive materials alone poses such a serious threat that it makes other forms of pollution seem relatively harmless.

It is questionable whether electricity from nuclear reactors serves us well when you factor in the billions of dollars of annual subsidies, life-cycle cost, and radioactive waste disposal. Nuclear power plants wear out in about 40 years. In 1956, the first commercial nuclear power plant was built at Shippingport, Pennsylvania, for $40 million. That reactor was retired several years ago and its decommissioning cost over $900 million, not counting the disposal of its nuclear waste. Over 100 nuclear power plants in the U.S. will be decommissioned during the next 20 years. By conservative estimate, it will cost over $300 million to retire each reactor.

Fusion or thermonuclear energy research has failed to produce a safe, practical alternative to the radioactive technology used in today's nuclear power plants. Researchers have said so many times that another 25 years of costly research is needed that they have lost their credibility. They have not, however, lost their federal funding, a fallout of the former nuclear arms race.

Wind power, geothermal energy, landfill gas, and water power are relatively clean energy sources. Hydroelectricity could produce up to 25% of U.S. electrical needs. Few nations have as many potential hydroelectric plant sites. Many rivers that were once deemed suitable for a hydroelectric plant have now become unacceptable because of the loss of migratory paths to fish spawning areas. Million of dollars are being spent to remove the dams that will restore these open waterways.

Wind power is cost effective and a growing contributor to the utility grid. Wind/PV combinations are popular among rural homeowners. Favored by utilities as a clean energy, wind power is very site-specific. Its noise, appearance, and hazards to wildlife are controversial political and environmental issues that must be resolved for each location.

Tapping natural geothermal sites for steam and manmade landfills for gas for electric production compete effectively with fossil fuels. Electricity from biomass combustion is being developed, but emission and soil depletion issues have yet to be resolved. Ocean and tidal power generation remains on the drawing board.

Hydrogen fuel cells are touted as the next major energy source to replace oil and natural gas. However, as the name implies, fuel cells consume fuel. Hydrogen fuel cells emit water and oxygen, but hydrogen fuel is not naturally occurring and so must be made from either non-renewable and polluting resources or from water. Extracting hydrogen from water requires electricity, which some say should come from PV. However, it costs less to simply use PV electricity directly.

PV has its detractors who say that the materials used in its manufacture are hazardous. Silicon can create problems if inhaled; the arsenic used in the limited production of solar cells for the space program is carefully controlled. Other PV materials are common to the semiconductor industry and, as in that industry, they must be carefully regulated and handled. However, once the chemicals have been made into solar modules, they are inert. Disposal of salvaged PV modules has not yet become an issue because most modules are still in use. One study concluded that 500 MW per year of thin-film modules could be safely recycled using current practices and recycling facilities.

Some people criticize large-scale PV land use because about 0.4 square mile (1 sq km) is required for every 20 to 60 MW of modules. That is the same amount of land required by coal power plants when strip mining is taken into account.

The subject of energy production and consumption is a complex issue with many political, economic, and social implications for present and future generations. Fossil fuels are being depleted rapidly and the scramble for remaining resources will continue to destabilize world politics and economics, leading to further confrontation and conflict. Meanwhile virtually limitless solar energy is globally available and waiting to be tapped.

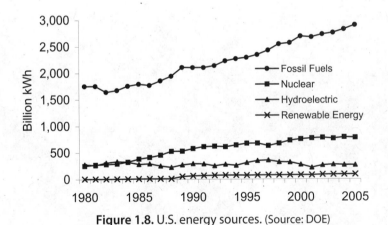

Figure 1.8. U.S. energy sources. (Source: DOE)

Future Trends in PV

The photovoltaics industry is growing at a healthy rate. Almost 10% of the installed PV modules are being used in cost-effective stand-alone applications, such as communications relay stations, water pumping and village power in developing countries, and telemetry devices. The future of off-grid PV does not require tax subsidies or another oil crisis.

The growing use of PV in specialty applications and consumer products assures continued growth of the industry.

Oil companies and multi-national corporations dominate PV manufacture due to the immense capital investment required and long lead time before profits may be realized. Approximately 15 American and over 100 foreign companies manufacture PV cells or modules. Some independents have a strong footing. On the other hand, major players such as Exxon's Solar Power Corporation pulled out completely, returning to their core competence: oil. Multi-megawatt utility-sized PV systems are being installed again as corporations return to PV. Meanwhile, the one-kilowatt to one-megawatt residential and commercial markets are growing steadily.

Amorphous PV production is increasing, yet single-crystal and semi-crystalline cell technology dominate stand-alone PV and grid-connected installations. Ribbon and concentrator PV remain special market/special use technologies. Tandem amorphous holds great promise but is years from the mythical 50¢ per watt panacea to global energy difficulties.

Figure 1.9. Recent developments include incorporating PV with roofing material. This 110,000-square foot, 329-kW system consists of amorphous PV modules bonded to a single-ply roofing membrane. Large commercial PV projects encourage manufacturers to increase capacity, which will ultimately reduce costs. (Solar Integrated Technologies)

Building-integrated photovoltaics (BIPV) is a rapidly growing market sector. Previously available products, such as PV roof shingles or glass panels, have a history of being either much more expensive than traditional panelized photovoltaics or more labor-intensive to install, thus increasing costs. They also had a lower output in watts per square foot than panelized PV, but prices are slowing dropping. PV is now being bonded to the material used for large-span roofs and installed the same way as traditional roofing. Architects and designers are increasingly replacing concrete, glass, and stone building surfaces with building-integrated PV.

Thin-film photovoltaics can be used to coat a surface, such as stainless steel or glass, which can be incorporated into building materials. These materials are less efficient than wafer technology, but they do have tremendous versatility in architectural applications.

Grid-connected photovoltaics is the fastest growing market in the United States, Japan, and Germany. Where off-grid PV applications continue to grow at a rate of about 10% per year, grid-connected distributed generation installations have more than doubled every year since 2000. The International Energy Association (IEA) Photovoltaic Power Systems Program reports the greatest proportion is grid-connected systems being installed in seven of the eighteen reporting countries (see Table 1.2).

While growth in industrialized countries is important, PV must reach markets in developing nations. Unfortunately, those who need PV the most can least afford it. As PV sales in industrialized countries grow, prices will drop for all. Barring a miracle technological breakthrough, mass production is the only significant factor that can drive prices down.

Asian PV companies will likely dominate the market by midcentury. As of this writing, the Japanese and German PV industries are well-established and subsidized by their governments. The U.S. PV industry will also benefit from similar government support to enable it to increase its foothold in the global market.

China is tooling up for PV world dominance. Fueled by extraordinary economic growth and driven to end its reliance on expensive foreign oil and fossil fuel pollution, China has become the fastest entrant to the PV market. Interestingly, over 95% of the raw materials used by Chinese PV companies come from outside China and over 95% of Chinese PV modules are exported. Foreign financing funds China's lopsided PV industry, which has yet to provide its own people with much-needed solar power.

Table 1.2. Worldwide Grid-Connected PV—2005

Country	Total Cumulative Installed PV (kW)	PV Installed in 2005 (kW)	Grid-Connected PV Installed in 2005	
			kW	%
Australia	50,581	8,280	1,980	19.3
Austria	24,021	2,961	2,711	47.8
Canada	16,746	2,862	612	17.6
Denmark	2,650	360	320	47.1
France	33,043	7,020	5,900	45.7
Germany	1,429,000	635,000	632,000	49.9
Great Britain	10,877	2,732	2,567	48.4
Israel	1,044	158	2	1.3
Italy	27,500	6,800	6,500	48.9
Japan	1,421,908	289,917	287,105	49.8
Korea	15,021	6,487	6,183	48.8
Mexico	18,694	513	30	5.5
Netherlands	50,776	1,697	1,547	47.7
Norway	7.252	362	0	0
Spain	57,400	20,400	18,600	47.7
Sweden	4.237	371	0	0
Switzerland	27,050	3,950	3,800	49.0
USA	479,000	103,000	70,000	40.5
Estimated Total	**3,676,800**	**1,092,851**	**1,039,917**	**48.8**

Extrapolated from IEA Photovoltaic Power Systems Program data

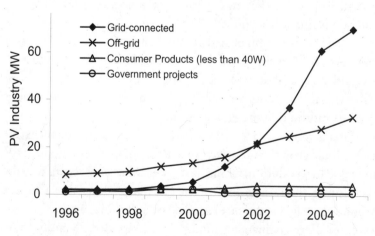

Figure 1.10. PV applications in the U.S. (Source: DOE, IEA)

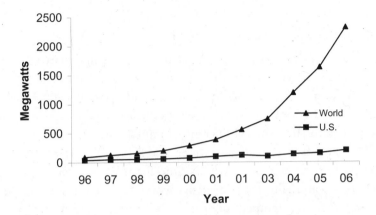

Figure 1.11. U.S. and world PV production. (Source: IEA)

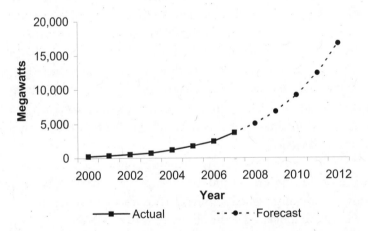

Figure 1.12. Long-term world PV demand. (Source: Prometheus Institute)

When surveyed, PV users almost always say they are glad they did not wait for promised lower prices. Price, tax credits, and major breakthrough rumors notwithstanding, PV has carved out a place in the global energy resource picture. Compared to bringing in a power line or operating and maintaining a fossil fuel generator, PV really shines. For remote or portable applications, PV is unsurpassed. As for reliability, its over 50 years of success in space plus more than five decades of use on earth are proof enough that photovoltaics is on its way to becoming as commonplace as the electric light.

The Boiling Frog Story

We are told that the ability to adapt to changing situations is good: Adaptation means survival. People who can change with changing circumstances are often called survivors. But there is another side to adaptability.

An experiment was made using a frog and a pan of water. The pan was placed on a stove and the water was heated. When the frog was dropped into the hot water, it instantly jumped out.

The same frog, same pan of water, and same stove were used in the second part of the experiment. This time the frog was placed in room-temperature water. The frog stayed in the water. The heat was turned on and slowly the water temperature rose. The frog stayed in the water. The frog's natural ability to adapt helped it adjust to the rising temperature. In fact, the frog adapted so well it was boiled.

When oil prices skyrocketed, there was outrage. People were shocked when gasoline prices rose from 25¢ to 50¢ a gallon. They swore they would never pay $1 a gallon. But pay they did. Prices are now over $4 a gallon.

At the same time, though less dramatically, the prices of natural gas, propane, and especially utility electricity were steadily climbing. We didn't notice the increase because it only hit us once a month in small increments and in small print. Not at all like the two-foot-high prices posted at the gas station.

We have adapted to a slowly changing condition: the rising cost of energy in a world where more and more people are more and more rapidly depleting already diminished resources. We are about to reach the point where we will no longer have the ability to leap to safety.

Notes

1. Y. Kuwano, "Genesis Project," Sanyo Electric Company, International Solar Energy Society, 1991.
2. National Renewable Energy Laboratory (NREL), "PV Facts"—www.nrel.gov/pv/.
3. John Perlin, *From Space to Earth: The Story of Solar Electricity* (Ann Arbor, MI: aatec publications, 1999), pp. 41–46.
4. *PV News*, Prometheus Institute.
5. U.S. Environmental Protection Agency: www.epa.gov/cleanenergy/powprofiler.html
6. "Construction is the Next World Trade Center Project," *USA Today*, 9 May 2002.

Chapter 2

GOVERNMENT, UTILITIES & PV

Times have changed since the first edition of *The New Solar Electric Home* was published. In 1983 it was virtually impossible to get a meeting with a utility manager or government official to discuss photovoltaics. There was little interest, no formal policy, and photovoltaics was considered a developing technology. Now PV is mainstream. Today the federal government has solar policies; state and local governments have solar programs; and utilities own megawatts of photovoltaic systems. Still, photovoltaics contributes less than 1% to the total amount of electricity produced in the United States.

Federal Policy & Legislation

There are two types of policy and legislation affecting photovoltaics—direct and indirect.

Direct policy and legislation include tax incentives specifying solar electricity. An example of direct legislation is the Energy Policy Act of 2005 that enacts a 30% tax credit for residential solar installations. Residential solar electric systems are eligible for a tax credit of up to $2,000.

The credit was available for systems installed after December 31, 2005, and before January 1, 2008. New legislation was introduced in 2006 and 2007 to extend and expand the federal tax credit.

The same Energy Policy Act requires federal agencies to acquire a specified percentage of their electricity from renewables (3% in 2007 rising to 7.5% by 2013), with double credit given for renewables production involving federal and tribal lands or buildings. The Secretary of Energy is required to procure at least 150 MW of PV over five years (2006 through 2010) for use on federal buildings: $50,000,000 to $60,000,000 per year has been authorized for this purchase.

Policies aimed at environmental issues indirectly affect photovoltaics. Federal regulations protect the environment by placing limits on or raising the cost of using the environment as a dumping ground for utility emissions and waste. These regulations do not necessarily say how the end goals are to be achieved, but they do make recommendations. Historically, projected dates for the enactment of environmental legislation are postponed because utilities lobby effectively against cleaner technologies, usually by simply saying they cost too much. Most recently, voluntary cooperation without legislative mandate has been encouraged by presenting case studies and scenarios where profits are maintained, expenses lowered, and projected environmental goals attained. The ultimate responsibility is passed on, either to state government, the utilities, or the private sector for voluntary compliance and implementation with no motivation other than public mandate.

The exception to states' authority or rights in regulating power plant emissions is what the White House Council on Environmental Quality has called "very narrow points of law."[1] For example, in 2003 federal agencies, including the Environmental Protection Agency, and several state governments sued each other over environmental issues. One issue concerned California's regulations to control carbon dioxide emission levels from power plants. A vanguard state in setting environmental policy, California's regulations are stricter than federal standards. On its behalf, the EPA referenced the 1970 Clean Air Act, stating that it precluded state regulation of carbon dioxide emissions linked to global warming.

But the issue is not California's law or states' rights. The issue is the ability of the federal government or an administration to redefine words

Figure 2.1. This 1 MW Sacramento utility-owned PV system, started in 1984, will grow to replace the adjacent nuclear power plant that citizens voted to shut down. By 2004, the PV system was 3.2 MW with another 8.3 MW of utility-owned PV distributed throughout the city. In addition, 6 MW of privately owned PV on homes and businesses make Sacramento a leading U.S. PV city. (California Energy Commission)

like "pollution" and "harmful." In spite of the scientific evidence that led to the Kyoto Accord, as well as a 2001 report issued at the request of the White House by the National Research Council and a 2002 report on global warming by the Environmental Protection Agency and the U.S. State Department, the Bush administration stripped the global warming section from an EPA report on trends in air pollution.[2] The administration contended that carbon dioxide is not a problem because CO_2 occurs naturally in the air. It does, but rising and uncontrolled levels of CO_2 in the atmosphere have been proven to be a causative agent in global warming. A similar argument could be made about cyanide, because the compound occurs naturally in apples, peaches, and other healthful fruits. But the consumption of cyanide in uncontrolled amounts is not recommended.

The Bush administration's Clear Skies Act was designed to replace the existing and more rigorous Clean Air Act; it is a two-stage plan that will not come into full effect until 2020. The cap on nitrogen oxide (NO_x) pollution levels will be lifted from the current 1.25 million tons by 2010 to 2.1 million tons by 2008, allowing a 68% increase in NO_x pollution. Sulfur dioxide (SO_2) pollution levels were supposed to be reduced to 2 million tons by 2012. The Clear Skies Act raises these levels to 4.5 million tons by 2010, an increase of 225%. Carbon dioxide reduction is now unspecified and voluntary. Mercury, limited to 5 tons per year by the Clean Air Act, is now capped at 26 tons per year, an increase of 520%. The new plan also creates loopholes that exempt new power plants from being held accountable under the Clean Air Act's New Source Review standards and existing power plants from being required to install clean-up technology (also known as "best available retrofit technology" or BART). Enforcement of public health standards for smog and soot has been delayed until the end of 2025.[3] It is important to note that in defining "electrical consumption" the standard had been simply per capita. This has been changed, and depersonalized, to a calculation relative to the GNP.

The Million Solar Roofs (MSR) program is a perfect example of federal policy that affects PV directly. It also illustrates the difference between Democrats, who generally want federal solar programs, and Republicans, who want to pass responsibility to state and local governments and individuals.

The Million Solar Roofs Initiative (www.millionsolarroofs.com) is a unique public/private partnership aimed at overcoming barriers to market entry for solar technologies. The press release detailing the June 1997 announcement by Clinton administration Secretary of Energy Federico F. Peña read:

> The initiative calls for the Department of Energy to lead an effort to place one million solar energy systems on the roofs of buildings and homes across the U.S. by the year 2010. This will be accomplished by using existing federal grant, procurement and other programs and working with local communities, businesses, state governments, utilities and other groups to spur sales of solar energy systems.
>
> "By putting solar cells on the roof, we're going to send solar sales through the roof," Secretary Peña said at an announcement of the initiative today in Washington. "We will marshal our considerable resources to reduce greenhouse gas emissions. And we will build on the increasing momentum in the U.S. to use renewable energy resources like solar."

The funding put in place by the Clinton administration was drastically reduced early in the Bush administration. This once highly publicized program received little federal funding or support. The program depends on partnerships similar to those of the Energy Star Appliance program, which is like a "Good Housekeeping Seal of Approval" for energy-conscious consumers. As of 2005 there were over 120 partners in MSR. The largest group is made up of public entities: states, cities, and utilities. In some cases very large and public commitments—such as the promise made by the Los Angeles Department of Water and Power (the single largest municipal utility in the United States and an MSR partner) to install the equivalent of 100,000 residential solar rooftop systems by 2010—has gone unfulfilled due to discontinued commitment.

.Since 1978, PURPA, the Public Utility Regulatory Policies Act, has allowed all photovoltaic electricity producers access to the electrical grid. This opened the door for net metering and has given photovoltaic power plants, both large and small, the opportunity to enter the power-generating marketplace. However, net-metering legislation has been relegated to the state level. So although PURPA permits PV systems to be grid-connected in all states, in those states that do not have net-metering laws, local utilities can, and usually do, require renewable energy users to install two meters: one for the retail purchase of utility electricity by the consumer and another for the sale of PV electricity at wholesale rates to the utility. For this reason almost no on-grid PV is installed in states that do not have net-metering legislation.

Renewable Portfolio Standards (RPS) are another form of legislation that has both direct and indirect effects on photovoltaics. These standards have been established in several countries to specify a percentage contribution of renewable energy to total electrical production. In the U.S. a 10% RPS implemented by 2020 has been proposed. Analyses by the Energy Information Administration (EIA) and the Union of Concerned Scientists show a total savings to consumers of $13.2 billion is possible between 2002 and 2020 based on that 10% RPS. However, critics label photovoltaics as too costly to play a major role in the economics of a RPS.

The Department of Energy is the only source of public funding for PV research and development. At one time the DOE was actively involved in R&D, but that role has shifted to the private sector. Now

Figure 2.2. The municipal-owned utility of the City of Ashland, Oregon, purchases solar electricity at the highest retail rate. In one year, this homeowner bought 4,139 kWh and sold 2,847 kWh to net $5.92 "profit," thanks to Ashland's pro-solar rate structure. This system has 32 100-watt PV modules mounted on an adjustable rack. (Owner: Andy Kerr; Design & installation: Electron Connection; Photo: *Home Power*)

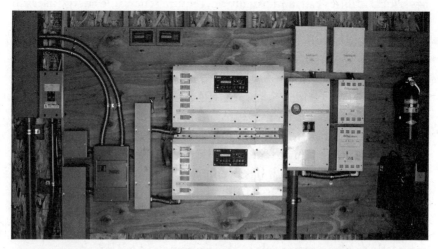

Figure 2.3. The PV array in Figure 2.2 feeds into two 30-amp charge controllers, two 4-kW inverters, and a 700-amp-hour, 48-volt battery bank. The wall-mounted power center uses garage space effectively. (Owner: Andy Kerr; Design & installation: Electron Connection; Photo: *Home Power*)

the DOE monitors and forecasts PV industry production and provides public information on PV activities in the U.S. In 2006, the DOE started the Solar America Initiative (SAI), another effort to cut federal subsidies for PV R&D while letting coal, oil, gas, and nuclear industry subsidies remain in place.

Federal Incentives for Photovoltaics

The Database of State Incentives for Renewable Energy (DSIRE) is a comprehensive source of information on state, local, utility, and selected federal incentives that promote renewable energy. Following is a list of federal incentives and programs from the DSIRE Web site (www.dsireusa.org) that apply to residential and commercial photovoltaic systems.

Energy-Efficient Mortgage (EEM)—This mortgage can be used by homeowners to pay for energy-efficiency improvements to new and existing homes. The EEM is federally recognized and can be applied to most home mortgages. Both government-insured (FHA, VA) and conventional (Fannie Mae) EEMs are available. All buyers who qualify for a home loan also qualify for an EEM, which gives the buyer benefits in addition to the usual mortgage deal. The lender uses the energy-efficiency of the house, as determined by a HERS (Home Energy Rating System, from the 1992 Energy Policy Act) rating, to determine exactly what these benefits will be. EEMs can be used to finance PV, solar water and space heating, and energy efficiency. To learn which lenders in your state offer energy-efficient mortgages, visit the Residential Energy Services Network at www.natresnet.org/.

Accelerated Depreciation—Under the Modified Accelerated Cost Recovery System (MACRS), businesses, including qualified home-based businesses, can recover investments in solar, wind, and geothermal equipment through depreciation deductions. The MACRS property class for solar, wind, and geothermal equipment is five years.

Renewable Energy Systems and Energy Efficiency Improvements Program—Rural Development grants can be used to pay up to 25% of eligible project costs. These grants range from $10,000 up to $500,000. Contact your USDA office for information (www.usda.gov).

Residential Energy Tax Credit—For the first time since 1985 home-owners are eligible for a federal energy tax credit. The Energy Policy Act of 2005 created a new 30% tax credit for residential solar installations. In 2006 and 2007, legislation was introduced to extend and expand the federal tax credit.

Solar and Geothermal Business Energy Tax Credit—The Energy Policy Act of 2005 increased the existing 10% tax credit for commercial solar installations to 30% for two years. In 2007, legislation was introduced to extend and expand the federal business tax credit.

Tribal Energy Program Grant—DOE's Office of Energy Efficiency and Renewable Energy's (EERE) Tribal Energy Program provides financial and technical assistance to Native American tribes for feasibility studies and shares the cost of implementing sustainable renewable energy installations on tribal lands. This program promotes tribal energy self-sufficiency and fosters employment and economic development on America's tribal lands. The program is managed by EERE (www.eere.energy. gov/power/tech_access/tribalenergy).

State Governments, Utility Programs & Incentives

Each state has its own policies pertaining to photovoltaics. Most states have net metering laws; some have state income tax credits or tax incentives and loan programs. Some of the more than 3,300 utility companies in the U.S. have their own solar rebate, loan, and incentive programs. Policies and standards are constantly changing, so check the Database of State Incentives for Renewable Energy (www.dsireusa.org) for the latest state, local, utility, and selected federal incentives, as well as regulations and policies. The database is easy to use and gives basic information on the programs available in your state. Links to individual state utility commissions and some local utilities will take you to detailed, up-to-date information.

Another source of information is your own utility company—the company you write a check to each month for your electricity. It might take some perseverance to find the right person to talk to because not all utilities have established a department or designated a specific person to handle PV inquiries.

The American Solar Energy Society is another valuable resource. ASES can put you in touch with people in your area who can provide information about local PV resources (www.ases.org; 303.443.3130).

A Glossary of Programs

There are three major types of programs offered by states, municipalities, and utilities. (See www.dsire.org for the most up-to-date information.)

- **Financial incentives (F)** include rebates, tax credits, and deductions that apply specifically to PV.
- **Outreach and voluntary programs (O)** are a mixed bag of educational and other PV-related benefits.
- **Rules, regulations, and policies (R)** are requirements that apply directly to photovoltaics (other than building codes, which will be discussed later).

Construction and Design Policies (R)—These include state construction policies, green building programs, and energy codes. State construction policies are typically legislative mandates requiring an evaluation of the cost and performance benefits of incorporating renewable energy technologies into state construction projects, such as schools and office buildings. Many cities have "Green Building" guidelines that require or encourage the use of renewable energy technologies. Some guidelines are voluntary measures for all buildings, while others are requirements for municipal building projects or residential construction. Local energy codes help us achieve higher energy efficiency in new construction and renovations by requiring that certain building projects surpass state requirements for resource conservation. Incorporating renewables is one way to meet code requirements.

Contractor Licensing (R)—Many states have rules regarding the licensing of renewable energy contractors. These requirements are designed to ensure that contractors have the necessary experience and knowledge to properly install systems.

Corporate Tax Incentives (F)—To promote renewable energy use, these incentives allow corporations to receive tax credits or deductions ranging from 10% to 35% against equipment or installation costs.

Direct Equipment Sales (F)—A few utilities sell renewable energy equipment to their customers as part of a buy-down, low-income assistance, lease, or remote power program.

Equipment Certifications (R)—Statutes that require renewable energy equipment to meet certain standards protect consumers from being sold inferior goods. Equipment certification benefits the renewables industry by reducing the number of problem systems and eliminating the resulting bad publicity.

Feed-In Tariffs (F)—A PV feed-in tariff is a special rate paid by the utility company or government to encourage people to install PV systems. A feed-in tariff PV system has a separate revenue meter. The system owner receives a monthly check for PV power sold to the grid.

Feed-in tariffs gained popularity in Germany, spread to other countries, and have been implemented in California as a performance-based incentive (PBI). In Germany the feed-in tariff started at twice the retail price paid to the utility for electricity. This gave PV system owners a 10-year payback on the purchase of their systems. Investing in PV systems became so popular in Germany that over 1,000 megawatts have been installed each year since 2005. The United States, with 3.5 times greater population, installed 85% less on-grid PV (150 megawatts) in 2007. American utility companies are generally opposed to feed-in tariffs and net metering because they feel that their control of the grid and their profits are threatened. Times are changing and customers are demanding that their utilities either provide clean renewable energy or allow people to do it themselves. Special rates for clean power and penalties for dirty power will eventually become commonplace.

Generation Disclosure Rules (R)—Generation disclosure requires utilities to provide customers with fuel mix percentages and emissions statistics. The related issue of **Certification** refers to the assessment of green power offerings to confirm that they are indeed utilizing the type and amount of renewable energy advertised (an example is the *Green-e* stamp). Both disclosure and certification help consumers make informed decisions about the energy and the supplier they choose.

Grant Programs (F)—States offer a variety of grant programs to encourage the development and use of renewable energy technologies.

Most programs offer support for a broad range of technologies. Grants are available primarily to the commercial, industrial, utility, education, and government sectors.

Green Power Purchasing/Aggregation Policies (R)—Municipalities, state governments, businesses, and other non-residential customers can play a critical role in supporting renewable energy technologies by buying electricity from renewable resources. A few states allow local governments to aggregate the electricity loads of the entire community to purchase green power and even to join with other communities to form larger green power purchasing blocks.

Industrial Recruitment Incentives (F)—These incentives focus on special efforts and programs designed to encourage renewable energy equipment manufacturers to locate within a state or city. Renewable energy industrial recruitment incentives include tax credits, grants, or a government or utility commitment to purchase a specific amount of the product for use by a government agency. Recruitment incentives are designed to attract industries that will benefit the environment and create jobs.

Leasing or Lease Purchase Programs (F)—Utility leasing programs are for remote power customers for whom line extensions would be very costly. Customers can lease the technology from their utility, and in some cases can opt to purchase the system later.

Line Extension Analysis (R)—Electric customers who request that electricity be brought out to a location not currently serviced by the grid are charged the cost of extending the power lines to their location. In many cases it is cheaper to purchase an on-site renewable energy system to meet their electrical needs. Certain states require utilities to provide customers with information on renewable energy options when a line extension is requested.

Loan Programs (F)—These programs offer financing for the purchase of renewable energy equipment. Low- or no-interest loans for energy efficiency and demand-side management are offered by some utilities. State governments also offer loans to assist in equipment purchases.

Net Metering Rules (R)—For consumers who have their own electricity generating units, net metering allows for the flow of electricity both to and from the customer through a single bi-directional meter. During times when the customer's generation exceeds use, electricity flows from the customer's system to the utility and offsets the electricity consumed at a later time. The excess electricity generated offsets the electricity that would otherwise have been purchased at the retail rate. Under most state rules, residential, commercial, and industrial customers are eligible for net metering, but some states restrict eligibility to specific customer classes.

Outreach Programs (O)—Increasing awareness and understanding of renewable energy technologies and providing technical assistance and training for their deployment are critical to building a strong renewables market. These programs include ongoing renewable energy awareness campaigns, state and local Million Solar Roofs Partnerships, and other government-sponsored programs and activities.

Personal Income Tax Incentives (F)—Many states offer personal income tax credits or deductions on the purchase and installation of renewable energy equipment.

Production Incentives (F)—These provide project owners with cash per kilowatt-hour payments based on electricity production, as is the case for the Federal Renewable Energy Production Incentive. Other incentives are based on a price per gallon of renewable fuels produced, as in state ethanol production incentives.

Property Tax Incentives (F)—Property tax incentives may be exemptions, exclusions, or credits. The majority of property tax provisions for renewable energy simply exclude the added value of the renewable energy equipment from the valuation of the property for taxation purposes. Some states allow local authorities the option of providing a property tax incentive for renewable energy devices.

Public Benefit Funds (R)—Public Benefit Funds (PBF) are typically state-level programs that support renewable energy resources, energy-efficiency initiatives, and low-income support programs. These funds are also frequently referred to as a System Benefits Charge (SBC) or Public Goods Fund. Such funds are usually supported by a charge to all customers based on electricity consumption. Twenty-five states have PBFs. The California Public Benefit Fund has proved to be the most successful program for expanding the use of PV in the U.S.

Rebate Programs (F)—Rebate programs for homeowners and businesses are offered at the state, local, and utility levels to promote the installation of renewable energy equipment. In some cases, rebate programs are combined with low- or no-interest loans.

Renewables Portfolio Standards/Set Asides (R)—These standards require that a certain percentage of a utility's overall or new generating capacity or energy sales be derived from renewable resources. For example, Arizona's Environmental Portfolio Standard requires that a minimum of 0.2% of its utilities' total retail sales come from renewables in 2001, increasing to 1% by 2005, and topping out at 1.1%, or up to 96 MW, in 2007. It further specifies that 60% of the renewable energy must come from photovoltaics. The term "set asides" refers to programs where utilities are required to include a certain amount of renewable energy capacity in new installations.

Required Utility Green Power Option (R)—A handful of states require certain classes of utilities to offer customers the option to purchase power generated from renewable sources. Typically, utilities may provide green power using renewable resources they own or for which they contract, or they may purchase credits from a certified renewable energy provider.

Sales Tax Incentives (F)—These incentives typically provide exemption from state sales tax for the cost of renewable energy equipment.

Solar Access Laws (R)—These statutes provide for easements or access rights. Easements allow the rights of a renewable resource property owner to be secured from an owner whose property could be developed in such a way as to restrict the use of that resource. Access rights automatically provide continued access to a renewable resource. Solar easements are the most common type of state solar access rule. Some states prohibit neighborhood covenants or municipal laws that preclude the use of renewables. Communities use many different mechanisms to protect solar access, including ordinances, development guidelines requiring proper street orientation, zoning ordinances that contain building height restrictions, and solar permits.

Utility Green Pricing Programs (O)—Green pricing is an optional utility program that allows customers to support a greater level of utility company investment in renewable energy. Participating customers typically pay a premium of $3 or more on their electric bill to cover the incremental cost of the additional renewable energy.

Voluntary Installer Certification Programs (O)—Certification provides a means by which to judge the skills and qualifications of solar practitioners, giving consumers increased confidence in the solar industry and rewarding practitioners for meeting high standards of training and practice. An increasing number of renewable energy associations, state energy offices, technical colleges, and other organizations are working together to develop voluntary training programs for those who install solar energy equipment.

Make Change Happen

Every municipal and utility policy change supporting PV began with a single individual lobbying for change.

Government, through its policies and laws, either recommends or controls how utilities operate. Assurances that utilities provide clean, safe, reliable, and affordable electricity are pivotal issues in elections throughout the United States. Our votes and letters make sure our lawmakers know how important environmental concerns are to us.

Notes

1. "U.S., State Clash Over Environment," *Los Angeles Times*, September 14, 2003. The 10th Amendment states: "The powers not delegated to the United States by the Constitution, nor prohibited by it to the States, are reserved to the States respectively, or to the people."
2. Jeremy Symons, "How Bush and Co. Obscure the Science," *Washington Post*, August 27, 2003.
3. "Facts About the Bush Administration Plan to Weaken the Clean Air Act," Sierra Club: www.sierraclub.org/cleanair/clear-skies.asp.4. See www.osti.gov/html/news/releases97/junpr/pr97060.html.

Chapter 3
CONSERVATION &
ENERGY EFFICIENCY

Conservation and energy efficiency are important first steps toward designing your PV system.

If you live away from utility power lines in an off-grid home, reducing your energy requirements by half will cut the size and cost of your PV array and battery bank in half.

On-grid PV users usually install systems to decrease their electric bills or to reduce their environmental impact. Decreasing your electrical consumption meets both goals.

When sizing a solar electric system for a 100% solar-powered off-grid home, solar production must equal the energy that is consumed.

When sizing a PV system for an on-grid home, you must know your energy consumption to determine what percentage of your energy needs you will power with PV.

In both cases you have to know how much electricity you use and what you use it for.

This is the first step in creating an energy budget.

How Much Electricity Do I Use?

Make a list, similar to the one in Table 3.1, of all the electrical appliances and tools you use or intend to use, the number of hours per day or week you use them, and how much electricity each consumes.

For a light bulb this is easy: A 100-watt light bulb uses 100 watts. To determine energy consumption for your appliances, look for a label on the back or underside that states electrical consumption. In some cases the number is in watts (W), in others it will be in amperes (amps, A). The power calculation is simple: The number of watts equals the amperage times the voltage.

For example, the label on the back of a printer reads "120 V, 60 Hz, 0.45 A." Hz is the abbreviation for "hertz," which is the frequency—the cycles per second—that alternating current (AC) flows rapidly back and forth. In most cases it is 50 or 60 cycles per second. All appliances in the United States operate at 60 Hz.

It is important to note that the amps or watts given on the power label indicate the appliance's maximum consumption. Some appliances have duty cycles that consume different amounts of electricity. Therefore actual consumption can be lower than indicated on the power label. Examples of this are a printer that is on but not actively printing or a refrigerator compressor that cycles on and off. To calculate the wattage of the printer, multiply the volts times the amps.

$$120 \text{ volts} \times 0.45 \text{ amps} = 54 \text{ watts}$$

Say the printer is in use one hour per day. To calculate the energy consumed, multiply power times the period of operation.

$$54 \text{ watts} \times 1 \text{ hour} = 54 \text{ watt-hours } (54 \text{ Wh or } 0.054 \text{ kWh})$$

If a 100-watt light bulb is on for 5 hours per day:

$$100 \text{ watts} \times 5 \text{ hours} = 500 \text{ watt-hours } (0.5 \text{ kWh})$$
$$\text{of electricity per day}$$

Table 3.1 details the power usage of a small home office that consumes 4,259 watt-hours (4.259 kWh) of electricity per day. Table 3.2 shows the power usage of that same small home office after energy-conserving principles have been applied.

Table 3.1. Energy Consumption of a Small Office

Appliance	volts	amps	calculation	watts	hours	watt-hours
Desk lamp	120			100	5	500
Computer monitor	120	0.5	0.5 × 120	60	8	480
CPU	120			50	8	400
Printer	120	0.45	0.45 × 120	54	1	54
Overhead lights	120			200	10	2,000
Coffeemaker	120			850	0.5	425
Radio	120			40	10	400
Total						4,259

Table 3.2. Energy Consumption of an Energy-Conserving Small Office

Appliance	volts	amps	calculation	watts	hours	watt-hours
Desk lamp	120			100	5	500
Computer monitor	120	0.5	0.5 × 120	60	8	480
CPU	120			50	8	400
Printer	120	0.45	0.45 × 120	54	1	54
Overhead lights	120			200	2	400
Coffeemaker	120			850	0.5	425
Radio	120			40	9	360
Total						2,619
% Savings						38.5%

Like any habit, bad energy habits can be changed. Modify your use patterns: Don't automatically flip on overhead lights during the day; turn off the radio when you leave for lunch. You will save almost 40% of your energy budget.

kW or kWh?

Power and energy are sometimes used to mean the same thing, but electrically they are very different. Electrical power is the flow of current at a voltage (amps × volts): for example, 2,500 watts or 2.5 kilowatts, which is also written 2,500 W or 2.5 kW. Electrical energy is power over a period of time (amps × volts × time). For example, 2,500 watts used for 3 hours is 7,500 watt-hours or 7.5 kilowatt-hours or 7,500 Wh or 7.5 kWh. In Chapter 1, we compared the power of full sun (1,000 W/m²) on 135 square miles to the peak capacity of all the electric power plants on Earth, which is 3,763,584,000 million kW. On the other hand, the energy of 3 sun-hours (3 kWh/m²/day) on 9,800,000 acres is more than the total U.S. energy consumption of 4,825 billion kWh/year. A much smaller example: Making toast with a 1,200-W toaster (1,200 W × 2 minutes = 40 Wh) uses less energy than a 60-W light bulb on for one hour (60 W × 1 hour = 60 Wh).

Your Electric Bill

Energy consumption and saving translate into dollars on your electric bill. Consumption is expressed in watt-hours or kilowatt-hours (1 kWh = 1,000 watt-hours) and the typical billing cycle is 30 days. The office in Table 3.1 would get a bill for 4.259 kWh/day × 30 days or 127.77 kWh. Billed at 13¢ per kWh, that comes to $16.61 (127.77 × 0.13). The office in Table 3.2, after minor conservation changes are made, would get a bill for $10.21 (2.619 × 30 × 0.13). While the fixed rate charges—taxes, transmission and distribution fees—remain the same, you have saved $6.40 or 38.5%.

If you have an analog kilowatt-hour meter, you can actually watch your house consume electricity.

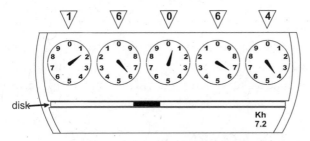

Figure 3.1.
Analog kilowatt-hour meter.

disk

Kh
7.2

Below the analog dials is a thin flat disk with a black reference mark on the edge. This disk rotates from left to right as power is consumed. The speed of rotation is proportional to the amount of power you are using. The number printed on the meter faceplate, usually at the lower right, is the Kh (kilowatt-hour) conversion number for that particular meter. In the illustration, the Kh number is 7.2. Using a stopwatch to count seconds, watch the black reference mark on the disk to count disk rotations. For example, count out two disk rotations and then record the number of seconds. Multiply the Kh factor times the number of rotations, and then divide by the number of seconds. For example:

(7.2 Kh × 2 rotations) ÷ 10 seconds = 1.44

Multiply this number by 3,600 (the number of seconds in an hour) to get the number of watts consumed.

$$1.44 \times 3,600 = 5,184 \text{ watts } (5.184 \text{ kW})$$

Your utility calculates your electric bill each month by reading the numbers on the analog dials on the utility meter (some utilities have digital meters). You can read these numbers yourself and calculate how many kilowatt-hours you have used. If the pointer is positioned between two numbers, use the lower number. If the pointer is directly on a number, look at the dial to the right to see if the pointer has moved past zero. If it has not passed zero, use the lower number. The meter in Figure 3.1 reads "16064." The number of kilowatt-hours you have used is 16,064 minus the previous meter reading. Note that the dials on some meters read from left to right, others read right to left. Compare your numbers to a recent utility bill to learn which direction your dial reads.

The reference mark on the rotating disk moves from left to right. This indicates that you are using electricity. If you have a PV system connected to the grid, and the reference mark is moving from *right to left* this means your system is producing more power than you are using. You are actually watching your electric bill go down!

Reading Your Bill

An important part of managing your home's electric consumption is reading and understanding your energy bill. Your cost per kWh includes both the basic utility rate per kWh plus additional charges such as transmission fees and rate adjustments, public purpose assessments, nuclear decommissioning charges, competition transition charges, energy cost recovery, fixed transition amounts, bond fees, service cost adjustments, and more. These charges are usually based on consumption per kWh and show on your bill as either "bundled" or "unbundled" (i.e., itemized) charges. There are other charges on your bill that are charged per utility meter, and then there are state and local taxes.

Your utility may also add a "fuel adjustment charge" to compensate for seasonal differences in the types and costs of fuel. The utility will tell you what is in their "portfolio," that is, what fuel they consume to make electricity. You are stuck with these charges unless you live in a state that has deregulated utilities, in which case you can change your electric supplier to one that has either more favorable rates or uses fuels you approve of, such as a higher proportion of renewables and less coal or nuclear.

Another basic charge you may pay is the "demand charge," which covers the utility's cost for maintaining the excess capacity it must have to meet peak power demands that occur from time to time. Utilities recommend that you shift your electrical loads to non-peak periods and run non-critical loads at night when overall demand is lower. You should practice load shifting and shedding (turning loads off) in your PV on-grid or off-grid home to reduce your system size.

If you consume very little electricity during peak periods (usually afternoons), consider switching to "time-of-use" (TOU) billing. The TOU rate is higher for peak periods, but much lower for off-peak (evening) consumption. It can be a windfall for some net-metered PV homes because PV systems produce the most power during peak periods. TOU is well worth investigating.

Conservation, Conservation, Conservation

Small changes can make a big difference. Let's compare three Los Angeles area neighbors. All three houses were built in the mid-1940s and are approximately the same size. The households have similar income levels, and all have home-based businesses. The homes are in the coastal climate zone with year-round average temperatures ranging from the mid-40s to the mid-80s. Household number one has monthly electric bills over $300. Household number two has monthly electric bills of about $125. The third household has winter monthly electric bills of $50 to $75 and summer bills of $25. Where are the kilowatts going? Why is household number three's bill so much lower?

Household number one keeps a small (4-cubic foot) second refrigerator outside on the west side of the house for cold drinks on the patio, and its window air conditioner runs at least 12 hours every day from March through October. Windows have been either painted shut or are the non-opening type. The wood shutters are kept closed so the owners, who work at home, keep lights on all day in at least three rooms. The homeowners say that they are in a constant battle with their children to turn things off and close the refrigerator door, but they really have no desire to conserve energy because "if I can afford to pay for it, I have a right to use it."

The second household has a different story. The homeowners think they are very energy conscious. They do everything their local utility advises them to do to conserve. The windows in their house have been replaced with new energy-saving models. They incorporated passive solar features when they remodeled in the early 1990s. They use energy-saving fluorescent lights. Other than computers and the radio, they use few small appliances simply because they choose not to.

Household three bought their house because it was bright, airy, and had a good-sized south-facing roof with the perfect pitch for a photovoltaic system. They do not think of themselves as particularly energy conservative. While their neighbors use more energy-efficient gas for heating and cooking, they opted to install a large electric oven. The home office is located in an addition not served by the gas furnace that heats the rest of the house so on cold winter mornings they turn on an electric space heater. The television is almost always on. But they do pay attention to the details. The house receives a lot of natural light so no lights are on during the day. The three computers and printers in the home office are turned off when not in use. They eliminate phantom electrical loads, such as the power drawn by the stereo system to maintain the presets, wherever convenient by either unplugging the device or installing a switch on the electrical cord. Their outdoor lights have motion detectors. They were pleasantly surprised to see their electric bill decrease after a refrigerator door gasket was replaced. Their electric bill would probably range from $75 to $125 per month if they had not installed a 2-kW photovoltaic system.

Utility Audits

Most utility companies offer on-line or mail-in home energy audits. The results of these audits tend to be painted with a very broad brush. Household number three attempted an energy audit three times. The mail-in energy audit came back from the utility company with the advice to not use room air conditioners, which they have never owned, and to replace the windows, which they had already done. The on-line audit came back from the utility company with a result listing their monthly electric bill as $339 per month. Their third attempt was a federal agency audit, which could not be completed because "zero" was not allowed on the form, as in "amount of heating oil consumed"–"zero." In most cases utility "audits" are a waste of time better spent doing a real energy audit.

Simple Steps to Conservation

Heating and Cooling Your Home

Heating and cooling consume a lot of energy. (Cooking and refrigeration will be discussed separately.) The first recommendation for an on-grid home is to not use electricity for heating or cooling. An electric space heater will cost $1,300 per year to operate. An equivalent gas heater (same BTUs) will cost $600, and a high-efficiency gas heater will cost $430. This advice applies to off-grid homes as well.

There are many ways to decrease heating and air conditioning costs. Insulation, low-E windows, and weather-stripping are recommended. Simple routine tasks such as changing or cleaning heating filters or vents will increase the efficiency of your existing equipment. Shading the house by planting trees, increasing the overhang of the eaves, or installing awnings over south and west windows are all very effective. When planting trees for shade near a solar home, keep two things in mind. First, the trees should be deciduous. When they drop their leaves, they allow passive solar heating to enter during fall and winter. Second, check the height that the tree is expected to reach. A PV panel that is only 10% shaded will have drastically reduced power production, often over 50%. You may have to either regularly trim overhanging trees or remove them altogether. Our home has a pergola that covers the entire west side. Its shade changed the west rooms from being unbearably hot in summer to being our coolest and most pleasant rooms, without shading the roof solar array.

The Importance of Delta T

The Greek symbol delta (Δ) denotes difference. ΔT means difference in temperature. The greater the ΔT, the greater the heat gain or loss. This is an important energy conservation and efficiency concept. Any

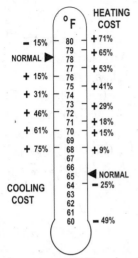

Figure 3.2. The relationship of temperature to heating and cooling costs. If your normal thermostat setting is 65°F/18°C, it costs 41% more to heat your home to 75°F/24°C no matter what fuel you use. Air conditioning to 72°/22°C from the normal 78°F/25°C will cost 46% more.

temperature difference that you can sense with your hand, about 20°F/ 11°C, indicates that a significant amount of energy is being consumed. The greater the temperature change, the more energy consumed, and the more it costs to make or prevent the change.

For example, say it is 70°F/20°C outside and 70°F/20°C inside your home. There is no difference in temperature, your heater is off, your air conditioning is off, and you are consuming no energy. When it is 50°F/10°C outside, you will have to consume energy to make up the 20 degrees difference or ΔT. If it is 0°F/-18°C outside, you must use much more energy to maintain your home at 70°F because the ΔT is 70. Each additional degree that you heat or cool your home translates directly into dollars. There is a 6% difference in cost between heating your home to 68°F as compared to 70°F. On the other end of the thermometer, there can be as much as a 15% increase in cost for two degrees of air conditioning.

One frequently asked question is, "Can I use PV to run an air conditioner?" The answer is yes, but there are a few things to take into consideration. One ton of air conditioning is equal to 12,000 BTUs or 2 kW. This is the peak AC output of twenty-five 100-watt solar modules on a sunny summer day in Houston. In an off-grid home, an air conditioner would require a larger 240-VAC inverter. If you must air condition, consider either a gas air conditioner or an evaporative air conditioner (swamp cooler). In an on-grid home your air conditioner will run the same way it always has—through your existing utility service panel.

Hot Water

The same rules apply whether you are heating water or heating air. Use gas and insulate both your water heater and pipes. An electric water heater can cost over $700 per year to operate. The equivalent gas heater (same BTUs) will cost $275, and a high-efficiency gas heater will cost $210. There is a solar option. Solar hot water systems are usually backed up by conventional water heaters to boost the temperature for household use or on cloudy days. Solar is ideal for heating the pool, where the water needs to be warm but not hot. Tankless water heaters are another possibility.

Large Appliances

The Cost of Cooking

This table from the *Consumer Guide to Home Energy Savings* compares the energy consumption of several cooking methods.

Table 3.3. Costs to Cook a Casserole

Appliance	Temperature	Time	Energy	Cost*
Electric oven	350	1 hour	2.0 kWh	26¢
Electric convection oven	325	45 minutes	1.39 kWh	18¢
Gas oven	350	1 hour ·	0.112 therm	07¢
Electric frying pan	420	1 hour	0.9 KWh	11¢
Toaster oven	425	50 minutes	0.95 kWh	13¢
Electric crockpot	200	7 hours	0.7 kWh	09¢
Microwave oven	"High"	15 minutes	0.36 kWh	05¢

*Based on gas costing 60¢/therm and electricity 13¢/kWh.
From Jennifer Thorne Amann, Alex Wilson, and Katie Ackerly, *Consumer Guide to Home Energy Savings*, 9th ed. (Gabriola Island, BC, Canada: New Society Publishing, 2007).

Although about 58% of American households cook with electricity, gas cooking continues to be popular. A gas stove costs less than half as much to operate as an electric, provided it is equipped with an electronic ignition instead of a pilot light. Electronic pilotless ignitions reduce gas use by about 30% compared to constantly burning pilots. They are more convenient because they are automatic.

There are small steps you can take to reduce cooking energy costs. When preparing foods like pasta or vegetables, heat the water to boiling with the lid on even if you remove the lid for cooking. Don't open the oven door when roasting or baking: Every time you do, the temperature drops 25°F. Invest in an external probe thermometer and a timer and you won't need to open the door. When designing your kitchen, don't position the ovens or stove next to the refrigerator and allow for adequate insulation in your floor plan.

When it comes to the refrigerator—shut the door. The refrigerator can be the largest single electricity consumer in an energy-efficient home, accounting for up to one-third of total consumption. The electricity consumed by an older refrigerator could power three newer models of the same capacity. Follow the manufacturer's recommendations for replacing and maintaining door gaskets and for cleaning the coils. Test the gasket by closing the door on a dollar bill. It should be held firmly in place.

After buying a new refrigerator for the kitchen, some people put the old one in the garage to keep beer and soda cold and to store Thanksgiving leftovers. Before you do this, remember that a new refrigerator is three times as efficient as the old one of the same size. This surprised one neighbor who kept a garage refrigerator: "Are you telling me I'm paying $50 a month so we can have more cold drinks?"

The kitchen is the heart of the home. It is also the room that can consume the most energy. When purchasing new appliances, apply the methods businesses use—payback analysis and life-cycle cost analysis. Treat these purchases like a business purchase.

Payback & Life-Cycle Cost Analyses
These analyses are based on simple calculations.

Payback Analysis
To determine simple payback, divide the installed cost by the annual energy savings.

For example, you replace four 100-watt, 1710-lumen incandescent light bulbs with four equivalent 28-watt, 1720-lumen fluorescent light bulbs for $10 each. The simple payback period is:

100-W bulb – 28-W fluorescent = 72 W

72 W × 4 bulbs = 288 watts savings

288 W × 5 hours use per night = 1,440 watt-hours or 1.44 kWh/day

1.44 kWh × 365 days = 525.6 kWh/yr savings
525.6 kWh × 9¢/kWh = $47.30 saved in one year
Cost of the new bulbs ÷ annual savings = ($10 × 4) ÷ $47.30
= $40 ÷ $47.30 = 0.85 years simple payback

Energy-efficient lights are initially more expensive than the bulbs they replace, but they pay for themselves in energy savings in a little over 10 months. If you subtract the replacement cost of the old light bulbs and factor in that the new lights last over twelve times longer, new fluorescent lights are definitely a good idea. Even if 100-watt light bulbs were free, using fluorescents would still save you a considerable amount:

5.5 years (expected life of a bulb) – 0.85 year (simple payback period)
= 4.65 years
4.65 years × 525.6 kWh/yr × 9¢/kWh = $219.96 savings

over the expected lifetime of new fluorescent lights.

If four little light bulbs can save you over $200, think of what you will save with more efficient large appliances!

Life-Cycle Cost Analysis

Another way to determine if a new appliance is a wise purchase is by "life-cycle cost analysis." The formula to calculate this is:

Life-Cycle Cost = Purchase Price
+ (Annual Energy Cost × Equipment Lifetime × Discount Factor)
$$LCC = p + (c \times l \times d)$$

Purchase price (p) should include sales tax, delivery, and installation charges. To determine the annual energy cost, use the yellow Energy Guide label as an estimate. You can easily calculate the appliance's actual operating expense—just multiply its energy consumption rate (in kilowatts) times the daily operating hours. Multiply this by 365 days, and then multiply the result by your electric utility rate.

Annual Energy Cost = Energy Consumption Rate (kW)
× Daily Operating Hours × 365 days/year × Utility Rate ($/kWh)

Equipment lifetime (l) is determined by the appliance quality, power quality, and how the appliance is operated and maintained. Table 3.4 lists average lifetimes and discount factors.

Table 3.4. Appliance Lifetime and Discount Factors

Appliance	Lifetime (years)	Discount Factor
Dishwasher	12	0.84
Water heater	13	0.83
Room air conditioner	15	0.81
Clothes washer	18	0.78
Refrigerator & freezer	20	0.76

Discount factor (d) is a simple way to express the time value of money or "real discount rate." To calculate the real discount rate, start with the nominal discount rate, which is the interest rate applied to future payments, and then discount to present value.

Real Discount Rate = [(1 + Nominal Discount Rate)
÷ (1 + Inflation Rate)] - 1
Discount Factor = 1 ÷ (1 + Real Discount Rate)^{year}

Table 3.4 and the example use a 5% nominal discount rate and energy prices increases that are 1% above inflation and an inflation rate of 3%. Compare the life-cycle costs of two refrigerators that are equivalent in capacity and features. The price of Refrigerator A is $750 and it costs $75 per year to operate at 9¢ per kWh. The price of Refrigerator B is $100 less and it consumes an estimated $24 more in electricity per year.

Table 3.5. Life-Cycle Costs of Equivalent Refrigerators

Refrigerator	Price	+	(Energy Cost	×	Estimated Lifetime	×	Discount Factor)	=	Life-Cycle Cost
A	$750.00	+	($75.00	×	20	×	0.69)	=	$1,785
B	$650.00	+	($99.00	×	20	×	0.69)	=	$2,016

Although Refrigerator A is more expensive initially, Table 3.5 shows that over its lifetime the energy it saves makes up for its higher upfront cost. If you live on-grid, the energy-efficient refrigerator consumes less utility electricity, shaving dollars and cents off your electric bill. Using less energy also means reducing air pollution and conserving natural resources. If you are off-grid, such energy efficiency allows you to install a smaller PV system.

Table 3.6. Life-Cycle Cost Analysis for Two Refrigerators

Assumptions & Factors	
Electricity cost (US$ per kWh)	9¢
Annual energy increase rate	1.0%
Inflation rate	2.0%
Nominal discount rate*	5.0%
Real discount rate **	1.942%
Appliance A	
Appliance price	$750
Energy consumption (W)	190
Daily operation (hr)	12
Annual operation (hr)	4,380
Annual energy consumption (kWh)	833
Annual energy cost (US$)	$75
Appliance B	
Appliance price	$650
Energy consumption (W)	250
Daily operation (hr)	12
Annual operation (hr)	4,380
Annual energy consumption (kWh)	1,095
Annual energy cost (US$)	$99

Year	Discount Factor***	Appliance A		Appliance B		B – A
		Lifetime Energy Cost	Life-Cycle Cost	Lifetime Energy Cost	Life-Cycle Cost	
1	0.99	$75	$825	$99	$749	-$76
2	0.97	$146	$896	$193	$843	-$53
3	0.95	$214	$964	$283	$933	-$31
4	0.93	$279	$1,029	$369	$1,019	-$10
5	0.91	$342	$1,092	$451	$1,101	$9
6	0.90	$405	$1,155	$535	$1,185	$30
7	0.88	$462	$1,212	$610	$1,260	$48
8	0.86	$516	$1,266	$682	$1,332	$66
9	0.85	$574	$1,324	$758	$1,408	$84
10	0.83	$623	$1,373	$822	$1,472	$99
11	0.81	$669	$1,419	$883	$1,533	$114
12	0.80	$720	$1,470	$951	$1,601	$131
13	0.78	$761	$1,511	$1,004	$1,654	$143
14	0.77	$809	$1,559	$1,068	$1,718	$159
15	0.75	$844	$1,594	$1,114	$1,764	$170
16	0.74	$888	$1,638	$1,173	$1,823	$185
17	0.73	$931	$1,681	$1,229	$1,879	$198
18	0.71	$959	$1,709	$1,266	$1,916	$207
19	0.70	$998	$1,748	$1,317	$1,967	$219
20	0.69	$1,035	$1,785	$1,367	$2,017	$232

* Interest rate applied to future payments.

** Includes inflation to discount future dollars to present value. RDR = [(1 + nominal discount rate) / (1 + (energy rate increase + inflation rate)] -1

*** Expresses the time value of money or the real discount rate, DF = 1 / (1 + real discount rate)^{year}

The EnergyGuide Label

The EnergyGuide label helps you compare the efficiency or annual energy use of competing brands and similar models of clothes washers, dishwashers, refrigerator/freezers, air conditioners, water heaters, pool heaters, and central home heating and cooling equipment. If you don't see an EnergyGuide label, ask a salesperson for the information.

The boxed number at the bottom of the EnergyGuide label gives the appliance's estimated yearly operating cost ($53 in the refrigerator example label below), which is based on the national average fuel cost when the appliance was manufactured. Your operating cost is based on your utility rates. If you have a tiered rate structure, the cost to operate the appliance is determined by your total electrical consumption. If you conserve energy in the rest of the house, the baseline rate will determine your annual cost. Baseline consumption is the lowest billing rate. The kilowatt-hours allowed at baseline are determined regionally by the utility and can vary seasonally. If your energy bills routinely go into the higher tiered rates, you could be paying over 30¢/kWh to operate that appliance.

Air conditioner energy guides have a Seasonal Efficiency Energy Rating (SEER). The national standard for energy-efficient air condition-

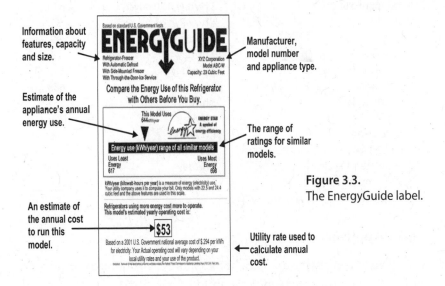

Figure 3.3.
The EnergyGuide label.

ers is currently a SEER of 12. Raising that rating to 13 would save U.S. consumers a total of one billion dollars annually.

Your utility may offer rebates or other incentives to encourage customers to purchase energy-efficient appliances or to turn in your older one. It is worth investigating.

Table 3.7. How Much Energy Does It Use?

Appliance	Time in use	kWh / year
Aquarium	24 hr/day	700
Central air conditioning	12 hr/day, 120 days/yr	2,700–3,780
Clock	24 hr/day	36
Clock radio	24 hr/day	44
Clothes washer (not including hot water)	2 hr/wk	31
Coffee maker	30 min/day	128
Computer	4 hr/day	520
Dehumidifier	12 hr/day	700
Dishwasher (not including hot water)	1 hr/day	432
DVD player	4 hr/day	30
Electric blanket	8 hr/day, 120 days/yr	175
Fan (furnace)	12 hr/day, 120 days/yr	432
Fan (whole house)	4hr/day, 120 days/yr	270
Fan (window)	4 hr/day, 180 days/yr	144
Hair dryer	15 min/day	100
Heater (portable)	3 hr/day, 120 days/yr	540
Iron	1 hr/wk	52
Microwave oven	2 hours/week	89
Radio (stereo)	2 hr/day	73
Refrigerator (frost-free 16 cubic feet)	24 hr/day	642
Refrigerator (frost-free 18 cubic feet)	24 hr/day	683
Television (color)	3 hr/day	264
Toaster oven	1 hr/day	73
Vacuum cleaner	1 hr/wk	38
Water bed (no cover)	12 hr/day, 180 days/yr	620
Water heater (40-gallon electric)	2 hr/day	2,190
Water pump (deep well)	2 hr/day	730

Source: EREN, U.S. Department of Energy

Small Appliances

Miscellaneous energy use in U.S. homes more than doubled between 1976 and 1995. It is projected to increase an additional 50% by 2010, accounting for almost all forecasted growth in residential electricity use. This growth is equivalent to the output of about fifteen 1,000-MW power plants.

One year we tried to list all the small kitchen appliances in the holiday catalogs. We gave up when we reached 100. How many do you use? How many do you need?

Phantom Loads

Many small appliances constantly "leak electricity." This is caused by "phantom loads," the electricity equipment consumes when off or in standby mode. The average U.S. home consumes 450 kilowatt-hours every year in phantom loads. This is the same amount of energy that a 12-cubic foot refrigerator uses in a year.

Low-voltage transformers—the little black cubes on the power cords of phones, answering machines, and other small electronics—consume around 5 watts all the time. Some TVs, stereos, and DVD players use up to 20 watts continuously. Fax machines are always on, but rarely used.

Table 3.8. Phantom Loads

Appliance	Phantom Watts	% of total energy use
Cable boxes	11.0	83%
Color TVs	4.2	21%
Compact audio systems	9.0	93%
Cordless hand vacuums	1.7	High
Cordless phones	2.3	High
Digital satellite systems	13.8	83%
Doorbells	2.2	High
DVD players	4.0	75%
Garage door openers	6.0	High
Hair/beard trimmers	0.9	High
Hand-held massagers	2.0	High
Microwaves	3.1	22%
Portable stereos (boom boxes)	2.2	88%
Projection TVs	2.0	6%
Rack audio systems	4.0	61%
Security systems	12.0	33%
Shavers, men's and women's	2.5	High
Telephone answering devices	3.2	High
Toothbrushes	2.2	High
TV/VCR/DVD combinations	9.1	45%
Video games	2.0	70%

Relatively small phantom loads can make a big difference. In an off-grid PV system, inverters with an energy-saving, no-load, low-power mode can be kicked into full power mode by as few as 7 watts. A clock or other seemingly insignificant electrical load can force your inverter on, causing it to consume the daily output of one or two 100-watt solar modules. The effect on the grid is considerable as well. It takes five large utility plants to power the estimated 5 billion watts of phantom load in the U.S.

Avoid buying devices that generate phantom loads. If you must use such a device, plug it into a power strip that you can easily turn off. Or just unplug them after use.

Two Examples of Conservation Design

Standby PV & Conservation

Since 1979, Bill and Jackie Perleberg of Golden, Colorado, have been living in a grid-connected home that is almost completely energy independent even though it is equipped with a surprisingly small solar electric system. Like other PV users they spent a lot of time thinking about energy and conservation before putting thought into action.

Figure 3.4. Bill and Jackie Perleberg's home located in the foothills above Golden, Colorado. Two 8-module passive solar trackers and a windcharger (not in photo) provide the power for this comfortable home.

The Perlebergs tried heating with wood, but decided it was too much work. Instead, they heat with waste oil, an efficient, smokeless and odorless high BTU "throw-away" fuel. Their existing electric furnace, electric fan, and duct system are used to circulate the waste oil heat. They don't like using this petroleum product, but they will have a steady supply as long as petroleum is used to lubricate engines. Different regions have different "waste fuels" available: corn cobs, sawdust, cow chips, and methane.

For heat and aesthetics, Bill and Jackie burn salvaged wood pallets in their fireplace. The seasoned pallets are free for the hauling, easy to handle, and require less processing than regular firewood. Bill fishes the nails out of the ashes with a magnet and sells them to a recycler.

They heat water with solar energy and heat diverted from the waste oil heater flue through a used commercial water heater tank with its burners removed. This modified tank serves as a preheater. Solar-heated water runs through coils on the outside of the tank and then flows to the conventional electric water heater. Excess electricity from their 560-watt, tracker-mounted PV array and 200-watt windcharger is diverted to three 12-volt, 500-watt water heater elements.

Their electric water tank has extra insulation and downsized heater elements. Bill replaced the two original 5.5-kW elements, first with two 2.5-kW units, then with 1-kW elements, and finally with 750-watt elements. Recovery time is longer, but they are never short of hot water and their utility electric demand rate is low. Instead of washing five loads of laundry in one day, they spread the chore over the week. Doing laundry daily fits their lifestyle since both work away from home. In Colorado, it is not always possible to dry clothes outdoors so they got an old gas dryer, removed the burners and ran a 5-inch diameter duct from the waste oil heater to the dryer for free heat. The dryer motor uses only 550 watts.

Bill disconnected the dishwasher's high current-drawing electric elements, and monitors water temperature with a gauge on the water heater tank to safely wash dishes at 140°F/60°C. The dishwasher motor draws only 700 watts.

They pay for utility electricity on the demand rate so heating element modifications keep the demand charge down. The demand charge is set by the highest 15-minute consumption each month. Because they control total kilowatts consumed during any 15-minute period, they pay the lowest demand rate. When shopping for appliances, they bring along a hand-held watt meter to measure consumption, which fascinates salespeople. Managing loads also keeps them within their 2,500-watt inverter rating.

Their Whirlpool 17-cubic foot refrigerator draws 200 watts. Their well pump is a 0.75-hp, 240-volt AC unit that draws about 1,000 watts. The combined pump and refrigerator loads determine the 1-kW consumption for their monthly demand rate. The remaining appliances draw less than 1 kW each and seldom are used at the same time.

After the Perlebergs installed their PV and wind energy system and reduced their energy consumption, the utility company suspected they might be tampering with the kilowatt-hour meter. The utility changed the meter several times during their first year with PV and wind, yet the Perlebergs' demand continued to drop. One day, the district manager and

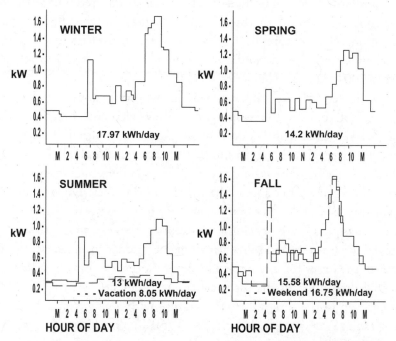

Figure 3.5. Seasonal characteristics of diversified daily electrical loads. These four charts show the pattern of electrical use for a family of four living in an all-electric home using 5,540 kWh per year. Note the use patterns based on time of day and compare them to your own use patterns. The average energy-conservative home will use less than half the total energy, but the time of use should remain similar.

a technician came out unannounced, pulled the meter from its socket, ran load tests, and checked the meter for accuracy and magnetism. Bill and Jackie were told that they couldn't heat water for what they were paying, much less heat their entire house and dry clothes. So Bill gave them a tour of their energy-efficient home. After that, the utility left them alone.

The Perlebergs' PV and wind system produces 120 VAC and 12 and 24 volts DC. The house was wired for AC and DC because low-voltage appliances were part of their original plan. The windcharger tower and PV trackers are earth-grounded for safety and lightning protection. Bill likes the extra PV electricity tracking provides, even though he says the tracker is slow in cold weather and does not tip back to the east until 10 a.m.

Their battery area is approximately 60°F/15.5°C with good ventilation. Still, battery compartment doors are left open during equalization charging to vent excess gas. Their first batteries were bought used and partially sulfated so Bill had to increase the charge cut-out from 14.5 to 14.8 volts DC to compensate for their lower efficiency. They bought Trojan T-105 batteries for their second set.

The Perlebergs' system produces 60 kWh per month.

16 each 35-watt modules = 560 peak watts
× 4 average annual peak hours of sun = 2,240 watt-hours
× 30% tracker increase = 2,912 watt-hours per day
at 70% combined inverter/battery efficiency
= 2,038.4 × 30 days = 61.1 kWh per month

Figure 3.6. Part of the battery bank used to store electricity for an energy-efficient home. The Perlebergs' home has four battery banks, each with ten 6-volt, 210-amp-hour golfcart batteries that can be switched from 12 volts to 24 volts. (Appropriate Systems, Inc.)

Their refrigerator uses 85 kilowatt-hours per month; the well pump, 8.33 kWh; and the hot water heater, 22.8 kWh. Bill and Jackie save over $1,200 per year through conservation alone.

For the past 20-plus years, their PV and wind system has been relatively trouble-free. Instead of upgrading their inverter, they put the well pump back on grid power. Golden has a lot of power outages, but Bill and Jackie don't notice because their PV power plant works just fine.

A Low-Consumption PV Home

The following is adapted with permission from Home Energy *magazine. Since 1987* Home Energy *(2124 Kittredge Street #95, Berkeley, CA 94704; 510.524.5405; www.homeenergy.org) has provided excellent information on all aspects of home energy conservation, efficiency, and use.*

Bob Hammond designed and built an off-grid PV home near Prescott, Arizona, that uses less than 2,700 kilowatt-hours per year. Through energy efficiency, conservation, and load management, Bob was able to live comfortably on a fraction of the 9,300 kWh per year consumed by neighboring homes. His efforts were rewarded with a beautiful home and an affordable PV system.

Completed in 1989, the 2,600-square foot, three-level house is wood-framed with a stucco exterior. The lower-level basement is earth-sheltered on the west and north sides. The house is about 1,200 feet from the Arizona Public Service grid, but a connection would have cost about $5,700. Bob opted not to hook up to the utility.

The 2,800-watt ground-mounted solar arrays feed a 24-volt DC bus. Three arrays feed into three charge controllers to charge 16 Trojan T-105 flooded batteries. The battery bank is rated 900 amp hours (22 kilowatt- hours). A 4-kW Trace inverter provides 120 volts AC to run household equipment and appliances. He uses 12- and 24-volt DC appliances so he installed a Vanner equalizer to balance battery cells. The PV system was the sole power source during home construction.

Loads grew from about 850 kWk per year in the early 1990s to 2,700 kWh, averaging a fairly constant 7 to 8 kWh per day regardless of the season. The PV system has grown to meet these needs.

A propane fuel generator is used approximately 25 hours per year. About 15 hours of runtime are used to charge the batteries during extended cloudy weather and the remaining 10 hours are used to exercise the generator once a month.

During over 20 years of operation, only two unscheduled outages occurred with total loss of power limited to about two hours. Both outages were caused by charge-controller failure that caused the batteries to overcharge, which, in turn, caused the inverter to shut down.

"Living in a stand-alone home is easy, as long as it's designed right," Hammond says. Located on a hilltop 20 miles north of Prescott, which has a fairly mild climate, south-facing windows provide sufficient heat during the winter. The 2"×6" stud exterior walls are insulated to R-26, the ceiling to R-40, and foundation footing, basement slab, and basement walls to R-5. Very little supplemental heating and no air-conditioned cooling are required. A propane stove is used occasionally on cold winter nights.

Cathedral ceilings and open architecture allow good air circulation and even heat distribution. Roof overhangs and decks block the summer sun to prevent overheating. Increased ceiling insulation, with a double radiant barrier, and passive ventilation also keep the house cool.

Figure 3.7. Bob Hammond's off-grid home near Prescott, Arizona, gets all of its power from the sun with a 1,240-watt solar array and 20 deep-cycle batteries.

Bob says the home's design allows temperatures on the second and third floors to moderate quickly according to the outdoor temperatures. It stays cool, for example, even on the hottest days. "I vent the house at night when it's cool outside, and close it up during the day," Bob says. "It stays very comfortable." The concrete thermal mass and earth-sheltering keep the basement 60 to 70°F/15.5 to 21°C year-round. Ample windows reduce the need for artificial lighting. At night, compact fluorescent lamps are used. Lighting is less than 4% of the total electric load.

Figure 3.8.
The Hammonds' home inverter (top) and battery bank (left).

The refrigerator is usually the biggest energy load in an off-grid PV home so Bob selected a 16-cubic foot, 24-volt DC Sun Frost. This energy-efficient refrigerator cost over $2,400, but only consumes 257 kWh per year, much less than conventional units. Bob used about $100 worth of propane per year for his oven, range, and water heater. The home does not have a washing machine or clothes dryer, which would increase electric consumption about 20%.

Water conservation in Prescott's arid climate is important. The home is not connected to a public water system and does not have a well. At first, household water was trucked in and stored in a 1,200-gallon cistern. Then Bob installed a rainwater roof collection system for everything except drinking water. One inch of rain provides over 700 gallons. The 0.5-hp pressure pump on the cistern consumes 47 kWh per year.

The 1.4 gallon per flush toilet is limited to three flushes per person per day and is not flushed after every use. A 3-gallon per minute low-flow showerhead and turning the water off while soaping cut consumption to 3 gallons per shower. Water consumption is 20 gallons per person per day compared to 80 to 100 gallons in a typical home.

Long pipelines supply water to the fixtures in this three-story house so a lot of water could be wasted before it comes out hot at the tap. To eliminate water waste, Bob installed a closed-loop circulating system that thermal siphons so hot water is immediately available. All hot water pipes are heavily insulated with rigid foam and fiberglass.

The home has all the usual modern appliances, entertainment equipment, office equipment, and power tools. The inverter delivers 4,000 watts continuous power. Load management has become a way of life and Bob says he never went without electricity.

Base and phantom loads contribute significantly to the energy consumption of an off-grid PV home. Remote-controlled equipment, such as televisions and video recorders, are always "on" even when turned "off." An easy solution is to turn them off using a power strip.

The batteries are usually fully charged by 10 a.m. Electricity is used directly from the PV array to power energy-intensive equipment and minimize battery cycling.

"Adapting to life in a stand-alone PV home was easy," according to Bob, "but private road maintenance, the high cost and limited choices

of food and other products in local stores, and limited local employment
opportunities did require major economic adjustments. People with or
without PV can benefit from energy-efficient design, conservation and
load management."

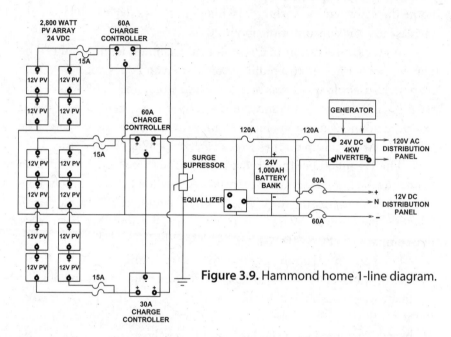

Figure 3.9. Hammond home 1-line diagram.

The Perlebergs and Bob Hammond represent only two of the millions
of PV households. Each is unique, based on where and how the owners
live. All practice energy conservation and energy efficiency. They act
on their desire to express their independence while making the world a
better place to live.

Chapter 4
LIVING WITH PV

Joel's PV-Powered Country Home

In 1972 I moved from California to the Ozark mountains of northwest Arkansas to live simply and become as self-sufficient as practical. I bought mountain property a mile from the nearest power line and built a pole-frame cabin. Sun shining through the cabin's large south-facing windows provided 50% of daytime heat in winter and wood provided the rest. The kitchen was equipped with a propane stove and refrigerator.

At first, I used kerosene lamps for light, but they were not very bright. I put metal pie plates behind the glass chimneys of the kerosene lamps to help reflect light and increase brightness, but soot and the possibility of fire still made kerosene lamps unacceptable. Coleman lamps were brighter, but not very pleasing. My cabin's first electric devices were two 16-watt, 12-volt DC fluorescent lamps powered by electricity tapped from my truck battery, which was kept charged by daily trips to and from work. On weekends, power consumption would discharge the battery so deeply that I had to roll the truck down my mountain road to get its engine started.

I had a well drilled but didn't strike water. If I had found water, I would have had to solve the problem of reliably lifting it up a 6-inch diameter, 400-foot deep hole. Instead of gambling on another drilling attempt, I installed a 500-gallon underground cistern to catch and hold rainwater. For pressure I installed a 4.6-ampere, 12-volt DC recreational vehicle water pump under my cabin in the crawl space and connected it to a pipe tapped into the cistern. Next I installed a 30-gallon propane water heater for hot running water in the kitchen and shower. A little electricity made a big difference in my comfort level.

I kept the propane water heater and stove pilot lights off to conserve energy and save money. When I wanted to cook or heat water, I started the flame with a spark igniter. This simple conservation measure kept my propane consumption to 130 gallons per year for cooking, heating water, and refrigeration.

I lived on a very modest energy budget while building and improving my home, barn, fields, fences, and pond. I worked my garden by hand and occasionally used a horse or secondhand gasoline rototiller. Wood, propane, sunlight, and electricity from a truck-charged battery fulfilled my other energy needs.

Figure 4.1. Some remote PV installations are really remote. A llama transports a solar electric system to a mountain site. (Photo: D. Stewart)

By 1977 I was ready to set up my energy system. Wind power seemed the logical choice for my mountaintop location. From a neighbor I got a windcharger, made from an automobile alternator, with three 6-foot birch plywood blades. Gusting wind smashed its free-spinning blades into the tower before I could complete the installation. My second windcharger had three 8-foot blades and was much sturdier. However, ice covered its blades and jammed the brake shortly after it was installed, and it was almost destroyed too.

I needed a power system that could withstand extreme wind, ice, and snow. I knew solar cells were reliably powering satellites in space and that terrestrial PV modules were being used to power mountaintop radio stations. With no moving parts and nothing to wear out, I decided PV was the ideal choice.

In 1978 I bought my first PV modules and stopped using my truck to charge the battery. When the first snowstorm arrived that winter, I was comfortable in my cabin watching the news on my solar-powered TV. The lead story was about the utility power outage affecting tens of thousands of homes throughout the region. Sitting snug and warm in my electrified cabin convinced me of PV's reliability.

In the 1970s PV was expensive, and my income was very modest, so I started with three used 10-watt solar modules designed to be deck-mounted on a boat. I fastened the modules to a piece of plywood that was mounted on the roof of my cabin and continued to use an automobile battery for energy storage. My PV system had no charge controller. Wiring was secondhand Romex 12/2 run the shortest possible distance or doubled up (paralleled) to minimize resistance. Inside the cabin I used standard outlets and plugs marked with a spot of red paint to indicate polarity. The system was primitive—and definitely not code-compliant—but it worked.

June 1979 PV production matched my modest 3-kilowatt-hour consumption. However, winter energy consumption was four times production because I needed more hours of lighting during the long winter nights. In addition, like most of the U.S., Arkansas has cloud cover at least half the winter. To make up for the difference between PV production and energy consumption, I expanded my battery bank and built a back-up generator.

Figure 4.2. Joel's Pettigrew, Arkansas, cabin. This small PV system used only four 33-watt solar modules and a 420-ampere-hour battery bank located in the crawl space to power lights, TV, stereo, radio, fans, computer, appliances, and tools.

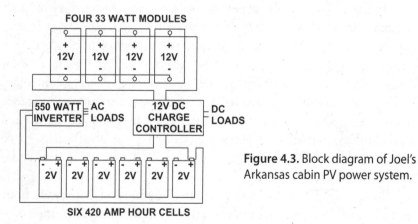

Figure 4.3. Block diagram of Joel's Arkansas cabin PV power system.

I bought used deep-cycle, flooded, lead-acid battery cells at salvage lead prices. The cells had been used by a telephone company for seven years but were still in good condition. I connected the six 130-pound, 2-volt cells in series for 4 kilowatt hours of storage (420 ampere-hours × 12 volts × 0.8 depth of discharge).

A 3.5-horsepower, horizontal-shaft gasoline engine turning a truck generator provided back-up power for battery charging. I used a flexible coupling to connect the engine and generator shafts, and put the homemade DC generator in a box 100 feet from my cabin to minimize noise and fumes. Power from the generator was fed through positive and negative cables to the batteries in the cabin. One-inch diameter insulated aluminum cables, purchased by the pound from a salvage yard, were laid unburied on the ground. The generator regulator was set to the maximum 30-ampere charge rate. Wire loss was 1.6 volts. The charger would run for 6 hours on 0.75 gallons of gasoline for 180 ampere-hours of charge. In 18 months I used only 8 gallons of gasoline.

Life was comfortable, but I wanted more PV to wean myself from fossil fuels. The more modules you buy the less each costs, so friends and I pooled our money to purchase several dozen. That purchase was the first of many quantity purchases, or PV bulk buys as they came to be known.

I bought four 33-watt ARCO solar modules to add to my array, quadrupling my solar production and eliminating the need for a back-up generator. I also bought a charge controller. The additional PV and controller were major expenses. The 40% federal tax credit for solar equipment was factored into the purchase.

Solar tax credits have always been controversial. Some people believe solar technologies need tax breaks and subsidies to compete. From 1976 to 1985, hundreds of thousands of people, including me, took $375 million in tax credits to buy solar thermal, PV, and conservation-related equipment instead of sending their money to Washington. Some people oppose government intervention and think the free marketplace should determine which technologies succeed or fail. However, all of the polluting energy production technologies are subsidized or depend heavily on tax breaks and other concessions. Taxpayers are paying twice: once to subsidize polluting energy production technologies, and then again, paying dearly, to clean up the resulting pollution.

In 1980, I moved from my mountain home to a nearby valley. I built another cabin and reinstalled my PV system. Moving it was easy. I disassembled the array and carefully wrapped the modules in blankets to protect them in transit. The 800-pound battery bank was secured with ropes in the truck bed. The whole system was moved in one trip.

The new cabin's metal roof was tilted 45 degrees (12:12 pitch) facing true south. My four 33-watt PV modules were fastened to the roof on a simple wood stand-off mount. The batteries were installed in a box in the cabin's crawl space. The sides and top of the battery box were insulated with 4-inch foamboard. Wire from the roof array ran through the cabin to a wall-mounted charge controller and meters and then to the batteries under the cabin. The 12-volt DC lights, water pump, and electrical outlets were wired as before.

My original 30-watt array was mounted on the ground and was used to charge a 12-volt battery that was equipped with a carrying strap. The battery was occasionally used to power a DC recreational vehicle pump to draw water from a nearby shallow well.

Newspaper stories and magazine articles were written about my simple cabin PV system. People wrote to me asking about PV and where they could buy modules. I began organizing more quantity purchases or bulk buys, helping thousands of people throughout the U.S. start using photovoltaics. And I began writing the first edition of *The Solar Electric Home*.

I added four more 33-watt solar modules and another set of used batteries to my system. The new modules were installed on a pole-mounted Zomeworks tracker. I also bought a 550-watt, 12-volt DC to 120-volt AC inverter to power a ⅜" drill, saber saw, sewing machine, vacuum cleaner, and computer.

Back to the City

Living in the mountains was a dream fulfilled. I had homesteaded two properties, built homes and barns, plowed and planted gardens and orchards with horses and mules, and eventually grew most of my food and animal feed. I was energy independent and unplugged from the grid, but I could not unplug from the world. I was painfully aware of the smog that stretched hundreds of miles from the industrial region of the Midwest to my home deep in the Ozark National Forest. No place could remain a refuge from pollution and resource depletion for long. Using PV was my small step in the right direction, but I felt that more should be done.

I sought the advice of William Lamb, my PV supplier in Los Angeles. Bill was a successful businessman who had returned from retirement to become the world's first PV distributor. He believed that PV could help solve the world's energy and pollution problems. Bill offered me a job, and, after ten years of off-grid rural living, I returned to the city to work full-time in the PV industry. Soon after, Fran and I met.

I designed and sold PV systems from my PV-powered office in the William Lamb Company building in Los Angeles. I rewired the office for 12 volts DC and retrofitted the AC fluorescent ceiling fixtures with DC ballasts. I had a DC stereo sound system and used a 12-volt black-and-white TV for my computer monitor. The computer was powered with a 550-watt inverter. I used a Specialty Concepts (SCI) charge controller to regulate PV power into four Trojan T-105 deep-cycle golfcart batteries for 4 kilowatt-hours of storage.

My office solar array included a mix of four 33-watt modules brought from Arkansas, three 28-watt modules, and one 35-watt module. The 8-module array was fastened to a fixed roof mount tilted 45 degrees. PV production averaged one kilowatt-hour per day. The battery bank should have been twice the size and capacity, but it was limited by the space available. As a result, the battery was deeply discharged daily in winter.

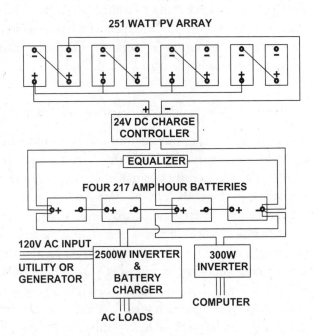

Figure 4.4. Block diagram of Joel's California office PV power system.

In 1983, when Heart Interface made the first inverters designed specifically for PV systems, I switched my office back to AC. The inverter consumed very little power while idling under no load and was over 85% efficient. Before Heart, 70% efficiency inverters with over 24 watts idling consumption were typical. I installed a 2,500-watt, 24-volt DC to 120-volt AC Heart inverter so I could power many different loads at the same time. I rewired the solar array and battery bank for 24 volts, but I did not add modules or batteries. The charge controller was upgraded to 24 volts.

The original AC fluorescent lamp ballasts were reinstalled and the 120-volt AC wall outlets were wired into the system. I used a separate 300-watt Heart inverter powered through a 24- to 12-volt Vanner equalizer as my computer's uninterruptible power supply (UPS).

During the 1980s, PV systems evolved from simple DC power sources for remote locations away from power lines to urban alternative energy systems that were cleaner and more reliable than grid power. My PV array provided enough power for my office except in winter or when I worked overtime. When consumption exceeded production, the inverter would switch automatically to utility power and then back to solar. I could have added more PV modules and batteries to make my office system 100% solar because PV is easily expandable. Instead Fran and I decided to power our home with PV.

Our PV-Powered City Home—Part 1

We were ready for urban PV, but custom inverters and utility interface equipment were expensive. A 2-kW utility-connected system cost $40,000 in 1986, and utilities did not make it easy for homeowners to connect PV systems to the grid. A home PV system required the same interconnection contracts, interconnect hardware, and costly insurance as commercial cogeneration power plants. The few people willing to pay the price were often treated in a hostile manner by their local utility, which felt threatened by customers producing all or part of their own power.

In 1992 the Clinton administration brought to Washington a renewed focus on environmental issues. Utility company interest in renewable energy grew with increased federal subsidies for grid-connected PV. Some utilities became PV-friendly, while others grew less hostile.

The real breakthrough for utility PV came when Trace Engineering, an offshoot of Heart Interface, introduced an efficient and reasonably priced inverter that connected safely to the grid, delivered grid quality power, and could be used with or without batteries.

Grid-connected PV made economic sense to us when PV net metering became law in California in 1996. Under net metering, interconnection was simple and we would get full value for the PV energy we produced, paying only the net difference between utility energy and the energy produced by our PV system. In 1998 another windfall reduced our PV system cost. Part of electric utility restructuring in California included a $3 per watt incentive to help buy-down the cost of installing PV. It was definitely time to again install PV on our home.

In 1984 Fran and I bought our house with energy conservation and PV in mind. It is a typical 1947 California bungalow with a clear south orientation. The neighborhood is pleasant, with stores and services within easy walking distance. The climate is mild so air conditioning and heating requirements are minimal. Our appliances were selected for quality, long-life, and energy efficiency. We consume 13 kilowatt hours per day in our home and another 5 kilowatt-hours in our well-equipped office.

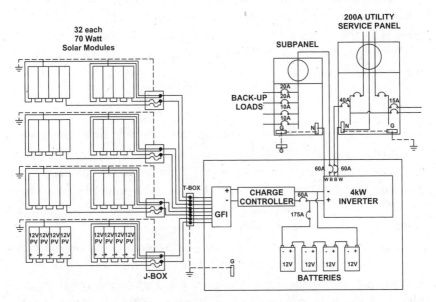

Figure 4.5. One-line diagram of Fran and Joel's home PV system. Their original PV system is on the left, the 2007 system expansion is on the right.

Hassle-Free Urban PV

The following story about our home PV system appeared in the December 1998/January 1999 issue of Home Power *magazine. The article has been modified slightly for inclusion here.*

The United States is an urban society with over 72% of Americans living on less than 2% of the land. PV can displace a significant percentage of polluting electric generators if it becomes an urban technology. With a little patience, planning, and help from an experienced contractor, city folks can install a PV system hassle-free.

Hire a PV Contractor

Experienced do-it-yourselfers can install a grid-connected PV system. They need the same skills and knowledge required to install a service panel and rewire a home. However, dealing with inspectors and your local utility require special skills.

I've built a lot of homes and even trained building inspectors. Fran and I have many years of PV experience. We also know that most inspectors prefer not to discuss codes and rules with owner-builders. That's why we hired Greg Johanson, owner of Solar Electrical Systems. Greg is a general and electrical contractor who has installed megawatts of PV; he has 8.8 kW of PV on his own home and 21.6 kW of PV on his office and shop.

Designing the System

We wanted the most PV we could afford that would fit on our home's low-pitched roof. We chose a 2.4-kW DC system, but the 225-square foot single-crystal array tilted optimally at 35 degrees would look like a billboard atop our home. Our office and garage, behind the house, have 5- and 18-degree south-facing roofs. A tilt-up array would still look bad and be a costly wind load problem.

We decided to use the low-profile, structurally engineered panel Greg and I had designed for PV Pioneer homes and churches in Sacramento. Although annual PV production at an average 10 degrees tilt in Los Angeles is 5% less than 34 degrees, we did not want to pay for extra structural engineering and hardware for a high-tilt mount. The low-profile array also put us in compliance with local building codes, which at the time prohibited roof panels and antennas. Our neighbors liked the low-profile panels and several have since gone solar, so we know we made the right decision.

The System

Our PV array has 32 Siemens 70-watt modules wired 4 in series to match inverter input voltage. Eight 4-module panels with standoffs are fastened to the roof plywood sheathing with wood screws. The mount meets local wind and seismic requirements. All wiring is in flexible and rigid conduit approved by the inspector.

We have a battery bank to protect our computers and for emergency power. Despite news reports to the contrary, Los Angeles has had relatively few power outages in the past 30 years. Most blackouts were less than a few minutes long. Our grid power was off 20 minutes during the 1994 earthquake. Initially, four Johnson Controls 12-volt, 86-ampere-hour sealed gel-cell batteries provided 3.3 kWh of energy storage (at 80% depth of discharge). After eight years, the battery bank had served its useful life and was replaced with four Dynasty 12-volt, 134-ampere-hour sealed batteries for 5.1 kWh of storage. If we need more autonomy, we can get locally manufactured industrial flooded batteries.

So many good things have been said about Trace Inverters that more would be redundant. We thank the engineers at Trace, who now work at OutBack Power, for helping make urban PV a reality. We installed a Trace Engineering Modular System and SW4048 sine wave inverter that can handle our largest combined loads. The modular cabinet looks good and is easy to install. It came prewired, UL-approved, and its tidy appearance impresses inspectors.

The PV system is wired into our home service panel. The office, garage, and specific house circuits for lights and refrigerator are wired into a subpanel that is connected directly to the PV system. If there is a utility power outage, these dedicated circuits get power from PV or battery storage. Solar power goes first to the house. Excess daytime solar power is fed into the grid and spins the utility kilowatt-hour meter backward at full retail value. Utility electricity powers the home when consumption exceeds PV production or when the sun is not shining.

Our system cost $19,742 in 1998 and qualified for a $5,835 California Energy Commission buy-down rebate. The net price was $13,907 or $7.15 per watt AC. The battery storage package costs another $2,709 but was not eligible for the grid-tie buy-down.

So Where's the Hassle?

If you want hassle-free PV, you have to understand inspectors. When we upgraded our service panel a year earlier, Fran told the inspector we planned to install PV. He was really interested and wanted to learn more. Our PV system would be the first in our town, so teaching was the key to opening inspectors' minds.

We put together a permit package that would educate inspectors. It included the scope of work, system description, design calculations, equipment specifications, parts list, wiring diagrams, drawings, plans, and elevations. Of course, we added the impressive California Buy-down Rebate Confirmation and attractive product literature, which helped to explain and endorse the technology.

We began the inspection "process" by applying for a homeowner's permit listing Solar Electrical Systems as our licensed electrical subcontractor. First Joel met with the electrical inspector and gave him a copy of the permit package, some photos, and additional information. Next, he met with the engineer responsible for inspecting signs, poles, towers, and other things stuck on roofs. The engineer liked the low-profile design.

Then Joel met with the construction permit engineer and hit a snag. The engineer couldn't care less about PV. All he wanted were site-specific structural calculations. Joel told him that our generic calculations included our roof type, but the engineer refused to look at them. So Joel politely asked to see his boss.

The building department director was a professional engineer (PE). We told him about the PV work we did for utilities, showed him lots of photos, and explained our structural calculations. He confirmed the calculations and even waived the construction permit because our design was under 3 pounds per square foot dead load. The three meetings took two well-spent hours because we ended up paying only $31.50 for an electrical subpanel permit.

Our equipment arrived on schedule. We installed the array on a Saturday during a light rain. Working in the rain is not recommended, but it was our only free day and the roof is nearly flat. Four guys worked three damp hours to get the array in place. We installed the wiring the next open, clear day. The inspector passed the job without a hitch. We mailed the final papers to the California Energy Commission and received our rebate check within a month.

A photo album of the step-by-step installation of our system begins on page 76. Photos of our system's 2007 expansion begin on page 86.

Net Metering

The next step was getting our net metering agreement. It is important for folks with PV to get full value for their home-grown energy while displacing polluting electricity. All U.S. utilities are required to allow qualified generating facilities (QF) to connect to the grid. Most states' utilities are required to net meter qualified residential PV systems.

We called our electric company, Southern California Edison (SCE), and promptly received an application by fax.

SCE recommends, but does not require, a lockable AC disconnect switch between the PV system and the grid. Other utilities may require lockable disconnects. It will be years before utilities and the PV industry agree on national interconnect standards, so consult with your local contractor.

Some utilities require homeowners to insure their grid-connected systems. They require this insurance even though, under the Price-Anderson Indemnity Act, they are not required to insure nuclear power plants. We told our insurance agent that our PV system was an electrical improvement approved by the building inspector, the utility, and the California Energy Commission. Our insurance rate remained unchanged.

Finally, we signed the net meter application, attached a one-line electrical drawing, and mailed it all to SCE. Three few weeks later, we got permission by mail to connect our PV system to the grid. SCE did not visit our installation until a year later, but your utility may require an inspection before allowing you to turn on your grid-interconnected system.

You Can Do It, Too

Our PV system performs flawlessly. The first full month it produced 292 kWh. Our electric bill dropped from $60 to $24. Yearly production is 2,400 kWh or about 50% of our home and office load.

The savings are nice. Inflation-free electricity for the rest of our lives is nice. What is most important is that everyone involved in this installation thinks positively about PV. The next PV installation in our town went even more smoothly, and now the city is looking for ways to encourage the use of PV.

Even though a lot of PV systems are installed in the U.S. every day, you may still have to do some trail blazing at the building inspection office when you go solar. An experienced contractor can be your guide. Have complete plans ready before you meet your building inspector and your utility. Follow the rules; don't fight them. If you run into a problem, calmly find the work-around. If you have any questions or need help, write to us. We installed a hassle-free PV system in the city and so can you.

Figure 4.6. Our office and adjoining garage roofs before PV.

Figure 4.7. Placing the solar modules on the 5° pitch office roof and the 18° pitch garage roof.

Figure 4.8. Attaching mounting feet to solar modules with stainless-steel self-tapping screws.

Figure 4.9. Caulking the mounting foot screws.

Figure 4.10. Fastening the mounting feet to the roof.

Figure 4.11. Installing solar array conduit, junction boxes, and wiring. Under the modules are interconnects with wire in flexible conduit.

Figure 4.12. (above) Completed rigid and flexible conduit, j-boxes, and homerun roof jack.

Figure 4.13. (left) Installing the subpanel and subpanel kWh meter hub in the garage.

Figure 4.14. (right) The subpanel with cover plate removed. Our home's main service panel is 15 feet from the subpanel. Wire in underground conduit connects the PV system and subpanel circuits to the home main panel.

Figure 4.15. (below) Roof homerun conduit and inverter output conduit in the garage.

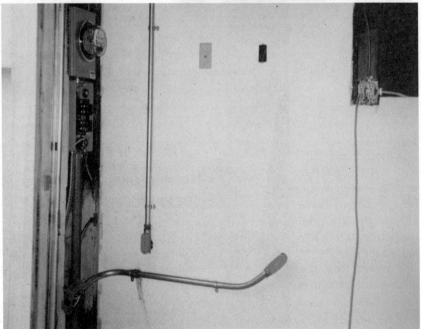

Figure 4.16.
Inverter and battery enclosure in the office. The enclosure holds a 40-amp charge controller and circuit breakers.

Figure 4.17.
Completed enclosure with a 4-kw inverter and 86-ampere-hour, 3.3-kWh battery bank. The inverter buzz and fan noise was distracting, so the equipment was moved from the office to the garage.

Figure 4.18. Inverter and battery enclosure in the garage. The homerun junction box was added after the equipment enclosure was moved from the office to accommodate the wiring change.

Figure 4.19. The closed equipment enclosure. Its neat, appliance-like appearance pleased the electrical inspector.

Figure 4.20. The low-profile solar array of thirty-two 70-watt solar modules provides power during the day. PV and the battery bank provide emergency power during utility outages day or night.

Our PV-Powered City Home—Part 2

In 2007 the time was right to expand our PV system. Our city now actively encourages people to go solar and no longer prohibits solar arrays that can be seen from the street. The state has also made a bold 10-year commitment to subsidize the installation of 3,000 megawatts of new PV by 2017.

We left our existing PV system—32 Siemens SP70 solar modules with the SW4048 and the UPS—in place and installed an additional 20 Siemens SP75 modules and a batteryless SMA America Sunny Boy 2.1-kW inverter. The new solar array matches the old array and is on a low-profile mount at the same 18-degree angle as the south-facing roof of our house. The modules are connected in series to provide 340 volts DC to the inverter. The inverter's 240-volt AC output feeds directly into our home electric service panel through a separate 15-amp circuit breaker. Other system equipment includes a solar array DC disconnect switch required by the *National Electrical Code* (NEC)®*, an AC kilowatt-hour meter required by a solar rebate program, and an AC disconnect switch. Our utility company still does not require a PV system output AC disconnect switch, but many electric utilities do. We included the AC switch so that other people could see how easy it is to comply with utility interconnection requirements.

Our roof shingles were 17 years old and in fair condition so we replaced the shingles under the PV array to avoid having to remove and reinstall the array in a few years when the shingles needed replacement. Pulling off the old shingles made installing the solar array mounting stand-off roof jacks and flashing easier. The solar array homerun wiring in flexible conduit was routed through the attic and then in rigid conduit to the DC disconnect switch next to the inverter.

Batteryless inverter systems operate more efficiently than our battery-based inverter PV system. At 80% DC-to-AC conversion efficiency, the new system's output is 1,500 watts DC STC × 0.8 = 1,200 watts AC.

The PV system cost $14,330 and qualified for a $3,110 solar rebate. The net system price was $11,220 or $9.35 per watt AC. The increased net price is the result of increased material and labor costs and a decreasing subsidy. The electrical permit cost was $365, but the city is considering waiving PV system permit fees in the future to encourage more people to go solar.

The PV system is still a good deal because electric utility rates have increased since we installed our first system and there are three rate increases scheduled for next year. PV now produces over 75% of our electricity. When we replace our 17-year-old 25-cubic foot refrigerator/freezer with a new energy-efficient model, we will be almost 100% solar.

We don't plan to move so the PV system's added value to our home is not important. However, having inflation-free electricity is important because it means greater economic security as we get older. What is most rewarding is knowing that our system is setting an example for other people in our city to go solar, making the world a better place for everyone.

National Electrical Code and *NEC* are registered trademarks of the National Fire Protection Association, Quincy, MA.

The Step-By-Step Installation
of Our Home PV System Expansion

Figure 4.21.
The solar array mount roof jacks are fastened onto the roof rafters with ⅜-inch stainless-steel lag bolts.

Figure 4.22.
Flashing is installed over the roof jacks and new shingles are laid. Unistrut rails to support the solar modules are fastened to the roof jacks.

Figure 4.23. The 75-watt solar modules were assembled into 4 module panels and prewired with flexible conduit before being delivered to the house. The panels weigh 75 pounds (34 kg).

Figure 4.24. The panels are put into position. Aluminum clips with stainless-steel bolts, nuts, and washers secure the panels to the Unistrut support rails.

Figure 4.25. (above) Positive and negative wires from each panel are connected in series in a weatherproof junction box. A separate earth-grounding wire (not shown) is connected to each solar module frame and run to the ground rod at the electric service panel.

Figure 4.26. (left) The inverter and DC and AC disconnect switches are mounted near the electric service panel.

Figure 4.27. The finished inverter installation. The solar array homerun is to the left of the inverter and feeds into the DC disconnect switch. Under the inverter to the right is the AC kilowatt-hour meter and AC disconnect switch. Most utility companies require the AC disconnect switch to be within 10 feet of the electric service panels with the billing meter so that utility electricans can shut off the solar power system when they are working in the neighborhood.

Figure 4.28. The new solar array is twenty 75-watt solar modules. The array in the background was installed in 1998 (see Figure 4.20). Our neighbor has a 2.5-kW PV system; on the next block, there are homes with 5-kW and 10-kW PV systems.

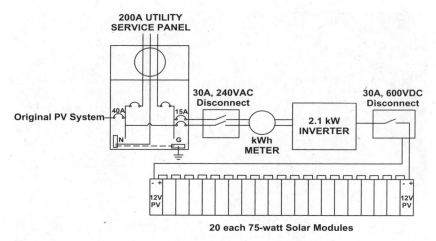

Figure 4.29. One-line diagram for Fran and Joel's home PV system expansion.

Figure 4.30. The new solar array as seen from the street. 20 solar modules × 75 watts DC STC at 80% DC-to-AC conversion efficiency = 1.2 kW AC. Estimated annual production is over 2,200 kWh.

Chapter 5
MAKING THE DECISION
Your Off-Grid Solar Electric Home

How an Off-Grid PV System Works

A stand-alone PV system has to produce all of the power consumed during the day and charge a battery bank to provide power at night and during cloudy periods. If the battery bank is fully charged during the day, the charge controller shuts off power from the array to the battery bank. At night, the charge controller disconnects the PV array from the battery bank to prevent the battery from discharging back through the solar array.

An off-grid PV system with generator backup operates in a similar way. Before sunrise, the home gets power from the battery bank. If battery voltage drops below a preset point, the generator is used to power the home and charge batteries. After sunrise when array voltage is greater than battery voltage, PV charges the battery bank and powers the home. On cloudy days, the house operates on stored power, although some solar current will continue to trickle into the battery bank.

If daytime loads exceed the inverter capacity, a PV/genset (combination PV system and generator set) can switch manually or automatically to the generator, which can power a battery charger or activate the

Figure 5.1. Sun, wind, and propane are all that power this off-grid home in northern California with no compromise on amenities. The PV system produces 8 kwh/day in winter and 15 kWh/day in summer. Two windchargers provide an additional 3 to 5 kWh/day. (Solar Depot)

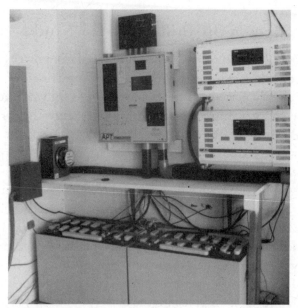

Figure 5.2. Battery bank, inverters, and power center for the home shown in Figure 5.1. A propane generator provides back-up power. The system cost about $45,000, which was half the cost of bringing utility lines to the house. (Solar Depot)

battery charger in a combination inverter/charger. When the daytime loads are small and the battery bank is fully charged, the charge controller automatically regulates the charge or it may shunt PV power to a secondary load, perhaps a DC water heater element. At sunset the PV array voltage drops below battery voltage and the charge controller disconnects the array from the battery bank until the next day. The house is then on battery power and generator backup.

When deciding whether to use PV to power your home, you need to know both sides of the story.

Advantages

1. There is a one-time cash outlay to purchase solar modules.
2. There is no monthly utility bill.
3. Installation of a PV system can be less expensive than a utility line extension.
4. Users aren't affected by electricity price increases or inflation.
5. Modules are reliable, sturdy, lightweight, and long-lasting.
6. PV can be used wherever the sun shines.
7. There are no moving parts to wear out or break.
8. Modular system allows for expansion as money permits and needs require.
9. DC appliances are available.
10. Solid-state, high-efficiency sine wave inverters make using AC appliances practical.
11. Battery technology has long been proven reliable.
12. Modules can be used in conjunction with commercial electricity, generators, wind or hydro power.
13. PV systems are quiet.
14. Users are not affected by commercial power outages.
15. Autonomous systems produce electricity with no on-site fuel consumption or pollution.

Disdavantages

1. The initial system cost is high.
2. Electricity is not produced at night.
3. Cloudy weather can significantly reduce power production.

4. Storage batteries must be serviced and replaced when they wear out.
5. An inverter must be used to power AC appliances.
6. The power output per dollar invested is low compared to a genset.
7. A generator or other power source may be needed if the PV array is undersized.
8. The manufacture of solar cells produces some environmental pollution.
9. Small solar arrays need seasonal tilt adjustment to maximize production.
10. You are solely responsible for maintaining the equipment.

Getting Started

Many PV people started on their path to energy independence with a fossil-fuel generator. We've seen this transition often. First they get a generator, which eventually breaks down. They make the repairs, which are followed by more breakdowns. Caught in the costly dilemma of replacing the generator—a noisy, expensive, high-maintenance, and unpredictable device they don't particularly like anyway—they begin to look for an alternative. So they buy a small PV system, sometimes under 1,000 peak watts. At first the PV system may be used to power a few 12-volt DC lights and a stereo. Next a small inverter is added to run a sewing machine or blender or vacuum cleaner. By this time the family has a pretty good idea what electrical equipment and appliances they want to use. From this point, it is just a matter of saving to buy the PV they need.

However, they have wasted time and money while suffering through a frustrating trial-and-error process. They own a generator that is seldom used and hard to sell because it is worn out. They have a collection of 12-volt lights and gadgets and an undersized inverter. Should they change over to standard 120-volt AC appliances? Probably yes, but then what do they do with the equipment they already have?

Thinking things out from the start would have resulted in a better PV system for about the same amount of money. That is what this book is all about. It will help you make long-range plans and either buy your entire PV system now or just those parts you need to get started. As things progress, you can add more modules and batteries without having to rebuild the whole system.

If we were to set up our off-grid system now, we would buy a few 12-volt DC lights and wire the house with standard AC wiring. We would buy the battery bank, regulator, and inverter that we would ultimately need and buy PV modules as we could afford them.

This long-range planning method has real economic advantages when it comes to investing in a PV system. It ensures that your batteries will last longer since you won't deep-cycle them as often as you would a small "starter-kit" battery bank. And you won't end up with a bunch of outgrown equipment collecting dust.

If you do decide to start with a generator, with the intent to eventually replace it with PV, an added benefit of proper system planning and design is that the generator will be run less and last longer. Its value will be higher if you sell it, and you can use the money to add more modules to your well-designed PV power system.

The Cost of Getting Started

The economics listed here are based on comparing PV to the cost of bringing a power line in to a home site. There are other factors that make PV a wise investment. A PV system is a high-value, low-service amenity that adds to your property's resale value and is property tax-exempt in many states. While it is impossible to place a dollar value on quality of life or lifestyle, your PV system allows you to have the best of both worlds. You can live in a pristine rural environment and still have all of the advantages of modern technology.

We must think about the cost of environmental pollution and health risks, nuclear waste storage, and an international energy supply network that requires armies to defend. It is not within the scope of this how-to book to examine "bottom-line PV." Bottom-line mentality is almost always short-sighted. Simply stated, short-term economics for non-renewable energy production only seems profitable. The long-term economics for PV is excellent.

While it would be nice to build a PV home from scratch, that is not necessary. Simply by adding a PV array to an existing home, with the proper interface equipment and using energy-efficient appliances wisely, you can begin using PV today.

Just how simple is it to go solar? PV modules can be mounted on readily available racks or on homemade frames. Inverters of all sizes are on the market. Interface equipment—such as automatic and manual switches and circuit breakers used to link emergency power systems to hospitals, banks, and offices—is easy to find. The list of energy-efficient appliances continues to grow.

Your PV system will pay for itself immediately compared to the cost of bringing in a power line. If you need to bring utility power to your home and are energy conservative, you can install a clean and efficient power system and never pay an electric bill again.

A Solar Sales Pitch

Let's take a closer look at these claims. What do they really say?

1. If you need to bring utility power to your home
2. and are energy conservative
3. you can install a PV system
4. to provide electricity for your home
5. that will pay for itself immediately.
6. You can install a clean, efficient power system
7. and never pay an electric bill again.

1. If you need to bring utility power to your home . . .

Let's look at the cost of utility power to get an idea of how expensive electricity really is.

The cost to have power lines brought to a new residence ranges from $4 per foot to over $16 per foot. Some utilities may run short lines for free. Free, that is, if you have an all-electric house and commit yourself to a high enough level of consumption.

How much utilities charge for line extensions varies. A Utah utility charged $35,000 per mile plus a $120 per month maintenance fee to be split among all the homes that hooked up to the line within the first five years. Recently, Southern California Edison quoted $32 per foot to extend a power line line to a new home site. Two years ago the same utility wanted a $3,000 deposit to begin installation of a half-mile utility line to a home in San Bernardino County near Los Angeles. The total cost was over $53,000. That's over $20 per foot for the "privilege" of paying an electric bill. Worse still is the fact that monthly bills in that locale have been increasing at a rate of over 10% per year. Paying over $50,000 dollars to get a monthly bill that doubles every nine years is incredible.

A Sandia National Laboratories study highlights another aspect of utility line extensions—the costs of replacement and maintenance. As our national electrical infrastructure ages, PV is less expensive to install than replacing and maintaining some existing power lines. In one example, the study reported, it cost $12,000 per mile to replace 40 miles of 50-year-old power lines to a water pumping facility. Plus there was a $200 per mile maintenance charge. This is not as cost effective as PV.

PV system owners in Colorado can see where the underground power line passes a few hundred feet from their living room window. In 1983 the cost to extend utility service to their home was quoted at $7,000.

It's the same story all over the world—except where government rural electrification programs force everyone to pay indirectly for power lines through taxation. The cost of extending electric service is very high.

In the United States the average cost to have a utility line extended to a new home is $7 per foot. That is what we will use in our example. Contact your utility company to find out how much they charge, what deposits and easements are needed, and what restrictions or requirements regarding electrical use and home occupancy are imposed. Also find out how long you will have to wait to have the line installed.

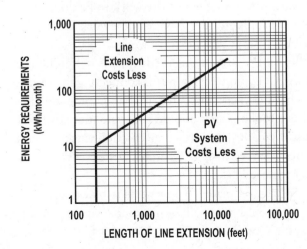

Figure 5.3.
Off-grid PV becomes more cost effective as the distance from the grid and/or the cost of grid electricity increases.

2. and are energy conservative . . .

This brings us to the next point: energy conservation. Much of this book is about energy conservation and for good reason. In a world with increasing population and finite resources, conservation is essential if we want a decent future for the inhabitants of this planet. The tips in this book may help you reduce your energy consumption to the point where you decide that PV is not for you simply because your electric needs are almost nothing. That's okay, too.

Energy conservation is an important factor in the "sales claim" calculation. Without energy conservation the cost of energy can be very high, not to mention environmental impacts or our secure future. With wise conservation your power consumption is reduced, and the size and cost of your PV power system is reduced.

3. you can install a PV system . . .

You can install a PV system. Those words divide us into two groups. The smaller group is the do-it-yourselfers, DIY in their jargon. The rest of us are the unskilled and all-thumbs who dread the words "some assembly required."

Our complex technological world requires specialization. To an extent this is good. However, how many times have you paid a specialist a high fee for some simple task? Granted, we sometimes hire others for convenience, to save time, and to get the benefit of their training and experience. In fact, we encourage novices to seek out experts. But we also encourage you to first learn something about the work to be done. The job may be beyond your capabilities, but do your homework so you can wisely select an installer and assess that person's work.

What does installing a PV system mean? It means designing the system and selecting the right equipment. It means connecting the equipment properly. It means having an understanding of the equipment. If you are handy and have a basic tool box, PV should prove no more difficult than any other new task. If you can do a little construction work—if you have wired a house, tuned a car, or installed a stereo system—you have, or can quickly acquire, the needed skills.

The development of the on-grid PV market has had a dramatic effect on off-grid PV systems in that installation options have expanded. You can still put a PV system together the old-fashioned way—piece-by-piece, component-by-component—but the on-grid market has resulted in excellent factory preassembled, prewired, and UL-approved equipment. You can purchase most of the electronics for your system as a single unit, ready to install.

If you really are all thumbs, hire someone to install your PV system. However, read this book carefully so you will know what you are getting. A well-designed turn-key PV installation is a pleasure. Even the DIYer can profit by bringing in an expert at some phase of the project, if only to check that everything is being done correctly.

4. to provide electricity for your home . . .

PV has been powering electrical devices for over five decades. You may have seen photographs of the first PV-powered repeater that was installed near Americus, Georgia, in 1955. Everyone has seen news photos of solar panels on space satellites. As a reliable electrical generator, PV is unsurpassed. And it can power your home.

Electricity produced by photovoltaics is direct current (DC) power. Most homes use alternating current (AC) power. Small cabins, boats, and recreational vehicles using DC can easily be converted to solar. AC applications require an inverter, which changes DC power to AC. Inverters and their use are explained in Chapter 13.

To provide PV power for your home, first look at the equipment you want to power and the work that you want the equipment to perform. Using the approach outlined in Chapter 16, create a list of equipment and power requirements. You must match your electric production to your consumption to ensure that you have enough power. Second, the size of your inverter and other system components is based on the work you want done. Also there may be more energy-efficient ways to perform the same task. This brings us back to the recurring theme of conservation. Perform the same task using less energy and your PV system will cost less.

The same cost savings hold true in other energy applications. For example, your car gets 25 miles per gallon, you drive 10,000 miles per year, and gas costs $4 per gallon. By driving a car that gets 45 mpg, you can save $711 per year. Same task, less energy, less money.

The same goes for home power systems. You could simply add a PV power system to a standard home. Or you could make the home energy efficient first and then install PV and save on the size of the system.

Figure 5.4. The Tarryall Resort is located at latitude 50° in Kenora, Ontario, Canada, north of Catherine Lake, and 3 miles from the nearest utility. The resort installed a 560-watt PV system in 1986 and gets most of its power from the sun. (Photo: Ron LaPlace)

Figure 5.5.
The Tarryall Resort equipment room. PV, inverter, and batteries reduced generator runtime from 24 hours a day to 27.2 hours per week; the payback time was 5.3 years. Today generator fuel costs much more and PV costs much less, so payback comes sooner. (Photo: Ron LaPlace)

5. that will pay for itself immediately.
To pay for itself immediately, a PV system must cost less than the cost of a utility power line extension. Again, this depends on the amount of electricity you use. If you have an energy-conservative home, your PV system will cost less.

6. You will own a clean, efficient power system . . .
Are PV power systems clean and efficient? If you live near a large, utility-sized PV power plant, pay a visit and see for yourself. The Sacramento Municipal Utility District (SMUD) PV power plant is particularly worth seeing. It is right next to the defunct Rancho Seco nuclear power plant, an interesting juxtaposition (see Figure 2.1, page 25).

7. and never pay an electric bill again.
Once a month millions of people wish that their electric bills would go away. Fail to pay your bill and it will—along with the electricity. Over 150,000 Americans who have gone off-grid with PV do not get a monthly electric bill.

Not everyone is ready for a total commitment to solar. Due to limited budget or limited space for an array, some people are able to produce only a portion of their power from the sun. Some use their PV to power an emergency or back-up system. Still others prefer to make the transition to PV gradually, by supplementing part of their utility power with PV or vice versa.

You may envision a gradual end to your dependence on utility power by going off-grid. Maybe you want to be totally energy self-sufficient by the time you retire in order to fix your expenses to match your fixed income.

Cost Comparison:
PV vs. Bringing in a Power Line

The system shown in Figure 5.6 was installed without a back-up generator. Five years later the battery bank was enlarged and upgraded to true deep-cycle batteries. The prices listed in the following tables are based on comparable equipment sold in 2006.

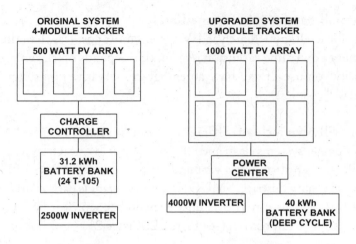

Figure 5.6. A cost-effective PV system upgrade.

Table 5.1. Initial Costs for the System in Figure 5.6

4 each 125-watt modules (500 watts)	$2,500.00
1 each 4-module tracking mount	$ 895.00
1 charge controller	$ 250.00
24 each batteries (31.2 kWh total storage)	$1,800.00
1 each 2,500-watt modified sine wave inverter	$1,200.00
Miscellaneous wiring & hardware	$ 750.00
Subtotal	**$7,395.00**
7% sales tax	$ 517.65
Installation charges	$1,500.00
Total	**$9,412.65**

How does this compare to the cost of running a power line one-half mile to the home site? At $7 per foot, one-half mile (2,640 feet) will cost $18,480. If you installed this system yourself, your solar electric power plant would be cheaper than the utility power line at $3 per foot. PV can be less expensive than utility service to a new home site.

If this system were upgraded to a pure sine wave inverter, a factory preassembled power center with controller and digital metering, and increased to eight 125-watt panels and industrial-grade batteries, it would cost $18,504.75, including sales tax and installation. This one-time cost is about the same as the cost of the power-line extension. The difference is that there will never be an electric bill to pay.

Table 5.2. Upgrade Costs (2006) for the System in Figure 5.6

8 each 125-watt modules (1,000 watts)	$ 5,000.00
1 each 8-module tracking mount	$ 1,200.00
1 each power center	$ 1,900.00
1 each industrial battery bank (40 kWh total storage)	$ 4,000.00
1 each 4,000-watt modified sine wave inverter	$ 2,575.00
Miscellaneous wiring & hardware	$ 750.00
Subtotal	**$15,425.00**
7% sales tax	$ 1,079.75
Installation charges	$ 2,000.00
Total	**$18,504.75**

Safety Considerations

PV systems are remarkably safe if you adhere strictly to safety codes. But electricity can be dangerous, so be aware. All PV systems should be installed according to the *National Electrical Code* and local ordinances. Follow the equipment manufacturer's recommendations. It is easy to have a safe PV system.

Insurance

Your photovoltaic system is an investment and should be adequately covered by standard homeowner's and liability insurance. It should be possible to simply add the equipment to your present homeowner's policy.

Coverage should include protection against hail, wind, falling trees or branches, fire, lightning, vandalism, and theft. If you hire a contractor to install your system, be sure you are insured against improper installation. If your insurance agent is not familiar with storage batteries and electrical equipment, have your PV supplier talk to your agent and explain that banks and offices have similar battery back-up systems.

Liability insurance to cover claims for personal injury, burns, electrical problems, and any utility interconnection should be considered. In some states, such as California, laws prohibit insurers from requiring customers to pay for additional liability insurance for PV. Make sure your homeowner's policy protects against a strong wind blowing your array into your neighbor's home. The best insurance in this situation, however, is to properly secure your array in the first place.

Loans, Grants & Rebates

Loans available for on-grid PV are also available for off-grid PV. Fannie Mae has financing available, typically up to $15,000 for off-grid energy efficiency upgrades.

If you can prove that your off-grid loads would otherwise have been powered by utility power, incentive programs may apply. For example, in some states if utility power runs to your property and you replace your AC well pump with a DC model and a PV system, if you can show that your off-grid loads are served at lower cost with PV, rebate or incentive programs may apply. In some states if a line-extension analysis shows PV to be more cost effective than utility power, the utility will rebate part of the cost of your PV system. In some states off-grid PV users are eligible for a state rebate program. There are programs such as the Renewable Energy Systems and Energy Efficiency Improvements Program (discussed in Chapter 2) that can help in cases of financial need. Rules change, so check www.dsireusa.org and local agencies for the most up-to-date programs and policies.

Tax credits and deductions available to on-grid homes and businesses also apply to off-grid homes and businesses. How these incentives affect the cost of your off-grid system are discussed in the next chapter.

Don't Wait—Go Solar Today

Government and industry reports in the early 1970s predicted that PV would be economical in 1986 at 50¢ per watt. No one realized the implications of that prediction. Not only did it encourage people to wait for lower prices that never came, it gave polluters an argument to continue their dirty technologies.

Despite long-held hopes for technological breakthroughs, at no time did it really appear that PV could ever be that cheap. Some companies tried to lower prices by selling below cost, hoping that lower prices would increase the PV market and bring about widespread public use of PV. Without incentive programs that didn't happen.

Hoping for breakthroughs is better than having no hope at all, but taking action and creating the life you want to live now is what really makes a difference.

   Industry and government are responsive to consumer and citizen
desires. If enough people want PV, incentives will be mandated. If we buy
and use PV, businesses will serve the market. Compare today's computers
and mobile phones with those sold a few years ago. People made demands,
competition developed, and government and industry responded. If we
express our desire for PV there will be more PV.

Figure 5.7. Located in a farming community in central California, just a short
drive from the Sierra Mountains and giant redwood trees, is a home that
gets most of its power from twenty-four 55-watt solar modules mounted
on passive solar trackers. Power is stored in deep-cycle batteries and used
through a 2.5- kW inverter. The family of four has all the modern conveniences: dishwasher, 19-cubic foot refrigerator/freezer, 21-cubic foot freezer,
automatic washer. Summer comfort is provided by evaporative cooling and
occasional air conditioning. For nine months, this energy-efficient home is
100% PV-powered, averaging 8 kWh/day solar production. Winter production is 4 kWh/day, and utility power provides backup. With one more tracker
array and expanded battery bank, this family can say goodbye to the utility
company bill and hello to complete energy independence.

Chapter 6
MAKING THE DECISION
Your On-Grid Solar Electric Home

Installing an on-grid PV system is a good investment. Your PV system will do more than produce electricity that is both pollution-free and inflation-free. It will increase the value of your home. Homeowners who invest in PV systems enjoy lower electric bills and can recover their investments if they sell their homes.

At first glance a PV system can seem expensive, but rising utility rates, utility and state financial incentives, and tax credits have made PV affordable. Your PV system is the only home improvement that literally goes to work and pays for itself.

How an On-Grid PV System Works

A Grid-Connected PV Home

The sun rises and its rays hit the solar modules. Electricity flows from the solar array through the inverter to the home's electrical service panel, which is the same distribution panel that was used before the PV system was installed. The family wakes up, gets dressed, drinks coffee, eats breakfast. They do all the same things in the same way that they have

Figure 6.1. Homes with more than average energy consumption can also go solar. This home gets almost all of its energy from a 12-kW solar array. The homeowners drive an electric vehicle. Their strong commitment to renewable energy and the environment is an inspiration to their community. (Light Energy Systems)

Figure 6.2. Seven 2.5-kW batteryless inverters get power from the array in Figure 6.1 above. See wiring diagram, Figure 17.5, page 352. (Light Energy Systems)

always done them before their PV system was installed. During the day, when the children are at school and the adults are at work, the electrical loads are at a minimum. The PV system feeds electricity into the utility electrical grid through the same electric meter that was there before the PV system was installed. Only now the meter is counting down instead of up, and the homeowners are "banking" credit for their solar electric production. This is net metering.

As the sun sets the family returns home and turns on the lights, TV, kitchen appliances, and computers. Electricity is supplied to the home at night by the utility grid, just as it was before the PV system was installed. These homeowners have opted not to include batteries for emergency back-up power. If there is a power outage, day or night, their home will be without electricity just like other homes in their neighborhood. The inverter will automatically disconnect their PV system from the electric grid to prevent PV-generated electricity from injuring utility technicians working on the outage.

The PV system operates silently and automatically. The big difference is that this family has lower utility bills because some of their kilowatt-hour consumption is being supplied by their PV system. While they are at work, so is their home PV power plant.

A Grid-Connected PV Home with Emergency Back-up Power

Before sunrise, the home gets energy from the grid. The batteries are used only for emergencies. During the day, PV powers the home. Excess electricity is fed back to the utility grid, which turns the meter backward.

If utility power fails, the inverter automatically disconnects from the grid. Selected loads on a separate subpanel continue to operate, getting power from PV and battery storage. When grid power returns, the inverter automatically switches back to net-metering mode.

At sundown, PV array voltage drops below battery voltage and the solar array is disconnected until the next day. At night power comes from the grid. If the grid fails at night, power comes from battery storage.

Designing and installing an on-grid PV system involves a different set of decisions than for an off-grid system. Off-grid you determine how much electricity you need and the equipment you need to make it. In an on-grid home you already have all the electricity you could possibly want literally at your fingertips. Just flip the switch. All you have to do is pay the electric company. Every month the utility company sends a bill and you write a check. "Renting" electricity has become completely automatic. Your bill keeps going up, but is the utility charging more or are you using more? Periodically little slips of paper that few bother to read are tucked into the envelope with the bill announcing another rate increase.

You say you hate smog and pollution, but you don't make the connection between the electricity you use and the smog you hate. The question is this: Do you really care?

Could investing in a PV system for your home be the answer? Can PV lower your electric bills? Can PV provide your family with pollution-free electricity? Can homeowners install PV systems themselves? Is PV for me, now? The answers are all "yes."

You need to know both sides of the story when deciding to install a PV system on your grid-connected home.

Advantages

1. There is a one-time cash outlay to purchase equipment.
2. A net-metered PV system will offset your electrical consumption from the utility grid and lower your bill.
3. Over the life of a PV system the kilowatt-hour cost is less than utility electricity.
4. A PV system will offset utility pollution.
5. Electricity produced by your PV system is inflation-free.
6. If a PV system is included in your home mortgage or home improvement loan, the value of the solar electricity it produces can be greater than the increase in the loan payment for the system.
7. PV systems consume nothing and have no emissions.
8. Modules are reliable, sturdy, lightweight, and long-lasting.
9. PV produces electricity wherever the sun shines.
10. PV has no moving parts to wear out or break.
11. Modular PV system design can be expanded as money permits and needs require.
12. You can continue to use any electrical appliance you want. There is no need to change your daily routine.

13. A PV system increases the resale value of your home.
14. PV systems are quiet.
15. PVers who install a battery backup or use PV in conjunction with a generator are not affected by utility power outages.
16. There may be federal, state, or local tax credits and deductions available to reduce first costs.
17. There may be rebates or other incentives available to reduce the cost of the system.

Disadvantages

1. The initial cost is high.
2. In the short term PV costs more per kWh than utility power.
3. Non-battery PV systems provide no power during outages.
4. Some utilities and building inspectors are overtly hostile to PV installations.
5. Power output per dollar invested is low in the short term.
6. The manufacture of solar cells produces some pollution.
7. Neighbors might object to the appearance of your PV array.
8. In states without net metering, you must sell the electricity to the utility at a wholesale rate through a separate meter.
9. You are responsible for maintaining the equipment.

Net Metering

What Is It & How Does it Work?

Net metering allows you to get full retail value for the excess electricity that your PV system produces during the day when you consume less power than you produce. Your regular utility electric meter runs either backward or forward and keeps track of the net difference between the electricity you generate and the electricity you use. Net metering, which is currently available in most states, makes your PV system much more cost effective. You are effectively "banking" excess midday solar electricity to use later in the day or at night.

Net-metering rules vary from state to state. This can affect how large your PV system should be. In some states, excess electricity produced during one month's billing cycle can be carried over to the next month.

This means you can increase the size of your system to offset electrical consumption in winter months when daylight hours are fewer and it is cloudier. If your state does not allow excess generation to be carried over from month to month or from year to year, the extra electricity that you produce may default to the utility, which means that they get it for free. In some cases the utility may opt to purchase net generation at wholesale rates. Some utilities will pay a premium for your "green power." This is why it is important to know how much electricity you consume on a daily, monthly, and annual basis. It can literally be a case of "use it or lose it." The goal is to optimize the advantages of net metering, as well as any utility rates—such as time-of-use (TOU) metering—that are favorable to PV.

What if My Utility Company Doesn't Allow Net Metering?

If your utility company does not allow net metering you can still have a PV system.

One option is to disconnect from the utility and go entirely off-grid, as over 150,000 households in the U.S. have done. Going off-grid may sound appealing, but may not be easy. You will have to fulfill 100% of your electrical needs 24 hours a day. Some utilities require an exit fee or that several years' worth of anticipated electric bills be paid in advance before they will allow you to unplug from the grid. In one case a homeowner had to prepay five years of electric bills and then produce his own power for five years before the utility would remove his meter.

A less drastic option is to stay connected to the grid and supplement your PV production with grid power. The equipment will be the same as that for a net-metered system except that you must have two utility meters. One meter will register electricity purchased at the retail rate. The second will record the electricity produced by your PV system and sold to the utility at its wholesale rate. This disparity in pricing is why few on-grid PV systems are installed in states that do not have net metering.

A third option is an autonomous off-grid PV system that powers specific loads. Your home will remain connected to the grid. The PV system will power loads on a separate subpanel that is not connected to the grid. This approach uses the same equipment as a grid-tie PV system with batteries except there is no bi-directional inverter connection.

The fourth option is to become a PV activist and work to change anti-solar laws and restrictive policies. You can do this in conjunction with any of the three options above. Contact your local politicians and utilities. Lobby for net-metering laws. Meet with like-minded people who are working to promote renewable energy and alternative power systems. Make a difference.

Getting Started

You have decided that the advantages of a PV system outweigh the disadvantages. You have read *The New Solar Electric Home*. You have made notes and lists. You are ready to go solar. Next find out what incentives are available and what, if any, obstacles stand in the way of installing your PV system. Start at the Database of State Incentives for Renewable Energy (www.dsireusa.org) to learn which programs are available in your area. Then contact the appropriate agencies and get the paperwork.

If you live in a net-metering state, contact your utility and get a net-metering agreement. If net metering is not available you still need to notify the utility. There may be two different utilities involved with your PV system. One is your electric service provider (ESP), who will handle the metering and billing agreements. A local distribution company (LDC) may separately handle how your generating system is connected to the grid. Usually the LDC and ESP are the same company. The LDC will spell out its requirements for you to safely connect your PV system to the grid. The requirements can include equipment such as meters and utility worker-accessible disconnect switches. In some states the LCDs cannot

- require you to pay for any meters beyond the existing bi-directional meter,
- impose any requirements or tests on your system if it meets national safety standards,
- require you to purchase additional insurance, or
- require you to buy electricity from them or their affiliates.

The laws regulating LDCs vary from state to state, so check with your state electrical or utility commission.

Most utility companies allow homeowners to install their own PV systems and connect them to the grid. This is called an "owner-installer."

Some electric utilities require PV systems to be professionally installed. Some states require certification for professional PV installers; others allow any electrician to install the system.

Your next step is to contact your local building, planning, and safety department (sometimes called the building permit office). All states require building permits and inspections to connect your electric generation system to the power grid. The PV system, its components, and installation must meet *National Electrical Code* (*NEC*) and building code requirements as interpreted or amended by your local building inspectors. The permit office will tell you what information you must submit to them to obtain a building permit and they will inspect the installation. All states allow owner-installers to be their own contractor and obtain permits, but they may require that a licensed contractor be involved in the project because most building inspectors think homeowners are incapable of installing a code-compliant system.

Covenants, codes, and restrictions (CC&Rs) in your neighborhood present a potential obstacle that could affect the installation of your PV system. Some communities have CC&Rs that do not allow equipment on a roof; some even specify no solar systems. Usually the intent is to restrict roof air conditioners, satellite dishes, and solar hot water systems. Under California's Solar Access Laws, anti-solar CC&Rs are unenforceable. California prohibits both municipalities and utilities from interfering in the installation of a PV system. Over 34 states have solar access laws. Check both your local CC&Rs and state solar access laws.

If there is an obstacle to the installation of your PV system, your first course of action is to be informed. Know what the laws are and what your rights are. That said, whether you are dealing with a neighbor who doesn't like the way PV looks or a legal restriction, it is always better to enter into negotiations with a smile on your face and good thoughts than to go into battle angrily waving a law book. Generally people like PV. It's clean. It's quiet. It increases property values. PV displaces peak loads so utilities don't get unfavorable press because of brown-outs. Be a PV ambassador. Ask people what they want from you to make your PV project happen. Start a dialogue. They might come around and decide to install a PV system too.

Start-Up Costs

This section is intended to help you get past PV system sticker shock. We can detail the dollars and sense for your PV system, but only you can put a value on the peace of mind that comes from providing your family with the electricity it needs with a technology that is clean, quiet, and safe. That transcends dollar value.

The following calculations are based on real systems and show how state and federal incentives make PV affordable. The calculations differ for each system. You will have to do the calculations for your system. Consult your tax professional about applying tax credits or, for businesses, accelerated depreciation. If unfamiliar with the laws, refer your accountant to your PV supplier or to www.dsireusa.org. Prices, rebates, and incentives change constantly, so make sure your information is current.

A Residential Net-Metered PV System

The first example is an average-size net-metered PV system on a home in coastal Southern California. It is Household Number Two in Chapter 3. The homeowners decided on a 2.5-kW PV system. They do not want battery backup for emergency power. They opted for a prepackaged system with twenty 125-watt solar modules, a grid-interconnect inverter, and a roof mount kit. The package price is $15,800, including shipping. They also need AC and DC disconnect switches, $200 each; wire and conduit totaling $500; and a 240-V, 15-A circuit breaker for their service panel, $15. The total for all the hardware is $16,715 (2006 retail prices). Sales tax is 8.25%, or $1,379. The building permit cost $300. The homeowners hired an electrician to install the system at a cost of $2,500.

Table 6.1. Contractor-Installed System Costs: Southern California

PV system kit	$15,800.00
AC and DC disconnect switches	$ 400.00
Circuit breaker	$ 15.00
Wire and conduit	$ 500.00
Subtotal	**$16,715.00**
8.25% sales tax	$ 1,378.99
Building permit	$ 300.00
Electrician	$ 2,500.00
Total installed cost	**$20,893.99**

The homeowner lives in the Southern California Edison service area and was eligible for a $2.50/watt rebate for a contractor-installed PV system. The state rebate calculation is based on the PTC (PVUSA test conditions) kilowatt rating of the solar modules and the inverter efficiency.

Table 6.2. PV System Rebate

Module California Energy Commission (CEC) PTC Rating	111.8 W
Inverter efficiency	94%
Rebate amount (20 × 111.8 × 0.94 × $2.50)	**$5,254.60**

Post-rebate the system cost is:

$$\$20,893.99 - \$5,254.60 = \$15,639.39$$

Prior to 2007, the rebate for an owner-installed system was 15% less than for contractor-installed systems. The 15% penalty was removed by the California Public Utility Commission. There are restrictions on owner-installed systems in other areas or states. Check with your rebate administrator or www.dsireusa.org for any restrictions on owner-installed systems in your area.

The 2005 Energy Policy Act allows a 30% tax credit, up to $2,000, on residential PV systems. The homeowners' federal tax credit is:

$$\$15,639.39 \times 0.3 = \$2,000 \text{ maximum}$$

The cost of their system is now

$$\$15,639.39 - \$2,000 = \$13,639.39$$

If they had installed the PV system themselves, they would have saved the $2,500 that the contractor charged to install the system. The system cost, after the rebate and tax credit, would be:

$$\$13,639.39 - \$2,500 = \$11,139.39$$

The homeowners' electric bill averages $125 per month, and their utility has a tiered-rate structure. Before they installed the PV system their monthly bill was:

Consumption	Rate	Cost
304 kWh at baseline	13.009¢/kWh	$39.55
91 kWh at tier 2	15.170¢/kWh	$13.80
213 kWh at tier 3	19.704¢/kWh	$41.97
114 kWh at tier 4	25.950¢/kWh	$29.58
Total		**$124.90**

Their PV system produces:

> 20 × 125 watts DC STC × 0.8 system efficiency
> × 5.6 sun-hours per day = 11.2 kWh AC/day

The billing cycle is 30 days. In 30 days the system produces:

> 30 × 11.2 kWh/day = 336 kWh

This displaces the 114 kWh billed at 25.95¢, the 213 kWh billed at 91.704¢, and the 9 kWh billed at 15.17¢.

> (114 × 25.95¢) + (213 × 19.704¢) + (9 × 15.17¢) = $72.92

This is the monthly dollar value of the electricity that their PV system has produced. Dividing the net cost of the PV system by the value of the electricity saved gives the simple payback period for the PV system.

> $13,639.39 ÷ $72.92 = 187.1 months or 15.6 years
> for the contractor-installed system

or

> $11,139.39 ÷ $72.92 = 152.8 months or 12.7 years
> for the owner-installed system

Add a conservative 3% inflation rate, compounded annually, to the utility rates and the monthly dollar value of their PV-produced electricity is:

Year	Monthly Amount	Annual PV Value	Cumulative PV Value
1	$ 72.92	$ 875.04	$ 875.04
2	$ 75.11	$ 901.29	$ 1,776.33
3	$ 77.36	$ 928.33	$ 2,704.66
4	$ 79.68	$ 956.18	$ 3,660.84
5	$ 82.07	$ 984.87	$ 4,645.71
6	$ 84.53	$1,014.41	$ 5,660.12
7	$ 87.07	$1,044.84	$ 6,704.96
8	$ 89.68	$1,076.19	$ 7,781.15
9	$ 92.37	$1,108.47	$ 8,889.62
10	$ 95.14	$1,141.73	$10,031.35
11	$ 98.00	$1,175.98	$11,207.33
12	$100.94	$1,211.26	$12,418.59
13	$103.97	$1,247.60	$13,666.19
14	$107.09	$1,285.03	$14,951.22

The packback period for the contractor-installed system is 12.8 years. The owner-installed system has a payback period of 10.6 years at 3% inflation.

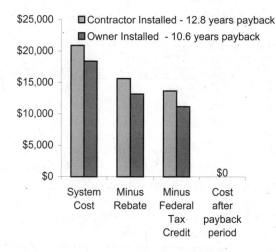

Figure 6.3.
Net-metered residential
PV system paybacks.

Residential Net-Metered PV in Other States

What about other states? The initial cost for the same residential net-metered PV system installed on a home in Atlantic City, New Jersey, in 2005 was:

Table 6.3. Contractor-Installed System Costs: New Jersey

PV system kit	$15,800
AC and DC disconnect switches	$ 400
Circuit breaker	$ 15
Wire and conduit	$ 500
Subtotal	**$16,715**
Building permit	$ 300
Electrician	$ 2,500
Total installed cost	**$19,515**

New Jersey has a sales-tax exemption for solar equipment. The New Jersey 2005 solar rebate for a PV system under 10 kW was $5.30 per watt DC up to 70% of the system cost.

2,500 watts × $5.30 per watt = $13,250

Subtract the rebate and federal tax credit and the net cost is $4,265.

Total installed cost	$19,515
State rebate: 70% of total installed cost	–13,250
Federal tax credit	– 2,000
Net cost	**$ 4,265**

Using sun hours from the Solar Radiation Chart in the appendix or the PVWATTS internet calculator (http://rredc.nrel.gov/solar/codes_algs/PVWATTS/), the system produces:

20 × 125 watts DC STC × 0.8 system efficiency × 4.6 sun-hours per day
= 9.2 kWh AC/day × 365 = 3,358 kWh /year

Using the utility rate of 10.7¢/kWh and a 3% inflation rate, the payback period for this system is 10.03 years.

In Ann Arbor, Michigan, the initial cost of the system would be the same as in New Jersey. The homeowner purchased the system from an out-of-state supplier so there is no sales tax. There are no rebates in Michigan, but the homeowner qualifies for a $2,000 federal tax credit. The net cost of this system is $17,515.

The average annualized sun hours in Michigan is 4.3. This PV system produces 8.6 kWh AC/day or 3,139 kWh AC/year. Using a current utility rate of 9.1¢/kWh and 3% inflation, this system produces $285.65 worth of electricity in the first year and has a simple payback period of 35.3 years.

The different payback periods for identical PV systems in California, New Jersey, and Michigan illustrate why solar power plants should receive subsidies that are, at the very least, equal to those currently given to polluting and resource-depleting coal, gas, and nuclear power plants.

A Commercial Net-Metered PV System

Now let's install the same PV system for a small business in California. The purchase cost, the installation costs, and the rebate remain the same. The system cost after the rebate is $15,639.39. The business owner is eligible for a 30% federal tax credit based on the post-rebate system price.

$15,639.39 × 0.3 = $4,691.82
$15,639.39 – $4,691.82 = $10,947.57

Businesses are also eligible to take five-year accelerated depreciation on their PV systems on their federal income taxes based on 95% of the system's price. The total depreciation on their federal income tax for a business in the 34% tax bracket, taken over a five-year period, is $5,317. In addition, state depreciation is $1,454. After five years their system cost is:

$10,947.57 (cost after tax credit) – $5,317 – $1,454 = $4,176.57

This system produces the same amount of electricity as the other one, 11.2 kWh/day, but the business owner is billed at a flat rate of 12¢/kWh. This comes to 4,088 kWh/year or $490.56 worth of electricity in the first year. Factoring in a 3% inflation rate, the payback period is 7.7 years.

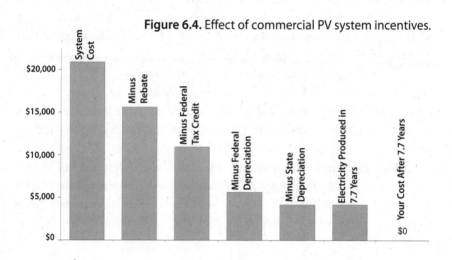

Figure 6.4. Effect of commercial PV system incentives.

A PV System & Your Home's Resale Value

The formula for calculating the dollar value of durable energy-efficiency improvements to a home is $20.73 for every $1 decrease in annual fuel bills. The Appraisal Institute developed this formula by studying the nationwide relationship between incremental cost of energy-efficient upgrades and the appraised value of a home.

The California home had a $124.89 average monthly electric bill or $1,498.68 annually. The PV system produces $72.92 worth of electricity per month or $875.04 annually. The value added to the appraised value of their home is:

$1,498.68 - $875.04 = $623.64 annual savings

$623.64 × $20.73 = $12,928 added to the appraised value

The net cost of the system was $13,439.39. Their cost recovery, not including the value of the electricity generated, is 96%. How does PV compare to other home improvements, as evaluated by *Remodeling* magazine?

Table 6.4. Home Improvement Cost Recovery

Project	Job Cost	Resale Value	% Cost Recouped
Window replacement (wood)	$11,040	$9,416	85.30%
Minor kitchen remodel	$17,928	$15,278	85.20%
Bathroom remodel	$12,918	$10,970	84.90%
Window replacement (vinyl)	$10,160	$8,500	83.70%
Two-story addition	$105,297	$87,654	83.20%
Major kitchen remodel	$54,241	$43,603	80.40%
Attic bedroom remodel	$44,073	$35,228	79.90%
Basement remodel	$56,724	$44,685	78.80%
Deck addition	$14,728	$11,307	76.80%
Bathroom addition	$28,918	$21,670	74.90%
Roofing replacement	$14,276	$10,553	73.90%
Master suite addition	$94,331	$68,458	72.60%
Family room addition	$74,890	$53,519	71.50%
Sunroom addition	$49,551	$32,854	66.30%
Home office remodel	$20,057	$12,707	63.40%

"Cost vs. Value 2006" from www.costvsvalue.com

A wood window replacement and minor kitchen remodel topped the list with about an 85% cost recovery at resale.

A PV system is the only home improvement that produces electricity. Add the dollar value of the electricity produced every day to the 96% cost recovery and installing a PV system makes economic sense.

The owner-installed system in California had a net cost of $11,139.39. The dollar value of the electricity produced by the PV system in the first year was $875.04. The cost recovery on their investment is 116%. The additional 20% return on the investment as compared to the contractor-installed PV system is the homeowners' "sweat equity."

The PV system installed on the home in Michigan had an equipment cost of $17,515. The value of the electricity produced in the first year was $285.65. The value added to the appraised value of the home in the first year is $5,920.49.

In New Jersey the net cost of the PV system was $4,265. The value of the electricity produced in the first year was $359.31. The value added to the appraised value of the home is $7,448.41. This is a 175.5% return on the investment.

Financing Photovoltaics

There are many ways to finance a PV system. The owner of the array in Figure 10.19 (page 178) took part of his retirement savings out of the stock market and put it into the Permaculture Credit Union as a certificate of deposit (CD). Then he took out a shared-secured loan at 3% above the interest his CD was paying and used the loan to buy his PV system. "I was able to support the credit union, put PV on my home, and finance my system at a low rate, all the while making my investments more socially responsible," the owner said. "This arrangement let me put the PV system to work now, rather than waiting for years until I could save the money for it. And the utility savings offset part of my loan payment."

Financing Your System with Your Home Mortgage

You can include the cost of your PV system in your home mortgage. The California homeowners take out a mortgage and include the net cost of the installed PV system ($15,639.39). The increase in their monthly payments for a 30-year fixed rate mortgage at 6% interest is $93.77. The 6% interest compounded over 30 years raises the total cost of the system to $33,755.78. Their PV system currently produces $72.92 worth of electricity per month. Allowing for a 3% annual inflation rate for utility prices, the cumulative value of the electricity produced at the end of 30 years is $41,630. The PV system has produced enough electricity to pay for the equipment and installation costs and the cost of the loan in 26 years—less than the term of the loan. The warranty on the output of the modules is 25 years and a 30-year system life is reasonable. The electricity that the PV system produces pays for the equipment. The mortgage interest is tax-deductible, adding even more savings over the life of the PV system.

A delicate balance between the interest rate, the term of the mortgage, and the value of the electricity will result in zero net cost. The example is based on a real house and real electric bills in California with a tiered utility-rate system. If your electricity costs more or less, the economics change. If interest rates are higher and utility rates are lower, use the simple payback method to amortize the cash cost of the system. Calculations have to be done on an individual basis.

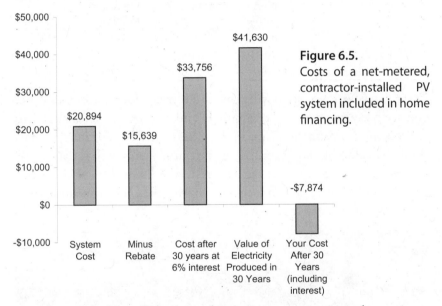

Figure 6.5.
Costs of a net-metered, contractor-installed PV system included in home financing.

In addition to adding PV to your home mortgage, a national movement has begun for cities to offer residents the ability to finance their PV system using increased property tax assessments to pay off the loan.

Safety Considerations & Insurance

Installing a PV system requires strict adherence to safety codes. Read chapters 7 and 8 to learn about electrical codes and electrical safety, and the sections in Chapter 5 on safety and insurance.

When connecting to the electric grid, you are dealing with 240 volts. Make sure that everyone involved knows how to work safely with electricity. You are responsible for complying with your utility's interconnection requirements, as well as making sure that your installation is code compliant.

Environmental Impact

Utility companies consume different fuel mixes to produce electricity. Installing just one kilowatt of PV in the three systems discussed in this chapter will reduce emissions over the 25-year warranty period of the solar modules by the following average amounts:

Table 6.5. Utility Emissions (in pounds)

	Southern California	Atlantic City, New Jersey	Ann Arbor, Michigan	mean
Nitrogen oxide	100	225	275	200
Sulfur dioxide	50	700	625	458
Carbon dioxide	88,400	99,050	137,700	108,383

Every day your PV system will provide electricity for your home. When you are away, your PV system quietly and efficiently makes electricity for you and deposits the excess in the grid "bank." In addition, you have the satisfaction of taking control of your utility bills with inflation-free electricity. How many other appliances in your home come with 25-year warranties? How many produce more power than they consume? How many literally pay for themselves?

Figure 6.6. Built in 1981, this Carlisle, Massachusetts, home was the first 100% utility-connected PV-powered home. The 3,200-square foot house has passive solar heating and cooling, super-insulation, internal thermal mass, earth-sheltering, daylighting, a roof-integrated solar thermal system, and a 7.5-kW PV system. The home needs no fossil fuel and exports surplus power to the utility. (Photo: Solar Design Associates)

Chapter 7
RESIDENTIAL BUILDING CODES & INSPECTIONS

When the first edition of *The Solar Electric Home* was published in 1983, PV systems were off the building inspector's beaten path and the *National Electrical Code* had not yet included photovoltaics. This is no longer the case. Now PV is a mainstream technology. When you install your PV system, whether grid-interconnected or not, code compliance is mandatory. The *National Electrical Code* should not frighten or inhibit you from installing a PV system. Electrical codes are tools you use for a safe and trouble-free installation. The building and safety codes and standards relevant to a PV system include:*

> *National Electrical Code* (NEC)
> International Residential Code (IRC)
> International Mechanical Code (IMC)
> Underwriters Laboratories Standards (UL)
> Institute of Electrical and Electronics Engineers
>   Standards (IEEE)
> American Society of Civil Engineers Standards (ASCE)
> Uniform Building Code (UBC)

*See Appendix C for contact information regarding individual codes and standards.

UL and IEEE standards that apply to your PV system are incorporated into the *National Electrical Code*. International Residential Code, International Mechanical Code, and American Society of Civil Engineers standards are reference models for state and local building codes. Most of the codes and standards books are available through your local public library. Local electrical and building codes may impose additional standards on your installation and can be referenced through www.dsireusa.org.

It is important that you contact your local building and safety department early in the process of designing your PV system. As soon as your system is designed—*and before you buy any equipment*—talk to your building inspectors. Show them your plans. Ask what information and documents are needed in your building permit package. Ask what they will want to see in your installation. All building codes are subject to interpretation by the building inspector. Find out in advance what your inspector wants to see.

Even the most experienced PV installation contractors sometimes fail to pass a final inspection because the inspector wanted something other than, or in addition to, the work that had been done. The first and best course of action in this case is to make the changes the inspector wants.

If your building inspector is not familiar with PV, he will draw on his experience with other electrical or structural systems, such as air conditioning, uninterruptible power systems, or skylights. You might have to show the inspector what you think is the appropriate section of the electrical or building code and give him the opportunity to consider its application to your installation.

Our home was one of the first net-metered PV homes in California, so our inspector had not seen one before. Two years before the net-metering law came into effect, when we were having other electrical work done, we told the inspector about the PV system we planned to install. He was very interested and supportive. When the time came to install our PV system and another inspector expressed concerns about the array being visible from the street (a violation of our local building codes at the time), the first electrical inspector interceded on our behalf and helped solve the problem. The point is that we had given this inspector information about PV so he wasn't just walking into his first PV system having to make immediate judgments.

Structural & Mechanical Requirements

A major purpose of the International Residential Code and Uniform Building Code is to ensure that building structures safely support all loads, including "dead loads,"[1] which are the actual weights of materials and construction.[2] Allowable loads (measured in pounds per square foot) vary depending on local seismic zone and climate conditions.

PV modules come with a 20- to 25-year factory warranty on their power output; they should continue to provide for your electrical needs long after that. Building codes and inspectors aside, no one wants to install, remove, and then reinstall a perfectly functioning PV system because the roof under it failed. Inspect your roof, or have it inspected, before you begin. Building codes usually allow two layers of shingles. Since most PV arrays weigh about 4 pounds per square foot (roughly the same as one layer of shingles), most inspectors will allow a PV system to be installed without a structural analysis or extra braces. The structural roof components must be able to support the PV system.[3] Make sure the existing roof provides an adequate base for your system. The inspector will examine the supports and sheathing to make sure they are sound and that there is no water damage or rot.

The structural design for roof-mounted PV panels should transfer dead loads directly onto the roof's structural members. You might be required to include a structural engineering report in your building permit package. Some commercial mounting kits provide the necessary information. If you design your own mounts and the inspector questions how and where they fasten to the roof or thinks they seem either flimsy or heavy, you may have to consult a structural engineer. Obviously this is something you should discuss with your building and safety department early in the design process.[4]

Roof-mounted PV modules and mounts are subject to the forces of nature. Wind and snow loads must be taken in to account. Wind loading requirements are determined by local conditions and local building codes. Design wind loads, pounds per square foot of uplift, are found in IRC Table 301.2b. The application of these design wind loads varies with geography and climate. The general requirement is that roof-mounted

---

[1] IRC 301.1, Design.   [3] IRC 909.2-3, Structural and Constructural Loads.
[2] IRC 301.3, Dead Load.   [4] IRC 301.2-4, Weights of Materials.

construction is "designed in accordance with ASCE 7-98,"[5] which provides minimum design load strength for structures. If the PV system
increases the uplift from wind loading over the entire roof, there may
be additional structural requirements for the roof and other framing
members.[6] Chapter 10 has more detailed information on wind loading
and on design requirements.

Snow loading is the downward force from the weight of the snow.
This is compounded by the fact that PV arrays tend to trap snow and may
make roof loads significantly greater than normal. Increased snow loads
decrease the allowable span and spacing of rafters. IRC 301.2[7] is a map
of design snow loads from 0 to 80 pounds per square foot. Local building
codes often are more stringent. If the face of your PV modules is glass, in
the absence of a suitable local building code specific to PV modules, the
building inspector may apply skylight or sloping glazing codes.[8]

Installation fasteners and roof penetrations must be weather-tight.
The inspector will check that roof penetrations are waterproofed, flashed,
or booted.[9] Requirements will vary with the type of roofing. Photovoltaic shingles and other building-integrated roofing materials that replace
traditional roofing materials must meet the code requirements and fire
rating as defined by UBC and UL for the type of material they replace. In
this case you have to rely on the PV manufacturer to supply the information necessary to answer a building inspector's questions and to recommend the product's proper application in terms of span and slope.

With respect to safety and access, building inspectors can judge roof-
mounted PV arrays the way they judge other roof-mounted mechanical
equipment. They may require work space around the PV system[10] or
that guards be installed if the roof pitch is steep or the system is close to
the edge.[11,12]

Other codes and local ordinances may apply. After the other inspectors had signed off our PV system, the aesthetics inspector wanted to reject
our array because 6 inches of aluminum frame on one panel might be visible

[5]ASCE 7-98, Minimum Design Loads for Buildings and Other Structures.
[6]IRC 802.10, Roof Tie-down.
[7]IRC 301.2, Map of Design Snow Loads (from 0 to 80 pounds per square foot).
[8]IRC 308.6, Skylights and Sloped Glazing.
[9]IRC 2107.2.7, Roof Penetrations.
[10]NEC 110-16, Working Space Around Electric Equipment (600 volts nominal or less).
[11]IMC 304.8, Guards.
[12]IMC 306.6, Sloped Roofs.

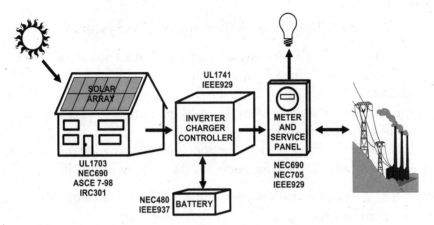

Figure 7.1. Codes that affect your PV system.

from the street and the color did not match our brown roof. We told him we would paint the edge of the offending panel brown if it really bothered him, and the other inspectors, who had discussed our plans with us throughout the installation, teased him into changing his mind.

Electrical Codes

The majority of electrical code requirements for a PV system are listed in Article 690 of the *National Electrical Code*. The *NEC* contains additional articles that also affect the installation of your PV system.

Table 7.1. Relevant Articles in the *National Electrical Code*

Article	Title
110	Requirements for Electrical Installation
200	Use and Identification of Grounded Conductors
210	Branch-Circuit Ratings
220	Branch-Circuit, Feeder, and Service Calculations
240	Overcurrent Protection
250	Grounding
300	Wiring Methods
310	Conductors for General Wiring
339	Underground Feeder and Branch-Circuit Cable, type UF
384	Switchboards and Circuit Boards
445	Generators
450	Transformers and Transformer Vaults
480	Storage Batteries
690	Solar Photovoltaic Systems
705	Interconnected Electric Power Production Sources
720	Circuits and Equipment Operating at less than 50 Volts

Installing a Code-Compliant PV System

The first thing that an inspector—structural, mechanical, or electrical—notes when walking onto a job site is how tidy or messy it is. A clean and orderly job site is the first indication of quality workmanship. Inspectors want to see neat wire runs, good connections, and get a general sense of an orderly installation.

The second thing an inspector does is evaluate the system. Look at your system and ask the same questions that the inspector will ask.

- Is the system well-designed and have you installed quality components?
- Have you used good common sense in the installation?
- Is the system properly sized to meet expected or required performance?
- Are equipment components appropriately rated and UL-listed?
- Are outdoor materials sunlight- and weather-resistant?
- Are components labeled properly?
- Are there any safety hazards?
- Are roof penetrations properly weather-sealed?
- Are the installation and interconnection code compliant?
- Is the grounding system code compliant?
- Are you familiar with the basic operation and maintenance of the system and components?

The Electrical Inspection Checklist

Use this checklist of system components and the sections of Article 690 (2008 edition of the *NEC*) that apply to those components to ensure that your PV system is code compliant.

Table 7.2. System Components and Corresponding Article 690 Sections

INSTALLATION		
Source circuits and Output circuits	Photovoltaic source circuits and photovoltaic output circuits are not contained with other conductors unless the conductors are separated by a partition.	690.4(B)
Array disconnect	Open circuiting, short circuiting, or opaque covering can disable the PV array for installation and service.	690.18

Inverters and Motor generators	Inverters or motor generators are identified for use in solar photovoltaic systems.	690.4(D)
Ground-fault protection	DC ground-fault protection shall be provided for roof-mounted DC PV arrays. (Residential roof-mounted PV systems cannot be grounded on the DC side per 690.41.)	690.5
Labels and Markings	Labels and markings are near the ground-fault indicator at a visible location.	690.5

CIRCUIT REQUIREMENTS

Source circuit voltage	The maximum photovoltaic source circuit voltage is calculated as the sum of the rated open-circuit voltage of the series-connected PV modules corrected for the lowest expected ambient temperature.	690.7(A)
Maximum system voltage	Maximum system voltage at lowest expected ambient temperature is less than module maximum voltage rating.	690.7
	Maximum system voltage to dwellings is less than 600 volts.	690.7(C)
	Circuits over 150 volts to ground are accessible to qualified persons only.	690.7(D)

CIRCUIT SIZING

Source circuit, Output circuit and Overcurrent protection	The maximum PV source circuit, output circuit, and over-current protection devices current are the sum of the parallel module-rated circuits multiplied by 125% ($\times$ 1.25).	690.8(A)
	PV source circuit, output circuit & overcurrent protection devices are sized for no less than the maximum current calculated in 690.8(A) multiplied by 125%. Note that the combined multiplication factor for 690.8(A) & (B) is 1.56.	690.8(B)
Inverter circuits	The maximum current of the inverter output circuit conductors and overcurrent devices are the continuous output rating of the inverter.	690.8(A)(3)
	Stand-alone inverter input circuit conductors and over-current devices are sized to the input current at rated output at the lowest operating voltage multiplied by 125%.	690.8(A)(4)
General	All equipment and devices are rated for 125% of maximum voltage unless specified by the manufacturer.	

OVERCURRENT PROTECTION

All circuits	Photovoltaic source circuit and output circuit, inverter output circuit, and storage battery circuit conductors and equipment are protected in accordance with Article 240.	690.9(A)
Transformers	Overcurrent protection has been provided for power transformers in accordance with Article 450.3.	690.9(B)
Source circuits	Branch circuit or supplementary-type overcurrent devices have been provided for photovoltaic source circuits and are no greater than the series fuse on the module listing.	690.9(C)
Overcurrent devices	Overcurrent devices are listed for use in DC circuits and have the appropriate voltage, current & interrupt ratings.	690.9(D)

Stand-alone systems	AC inverter output has a means of disconnect.	690.10(A)
	Circuit conductors are sized based on the output rating of the inverter and in accordance with Article 240.	690.10(B)
	There are no multi-wire branch circuits and the equipment is adequately labeled.	690.10(C)

DISCONNECTS

All conductors	A means of disconnect has been provided between the photovoltaic system output and other building conductors, except for the grounded conductor.	690.13
	The system disconnect is in a readily accessible location either inside or outside the building, but specifically not in the bathroom.	690.14(C)
Labeling	The PV system disconnect is labeled and is suitable for the prevailing conditions.	690.14(C)
Number	There are no more than six grouped disconnects for the system.	690.14(C)
All equipment	A means of disconnect has been provided for inverters, batteries, charge controllers, and any other source of power.	690.15
Fuses	A means of disconnect is supplied to disconnect a fuse from all power sources if the fuse is energized from both directions.	690.16
Disconnect switch and Circuit breaker	A manually operable switch or circuit breaker has been provided to disconnect underground conductors and has an appropriate interrupt rating.	690.17
Labeling	If the terminals of the disconnect are energized in the open position the hazard has been appropriately labeled.	690.17

WIRING

Wiring system	Appropriate code-compliant wiring has been used. Fittings specifically intended for use for PV are allowed. Devices with integral enclosures have enough wiring slack to allow replacement.	690.31(A)
Single conductor cable	Types SE, UF, USE, and USE-2 are allowed in source circuits. All cables exposed to sunlight are identified as sunlight-resistant.	690.31(B)
Tracking array	Tracking arrays can use flexible cords and cables to connect moving parts. This cable must be suitable for extra-hard usage and listed for outdoor use—water-resistant and sunlight-resistant.	690.31(C)
Module interconnections	Module interconnection cables are outdoor-rated, with a temperature rating or 194°F (90°C) or greater and meet the requirements in 690.8.	690.31(D)
Component interconnections	Connectors are the locking or latching type and are polarized and non-interchangeable with receptacles in other electrical systems on the premises. Male and female connectors can only be used in systems with the same voltage, and AC plugs and DC plugs must be different.	690.33

Grounding	The grounding member is the first to make contact and the last to break contact.	690.33(D)
Junction boxes	Junction, pull, and outlet boxes are accessible and are secured by removable fasteners and connected by a flexible wiring system.	690.34
50 volts or less	Conductors in systems operation 50 volts or less are not smaller than #12 AWG copper or equivalent.	720.4

GROUNDING

DC conductor	The DC conductor is solidly grounded at a single point for a two-wire system with a voltage of over 50 volts (except roof arrays, which must be ground fault protected per 690.5).	690.41
Connection	The ground connection can be at any point, but should be as close as practical to the PV array.	690.42
All equipment	Non-current carrying metal parts of module frames, equipment, and conductor enclosures are grounded regardless of voltage.	690.43
	If there is no ground fault protection, the grounding conductor is sized to 125% of the PV source and output circuits. If there is ground fault protection (as is required in Article 690.5), the grounding conductor is sized in accordance with Article 250.122.	690.45
	All exposed surfaces in the PV system are grounded with equipment grounding conductors that must remain grounded if any equipment in the PV system is removed for service.	690.48
Electrode system	The DC and AC grounding electrodes are bonded to a single grounding electrode system (ground rod) and the DC and AC grounding conductors are bonded to a single electrode.	690.47(C)

MARKINGS

Modules	PV modules are marked with open-circuit voltage (V_{oc}), operating voltage (V_{op}), maximum permissible system voltage (V_{max}), short-circuit current (I_{sc}), operating current (I_{op}), and maximum power (P_{max}).	690.51
DC disconnect	There is a label at an accessible location indicating I_{op}, V_{op}, V_{max}, and I_{sc}.	690.53
Other power sources	There is a label at the point of interconnection indicating maximum AC operating current and the operating AC voltage.	690.54
Batteries	Batteries are labeled with V_{max}, equalization voltage, and polarity.	690.55
Stand-alone systems	Off-grid PV systems have a permanent plaque indicating the location of the disconnect in an accessible location.	690.56(A)
Utility interconnect systems	There is a permanent plaque indicating the locations of the service disconnect and the PV system disconnect.	690.56(B)

CONNECTION TO OTHER SOURCES

Inverters	Inverters are listed and identified for interactive operation.	690.60
	The inverter de-energizes when the interactive source of power is lost.	690.61
	A grid-connected PV system can have stand-alone capability when the interactive source of power is lost.	690.61
Interconnections	There are no unbalanced interconnections.	690.63
Connections	The PV power source can be connected to either the supply or load side of the service disconnect.	690.64(A,B)
	PV connected to the load side of the service disconnect has (a) a dedicated branch circuit or fusible disconnect, (b) amp ratings of breakers that do not exceed the busbar rating (120% for homes), (c) the interconnection point is on the line side of any ground fault protection, and (d) back-fed breakers are identified.	690.64(B)

BATTERIES

Installation	Appropriate racks, trays, and ventilation are used.	480.8, 480.9, 480.10
Interconnections	Flexible cables identified in Article 400 in sizes 2/0 AWG and larger and identified for hard service use are used.	690.74
Voltage	Batteries for residential applications have the cells connected so that the operating voltage is less than 50 volts, nominal.	690.71(B)
	Live parts are not accessible during routine maintenance.	690.71(B)
Fuses	Current limiting fuses are installed on battery output circuits.	690.71(C)
Cases	Lead-acid batteries with more than 24 2-volt cells in series are in non-conductive cases.	690.71(D)
Disconnects	Series disconnects are provided for battery strings over 48 volts.	690.71(E)
	A disconnect, accessible only to qualified persons, has been provided for the grounded circuit conductor in battery systems over 48 volts.	690.71(F)

CHARGE CONTROLLER

Charge controller	A charge controller is used in any system where the charge rates are greater than 3% of battery capacity. Adjustment of the charge controller is accessible only to qualified persons.	690.72(A)
Diversion charge controller	Diverted power has an independent means of charge control and DC diversion loads, conductors, and over-current devices are rated for at least 150% of the controller current rating.	690.72(B)

Avoid Installation & Inspection Problems

A final review of common problems areas will help both your installation and inspections go smoothly.

- Are all structural attachments of PV arrays to rooftops and other structures secure?
- Are all roof penetrations adequately sealed and weather-proofed?
- Is all wiring safe?
- Are all conductors sufficiently rated and insulated?
- Is all overcurrent protection properly placed and rated?
- Are all disconnects properly placed and rated?
- Are the batteries installed safely, with proper containment and ventilation. Are you prepared to discuss maintenance?
- Have you used all UL-listed equipment?
- Is each UL-listed component used for the purpose it was designed for?
- Is everything properly grounded?
- Are all major system components and disconnect devices properly labeled?
- Do you have adequate documentation on system design and operating and maintenance requirements?

If you can answer "yes" to all of these questions, you are ready for the building inspector.

IMPORTANT
You and your installer are responsible for the safe design, installation, operation, and maintenance of your PV system.

Chapter 8
A SHORT COURSE IN ELECTRICITY & ELECTRICAL SAFETY

Electricity

Electricity is a fundamental property of charged particles. **Current** is electricity in motion. Electromotive force (**voltage**) is the electric pressure that causes current to flow in a circuit. The rate of flow of the current is measured in **amperes** (**amps**).

If we compare a wire or conductor of electricity to a water hose, voltage would be water pressure and current would be the rate of flow. Filling a bucket with water using 12 pounds per square inch of pressure at a flow rate of 2 gallons per minute is analogous to filling a battery with electricity using 12 volts at 2 amperes. We can also compare the diameter of a hose with the diameter of a wire. A large diameter hose will let water flow with less resistance; a large diameter (cross-sectional area) wire will let current flow with less resistance. **Resistance** is a property of the conductor and is measured in **ohms.**

There are two types of electric current: **direct current** (**DC**) and **alternating current** (**AC**). Direct current is one-way flow of electrons from minus to plus (although by convention, current is said to flow plus to minus). The most common uses of DC are in flashlights, portable radios, and the electrical system of cars and trucks. All of these devices are energized by the direct current from batteries, rectifiers, or DC generators.

PV modules produce DC electricity, which can be used to charge storage batteries or directly power DC electrical loads such as a DC fan or well pump motor. DC can also be fed into an inverter and converted to AC to power conventional appliances.

On-grid homes use alternating current—the flow of electrons first in one direction and then reversed in the other direction at a **frequency** of 60 cycles per second or 60 **hertz (Hz)**. In our homes we use AC current from the electrical grid. Typical house current in the United States is between 110/220 and 130/260 volts and cycles 60 times per second (60 Hz). This is commonly called 120/240 VAC. Most household appliances operate at 120 or 240 volts AC.

DC-to-AC PV systems can be paralleled to the utility grid to offset your on-grid PV home's electrical consumption. Only inverters that produce a sine wave synchronized in shape, frequency, and amplitude to the electrical grid can be used in an on-grid PV system. AC volts and amps are measured with **root-mean-square (RMS) meters** that measure the average of the peak cycles.

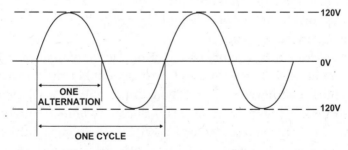

Figure 8.1. Alternating current.

Watts (W) are a measure of the amount of electric power needed to do a specific unit of work called a **joule** in a specific amount of time. One watt is one volt of current at one amp flowing through a resistance of one ohm and doing one joule of work in one second. W is the symbol for both watts and power in general. **Horsepower (hp)** is another term for an amount of work done by a motor. One horsepower equals 746 watts.

Electrical energy is measured in **watt-hours (Wh)**, the number of watts used or produced in one hour. For example, a 100-watt light bulb that is on for 1 hour consumes 100 watt-hours. If the same 100-W bulb is on for 10 hours, its consumption is 1,000 watt-hours or 1 **kilowatt-hour (kWh)**.

Table 8.1. Electrical Terms

Term	Unit	Abbreviation	Description
Power	horsepower	hp	746 watts
Energy	joule	J	A unit of work = 9.48×10^{-4} BTU or 2.78×10^{-7} kWh
Resistance	ohm	Ω	1 volt / 1 amp
Electric potential	volt	V, E	1 joule / 1 coulomb
Electric power	watt	W	1 joule / 1 second
Electric energy	watt-hour	Wh	1 watt / 1 hour; 2.778×10^7 J
Electric current	ampere	amp, A, I	An internationally agreed upon unit of current 1 coulomb / second
Electric charge	coulomb	C	The amount of electricity transferred in one second by a current of one ampere The charge equivalent of 6.281×10^{18} electrons

All of these electrical terms are defined in relationship to each other. This relationship is expressed in the formula:

$$E = IR$$

Volts (E or V) = Current (I or A) × Resistance (R)

This formula is called **Ohm's Law** and relates work and current to resistance. We used this formula in Chapter 3 when we converted amps to watts to calculate how much energy appliances consume.

If there is electron flow (current) and voltage, work (watts) is performed.

$$W = IE$$

Watts = Current (I or A) × Volts (E or V)

And since E = IR:

$$W = I \times IR \text{ or } W = I^2R$$

The relationships between these formulas and definitions are illustrated in the following chart.

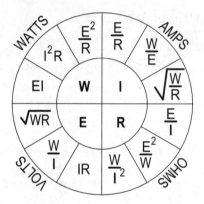

Figure 8.2. Ohm's Law chart.

Wire Resistance & Voltage Drop

The formulas in the chart can be used to calculate the size wire needed for your PV system.

Table 8.2. Copper Wire Resistance
(ohms per 1,000 feet of wire)*

Wire Gauge (AWG)	Resistance
18	6.385
16	4.016
14	2.525
12	1.588
10	0.999
8	0.628
6	0.395
4	0.249
2	0.156
1/0	0.098
2/0	0.078

*Single-stranded uncoated copper wire at 68°F/20°C.

Resistance increases as a function of temperature. The formula for calculating resistance of wires at different temperatures is:

$$R_2 = R_1 [1 + \alpha (T_2 - T_1)]$$

α is a constant characteristic of the metal. The constant for copper (α_{Cu}) is 0.00323. The constant for aluminum (α_{Al}) is 0.00330.

From the table we determine that #10 copper wire has a resistance of 0.999 Ω, or about 1 Ω per 1,000 feet or 0.1 Ω per 100 feet. If you connect a battery to a 100-watt, 12-volt light bulb, as illustrated in Figure 8.3, the total path length is 50 feet (25 feet to the bulb and 25 feet back).

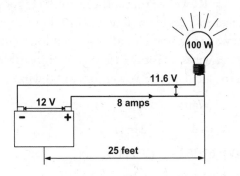

Figure 8.3.

Using Ohm's Law calculations, the resistance is:

$$1 \,\Omega \times 50 \text{ ft} \div 1,000 \text{ ft} = 0.05 \,\Omega \text{ wire resistance}$$

and the current is:

$$I \text{ (or A)} = W \div E = 100 \text{ W} \div 12 \text{ V} = 8.33 \text{ A}$$

Voltage drop is the loss of electrical pressure as the current flows through an impedance or resistance. To calculate the voltage drop we again refer to the Ohm's Law chart and use the formula:

$$\text{Volts or } E = IR = 8.33 \text{ A} \times 0.05 \,\Omega = 0.416 \text{ V}$$

The amount of voltage loss is 0.416 volts. So

$$12 \text{ V} - 0.416 \text{ V} = 11.59 \text{ volts}$$

are delivered to the bulb. The power actually available to the bulb is

$$11.59 \text{ V} \times 8.33 \text{ A} = 96.54 \text{ W}$$

The *NEC* recommends, but does not require, 3% or less voltage drop. More than 3% voltage drop should be avoided. Since 0.416 voltage loss divided by 12 volts equals 3.47% voltage drop, the #10 wire should be changed to #8 wire for 2.17% voltage drop. A #6 wire would allow even more power to the bulb.

$$0.628 \ \Omega \times 50 \text{ ft} \div 1000 \text{ ft} = 0.0314 \ \Omega \text{ wire resistance}$$

$$E = 8.33 \text{ A} \times 0.0314 \ \Omega = 0.261 \text{ V}$$

$$0.261 \text{ V} \div 12 \text{ volts} = 0.0217 \times 100 = 2.17\% \text{ voltage drop}$$

If we look at the formula $E = IR$ we see that if the current (I) goes up so does the voltage. If the resistance goes up, as it would in a thinner wire, the voltage must go up. If you want to carry a lot of current at lower voltages, you need shorter and thicker wires, as is the case with battery cables. We can also use this formula to calculate the power loss in #8 wire.

$$\text{Watts} = I^2 \times R$$

$$W = 8.33^2 \text{ A} \times 0.314 \ \Omega$$

$$W = 69.388 \text{ A} \times 0.0314 \ \Omega$$

$$W = 2.178 \text{ power loss}$$

Series & Parallel Circuits

The characteristics of a combination of circuits can be expressed as a single equivalent circuit. If the circuits are attached end to end, they are in series. In the example below there is a single path for the current, I (amperage), through the resistors or loads.

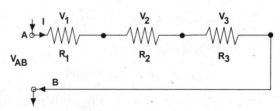

Figure 8.4. Resistors in series.

The resistance of the circuit is the sum of all three resistors or:

$$R = R_1 + R_2 + R_3$$

In a solar module the cells are connected in series. If there are 36 cells at 0.5 volts each, the voltage for the module is the sum of all the voltages or 18 volts. In each case the current (I) remains the same:

$$V = V_1 + V_2 + V_3$$

If the connections are made between two points in the circuit (as in the example below), the circuits are in parallel.

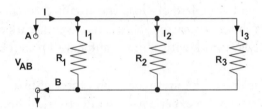

Figure 8.5. Resistors in parallel.

In this case voltage is the same across each element of the circuit. There are three different paths for the current, so the total amperage is the sum of the amperages for the three circuits or:

$$I = I_1 + I_2 + I_3$$

Going back to the Ohm's Law chart, I = V ÷ R, so—

$$I = (V_{AB} \div R_1) + (V_{AB} \div R_2) + (V_{AB} \div R_3)$$

An example of circuits in parallel is the wiring in your house. Each load operates independently of the others and each load draws its own current. The total circuit current is the sum of all the currents of all the appliances on that circuit. If this sum of all the currents (amperages) on a circuit exceeds the amp rating of the fuse or circuit breaker, it will interrupt the circuit. If no fuse is used, resistance can cause the wire to get hot and cause a fire.

Electrical Safety

The *National Electric Code* (*NEC*) is a comprehensive list of practices to prevent electrical hazards such as fire and electrocution. Follow the electrical code to ensure that your PV installation is safe.

Electricity is an invisible, constant part of our lives; usually not much thought is given to its potential dangers. Working with power tools or electrical circuits has the risk of electrical hazards. Coming in contact with electrical voltage can cause current to flow through your body, resulting in electrical shock and burns. Serious injury, even death, may occur. If you touch a live wire or any live component of an energized electrical device and complete or close the circuit through your body, you will receive a shock, a burn, or you will die. Never work on electrical circuits alone. Always work with a buddy and have a plan for what to do in case of shock or injury.

Electricity is one of the most common causes of fires and thermal burns in homes. If you do experience a small electrical fire, use only a Class C or multi-purpose (ABC) fire extinguisher, otherwise you will only make the problem worse. Water on an electrical fire can give current a path to flow and result in electrocution.

Protect yourself by identifying the hazards.

- Inadequate wiring
- Exposed electrical parts
- Overhead power lines
- Wires with bad insulation
- Electrical systems and tools that are not grounded or not double-insulated
- Overloaded circuits
- Damaged power tools and equipment
- Not wearing proper protective clothing
- Using the wrong tool
- Defective ladders and scaffolding
- Ladders that conduct electricity
- Water on or near equipment, workers, or the work location

All of these hazards are very dangerous.

In order to eliminate hazards, you must begin by creating a safe work environment and then work in a mindful manner. A safe work environment means avoiding contact with electrical voltages and currents. Before equipment is inspected or repaired, even on low-voltage circuits, the current must be turned off at the switch box. The equipment must be securely locked and tagged to warn everyone that work is being performed. Test circuits and equipment to make sure they have been de-energized. Treat all conductors—even de-energized ones—as if they are energized even if you have locked them out and tagged them.

Use tools correctly and only for their intended purposes. Follow the manufacturer's safety instructions and operating procedures. When working on a circuit, use approved tools with insulated handles. Remember: These tools cannot protect you if you use them to work on energized circuits. **Always shut off and de-energize circuits before starting work.**

Portable electrical tools are classified by the number of insulation barriers between the electrical conductors in the tool and the worker. Equipment that has two insulation barriers and no exposed metal parts is called double-insulated. When used properly, double-insulated tools provide reliable shock protection. Power tools with metal housings or only one layer of effective insulation must have a third ground wire and three-prong plug. Removing the ground prong or using a two-prong adaptor is hazardous.

Any electrical hazard becomes much more dangerous in damp or wet conditions. Assume your work location is damp, even if you do not see water. Wooden ladders can soak up water and become conductive. Metal ladders can conduct current. Use fiberglass ladders.

More than half of all electrocutions are caused by contact with overhead lines. When working in an elevated position, on a roof or near overhead lines, stay away from unguarded or uninsulated wires and metal that may conduct current. Keep yourself and anything you touch at least 10 feet away from high-voltage transmission lines.

Use multiple safe practices. There is always a chance that a circuit may be wired incorrectly. There may be unseen dampness. Wires may contact other "hot" circuits or metal frames or enclosures and boxes. Someone else may do something that could put you in danger. Take all possible precautions and then some. Thousands of homeowners have heeded this advice and safely installed their own PV systems. You can too.

For more information on working safely with electricity, contact the National Institute for Occupational Safety and Health (800.356.4674; www.cdc.gov/niosh). DHHS (NIOSH) Publication No. 2002-123 is a basic training manual on electrical safety. OSHA Occupational Safety and Health Administration standards are located in the Code of Federal Regulations (CFR). The full text of these standards is available for free on OSHA's Web site (www.osha.gov).

Chapter 9
SOLAR MODULES

Solar modules are the building blocks of your PV system. Modules are also the most expensive part, accounting for about 50% of the cost of an off-grid system and up to 85% of the cost of an on-grid batteryless system. The word "module" describes an important feature of photovoltaic system design. You can increase your PV system size and power output modularly as your electric loads increase or your budget allows.

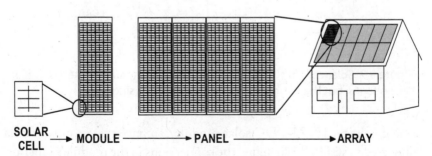

Figure 9.1. PV modularity makes installing and expanding your solar power system simple.

How Solar Modules are Made

A PV module is an assembly of hermetically sealed solar cells. Single-crystal and polycrystalline silicon solar cell modules are used most often for residential PV systems because they are designed for this application, are readily available, competitively priced, and have twice the power density (watts per square foot) that amorphous solar modules have. Higher power density makes crystal cells preferable when the space available for the PV array is a consideration in the design of the system.

Cells are connected in series to produce useful voltage. Single-crystal and polycrystalline cell solar modules can have 36 cells (of about 0.5 volts each) connected in series to produce at least the 14.5 volts needed to charge a 12-volt battery. A 24-volt PV system requires two 36-cell modules in series to produce the operating voltage. A 48-volt system needs four 36-cell modules in series. Most grid-tie inverters operate at up to 600-volts DC input and require several 36-cell modules connected in series. The total voltage of a module is the sum of individual series cell voltages; the current in the series cell string is the same as that of a single cell. Most 36-cell modules produce between 16 and 18 peak power volts.

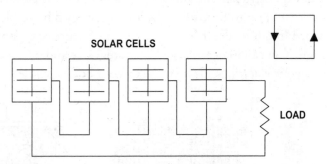

Figure 9.2. Solar cells in series. The total voltage is the sum of the individual cell voltages, but the current is the same as that of a single cell.

A dual voltage, 12/24-VDC solar module can have two groups of cells connected in series. By changing the connections in the module's junction box, the two series strings can be connected either in parallel (for 12 V) or in series (for 24 V) at the time of installation.

Table 9.1. Specifications for a Typical 12/24-Volt Module		
Electrical Ratings	24 V	12 V
Typical power (W)	100	100
Minimum power (W, P_{min})	95.1	95.1
Operating voltage (V, V_{max})	34.4	17.1
Open-circuit voltage (V_{oc})	43.2	21.6
Current at rated operating voltage (A, I_{max})	2.9	5.9
Short-circuit current (A, I_{sc})	3.0	6.0

Both configurations obey Ohm's Law (see Chapter 8). Power (watts) remains the same, and the current (amps) and electric potential (volts) are inversely proportional to each other. In some 72-cell solar modules all of the cells are connected in series. These are nominal 24-volt modules. In all cases, the primary voltage of the battery bank or the DC load or the inverter determines the number of solar modules to be connected in series.

Module assembly begins at the factory when cells are tested and matched by voltage and current. Next, two flat wires are soldered from the top of each cell to the bottom of the next cell to connect the cells in series. The two cell interconnection wires create redundant electrical paths that provide backup should one conductor or solder connection fail. Some modules have three cell interconnection wires for more conductivity, but there is a limit to the number of wires and cell grid lines. At some point they will block light from reaching the cell's photoactive surface and reduce overall power. Some cells have back contacts only and no front grid.

Solar cells are very fragile and must be protected from the weather and physical damage. After the cell connections have been made, the cells are sandwiched between layers of plastic, usually ethylene vinyl acetate (EVA), and a glass front and a weatherproof plastic back sheet. This assembly is then laminated using heat and vacuum to hermetically seal the cells and their connecting wires. The glass front sheet is tempered for strength and is lower in iron than regular glass, which allows more light to enter and strike the cell. The protective back sheet is usually made of white plastic Tedlar®,* which reflects light and keeps the cells cooler.

Impact-resistant or bullet-proof plastic is sometimes used to protect cells from vandalism. Although plastic may not shatter like glass, plas-

---

*Tedlar is registered trademark of E. I. duPont de Nemours and Company, Wilmington, DE.

tic solar modules are not "unbreakable." Plastic can scratch and attract dirt, and it may not protect cells from impact shock fractures. Crystal solar cells encapsulated in plastic or large spans of glass can break if the module is flexed during handling. Tempered glass solar modules are built to withstand the impact of a 2-inch (52-mm) steel ball dropped from a height of 51 inches (1.295 meters). This is similar to the terminal velocity of hail. Modules can also withstand over 30 pounds per square foot of snow load.

A frame is assembled around the laminated module to add strength and protect the edges. Anodized aluminum is the most common module frame material, but steel, plastic, and even wood can be used. Anodized aluminum frames and stainless-steel fasteners are recommended because mismatched metals and fasteners will cause electrolytic corrosion. Hard-anodized aluminum frames are very long lasting, but eventually they will pit and oxidize, especially in marine environments. Wood can be attractive, but requires extra maintenance, such as painting, and it does not provide an earth-grounding path. Steel, if used, should be stainless, galvanized, painted, or epoxy-coated. Coated and painted metal frames and anodized aluminum must be sanded or scratched down to bare metal to ensure good conductance to an earth ground. Stainless-steel star washers are useful because they scratch into the frame surface for grounding conductivity. This practice is not Underwriters Laboratories (UL) approved, however, and may not pass inspection.

In addition to protecting the glass laminate, the frame is part of the mounting and grounding systems. Predrilled frames should be easy to re-drill because mounting holes do not always line up during installation. Some manufacturer warranties do not allow drilling. Box and channel frames are strong but may require special insert fasteners and tools.

Some modules are made with dark frames and dark back sheets for uniform appearance. But because of added heat absorption, black modules have higher normal operating cell temperatures (NOCT), which means power loss. Typical crystalline-cell module temperature-induced power loss or gain is 0.5% per degree above or below 77°F/25°C. Black-framed modules with no ventilation can reach 178°F/81°C and can lose as much as 20% of their energy production from heat-induced voltage drop.

Glass-front modules with grounded metal frames have survived direct lightning strike tests with little or no damage. The glass insulates the cells, and the grounded metal frame provides a good current path to earth. Plastic and ungrounded modules did not survive field and laboratory lightning tests. Charge controllers, batteries, and inverters can be damaged by lightning, so lightning arrestors must be used if your region is prone to lightning.

Unframed laminates are sometimes used instead of framed modules in applications where the unprotected edges will not be hit accidentally. Laminates can be bonded to a support structure or mounted with silicone glue. Unframed laminates cost slightly less than framed modules, but have higher shipping and handling costs and increased breakage, which often offset any possible savings.

Some UL-listed solar modules have junction boxes (j-boxes) to protect the positive and negative wire connections. The j-box may also enclose bypass and blocking diodes. Easy-to-remove "knock-outs" allow wire conduit or watertight cable connectors to be fastened to the j-box. Otherwise, holes must be drilled into the j-box to make connections. Most modules now have factory-installed wires and connectors with "quick-connects" for faster installation. Quick-connectors have been used since 1990 and are expected to withstand 30 years of exposure to sun and weather. Locking or latching module interconnectors are becoming required equipment.

Module size has increased with PV manufacturers' ability to make bigger cells. In the 1970s, the largest cells were 2 inches (50 mm) in diameter. By 2000, 6-inch (152-mm) square cells were made from polycrystalline ingots, and 8-inch (203-mm) crystal wafers are common. Some modules are over 25 square feet and weigh over 100 pounds. Large modules require fewer array interconnects, but reduce design modularity, require more people for handling, and cost more to ship.

Automation and quality control have improved PV module reliability and reduced costs. Many PV factories are certified by the International Organization for Standardization (ISO) for quality standards. Technical standards set by the Institute of Electrical and Electronics Engineers (IEEE) and Underwriters Laboratories (UL) safety criteria have become standard. UL-listed modules are given repetitive cycle testing between -40°F/-40°C and 185°F/85°C at 85% humidity and are guaranteed to perform to specifications in extreme weather conditions. Field testing of well-designed PV systems is ongoing. Long-term field tests have found 0.5 to

2% per year module power degradation. However, 25-year warranties at 80% of factory-rated power attest to built-in longevity and are emphasized in competitive marketing literature.

How to Read a Solar Module Spec Sheet

The information you need to size your solar array should be given on the solar module specifications sheet. The information in this section refers to the sample spec sheet shown on pages 159–60.

Your Application
First, determine if the solar module you are considering is designed for your application. Most glass-encapsulated, aluminum-framed 50- to 300-watt solar modules are designed for general use, which includes powering DC loads directly, battery charging, and generator- and utility-interconnected applications. Modules made for marine or vehicle use may not be appropriate for residential applications.

Appearance
Most spec sheets include a black-and-white or color photograph of the module. In some applications, the color and shape of the cell may be important. Single-crystal cells range from round to square and dark gray to almost black. Polycrystalline cells are usually square and range from blue to very dark blue. The color varies with the anti-reflective treatment used. Some people prefer the traditional round-cell "polka dot," while others want a more uniform appearance. Some people like the blue, variegated "jack frost" appearance of multi-crystalline cells.

The sample spec sheet shows two models of its module. Although some find all-black modules aesthetically appealing, remember that they have higher operating temperatures and associated power loss.

Power Density
The photos also help you visualize module power density or watts per square foot. Watts per square foot and cost per watt help you determine which module is your best choice. A module made of round 6-inch cells has a lot of inactive area compared to a module made of closely packed 6-inch square cells, but power density may not be as important to you as

the price difference. Most 36-cell modules are rectangular with 4 rows of 9 cells. Thus, a module with 6-inch cells will be approximately 2 feet by 5 feet. Once you determine your PV array watts, your roof size will determine which array of modules will look and fit best.

Size, Weight & Cost

Next, look at the module's physical size, weight, electrical characteristics, and cost. Small modules may not be cost-effective and very large modules may not offer the modularity you need. Weight affects shipping costs and ease of installation. The S125 module in the sample spec sheet weighs 25.0 pounds (11.4 kg). If each carton contains one module, shipping weight and size can be extrapolated by adding about 5 inches to each module dimension and about 10 pounds to the weight. UPS and other carriers charge extra for oversized packages. Always confirm shipping method and charges before you buy your modules. Modules shipped on a pallet require a forklift or lift-gate for unloading and handling.

Weight and size are important array design parameters. Most solar modules weigh 3 pounds per square foot. The S125 module weighs 2.5 pounds per square foot. The allowable limit on dead loads for roof-mounted equipment is usually 4 pounds per square foot; over that weight structural analysis and strengthening are required. Some inspectors require structural engineering calculations for wind loading, seismic loading, roof framing, array mounting structure, and fasteners. The added cost for a structural engineering report for modules made of heavy glass and mounting hardware that add up to 4 or more pounds per square foot often offsets other savings.

Take weight and size into consideration when planning the installation. The S125 module assembled with aluminum rails into a prewired 6-module panel would weigh approximately 200 pounds. That may be too heavy or awkward for two people to handle. Modules can be individually fastened to a mounting structure and then field-wired. If you use this method, be prepared to drill out any mounting holes that do not match. Some mount manufacturers offer racks with clips to hold modules from the top instead of bolts through module frame holes. Always plan your installation for ease and safety, especially when installing conduit and wiring to premounted modules.

Standards & Warranties

This spec sheet includes the power output warranty and confirms that the manufacturer uses International Organization for Standardization (ISO 9001) quality standards. The module is listed with Underwriters Laboratories (UL), certified by Factory Mutual (FM) and TUV for safety, meets European Community (CE) standards, and meets Jet Propulsion Laboratory's (JPL) comprehensive standards. Another testing service, ETL (Electrical Testing Laboratory), now called Intertek Testing Services or ITS, tests to UL standards. Building inspectors require the UL label on electrical products, including PV modules to ensure that the products meet safety requirements. FM indicates that the module is safe for use in potentially explosive areas, such as oil and gas fields.

The Jet Propulsion Laboratory, Pasadena, California, established a series of environmental and performance standards known as the JPL Block V tests. The JPL PV program was terminated in 1984, but Block V, with the addition of fire tests, has survived to become the standard for Underwriters Laboratories module testing in the U.S. UL-1703 testing standards include thermal cycling, thermal shock, impact, wind load, humidity, freezing, corrosion, and electrical isolation. Modules that pass UL testing are strong and durable and should last for decades. Japan and the European Union have similar tests established by the International Electrotechnical Commission (IEC).

Solar modules are tested and rated to standard test conditions (STC): 1,000 watts per square meter irradiance at 77°F/25°C cell temperature and air mass 1.5. These conditions are approximately the sun's full intensity at noon on a very clear day. A module's peak power rating is the highest combination of load current and load voltage, plus or minus 5% or 10% to allow for production tolerances.

International power and energy standards are being developed, but it may take several years before they are universally accepted. Some U.S. utilities and rebate programs use PVUSA Test Conditions (PTC), which were developed at the Davis, California, PVUSA test site. PTC is based on field measurements corrected to 1,000 watts per square meter sun, 68°F/20°C ambient temperature, and 1 meter per second (2 miles per hour) wind speed at 10 meters (39.37 feet) above grade.

PTC ratings are closer to real-world performance because they factor in the effect of increased temperature. In full sun most solar modules are heated to 86°F/30°C above ambient temperature. Most crystalline modules lose approximately 0.5% power per degree above 77°F/25°C cell temperature. At 68°F/20°C ambient temperature this module could be 122°F (50°C) and power would be reduced by 12.8% to approximately 109 watts.

The small charts, drawings, and boxes on the back of the spec sheet contain important data. The dimension drawing shows that this module is 56.33 inches long (1,429 mm) by 24.75 inches wide (627 mm). From this we calculate that the module is 9.68 square feet (0.896 square meters). Power density is 125 W divided by 9.68 square feet, which equals 12.9 watts per square foot (145.3 watts per square meter). Module sun-to-power efficiency is also based on module area.

$$\text{Efficiency} = [\text{module watts} \div (\text{module m}^2 \text{ area} \times 1{,}000 \text{ W} \div \text{m}^2)] \times 100$$
$$= [125 \div (0.896 \times 1{,}000)] \times 100 = 13.95\%$$

Frames & Mounts

The module frame depth is 2 inches (50.7 mm). The locations of ten 0.26 inch (6.6-mm) mounting holes are shown. If you build your own mounting structure you will need to get a more detailed drawing from the manufacturer.

The module's j-box extends 0.6 inch beyond the frame, which could interfere with some mounting structures. The j-box location is specified, but its dimensions are not given. The spec sheet indicates that the module is available with plug-in connectors. This can make connecting modules in series simple. Many installations require electrical junction boxes, conduit, and wire, so plug-in connectors may not be suitable. J-box connector holes or knock-out dimensions are not specified on this spec sheet. Be sure to get these dimensions before selecting any module.

UL requires that manufacturers include a detailed instruction manual with each solar module with recommendations for wire size, series fuses, and detailed information about the junction box (including installation diagrams) and connectors. The manual is usually available on the manufacturer's Web site.

Qualification test parameter data offers other important information used in system design. Wind loading of 30 pounds per square foot means this module can withstand hurricanes. This module can also survive

1-inch hailstones traveling at 54 miles per hour. It can be flexed 0.74 inches (1.2 degrees diagonal distortion), but it must be mounted on a level plane. Be careful not to twist modules and panels out of shape when moving or lifting them to their mounts to prevent cracking the solar cells.

Maximum power current (I_{max}), short-circuit current (I_{SC}), maximum power voltage (V_{max}), and open-circuit voltage (V_{OC}) are listed on the spec sheet. This information is useful when designing your PV system and sizing your charge controller, and can be used to test a module straight out of the box. On a cloudless day you can measure open-circuit voltage to check wiring and polarity. Bright sunlight may vary from 700 to 1,500 watts per square meter intensity. Outdoors on a bright day (800 watts per square meter or 80% of full sun conditions) at 77°F/25°C, the S125 module will put out 6.4 amps (80% of I_{SC}). Current ratings are also used to calculate wire size and overcurrent protection device ratings. To avoid shock, burns, and falls from mounts, roofs, and ladders, take extreme care when you make field tests.

IV Curves

The current-voltage characteristics charts, or IV curves, contain important information, although it is often printed in type too small to be useful. Maximum power is the highest combined current and voltage point called the "knee" of the power curve. At full sun and 140°F/60°C, it looks like the example module's voltage drops below 14.5. This means that this module may not fully charge a battery in very hot climates. Note the current output curve at different light levels. For reference, you hardly cast a shadow at about 400 watts per square meter of daylight or 40% of full sun. Overcast is about 100 watts per square meter. We can see that this module has a very low threshold and could actually produce about 0.25 amps during a daytime rain shower.

The example IV curve notes the difference between the reference temperature of 77°F/25°C and the normal operating cell temperature of 116.6°F/47°C, as well as a very high temperature of 167°F/75°C. As temperature increases, current increases slightly but voltage at the knee of the curve (peak power point) has fallen off about 12% to approximately 15.3 volts. In summer when the ambient air temperature is above 80°F/26.6°C, this module may reach 133°F/56°C or more. Wind blowing across the module front and back can keep cell temperature down a few degrees and slightly improves power performance.

Peak volts times peak current equal peak watts or maximum power (P_{max}). Maximum power is 125 watts plus or minus 10%. You need voltage, current, and power temperature coefficients to calculate module power under real world temperature conditions. This information is missing from the spec sheet but should be available from the manufacturer or distributor. The temperature coefficients for the S125 module are:

$V_{OC\,(STC)}$	- 0.08 V /°C
$I_{SC\,(STC)}$	0.065% /°C
$W_{max\,(STC)}$	- 0.5% /°C

You need a thermometer to measure cell temperature, an irradiance meter, or a calibrated and metered solar cell, like the Daystar Solar Meter, and a voltage and current multi-tester to test modules. It is always a good idea to test your modules for V_{OC} and I_{SC} right out of the box, before they are installed, to make sure that they perform to the manufacturer's specifications. You can also use these tools to test and maintain your PV array.

Maximum power voltage (V_{max}) generally changes with temperature. Short circuit voltage (I_{SC}) changes with light intensity. For example, if you measure 800 W/m² irradiance (80% of full sun), the S125 solar module should measure 6.4 amps (80% of I_{SC}). If the solar cells are 55°C, then:

55°C cell temperature – 25°C STC = 30°C difference in temperature
30°C × (-0.08) voltage coefficient = (-2.4) volts
17.4 peak volts - 2.4 volts = 15 volts
15 volts × 7.2 amps = 108 watts
108 watts at 80% full sun = 86.4 watts

What to Know Before You Buy Your Modules

Solar modules are the most important and costliest part of your PV system. Buying solar modules is a major purchase. Here are some questions to ask your module supplier.

1. What are the module electrical and mechanical specifications? Get the spec sheet and installation instructions.
2. What are the physical dimensions of the modules? What array layout can you use?
3. Are the modules UL-listed? Even if your system is off-grid and not subject to building inspection, UL-listing assures that the modules meet strict electrical and safety standards.
4. If you are applying for a rebate, is the module on the rebate list of eligible modules?
5. What is the price of the module? What is the cost in dollars per watt?
6. How much does shipping to your door cost? Is shipping insurance included?
7. Are the modules delivered in boxes or on pallets? If they are being delivered on pallets, do you have, or can you arrange for, a delivery dock, forklift, or some other way to unload the truck?
8. What is the module warranty? Get the warranty in writing.
9. Are the modules new and unused?
10. What mounting structures are recommended by your supplier? Why?
11. Are the solar modules in stock for immediate delivery? If not, what is the lead time?
12. Can the supplier provide references?
13. Is there a quantity discount or discount for immediate cash purchase?
14. What incentives or services does the supplier give to encourage you to buy from them?

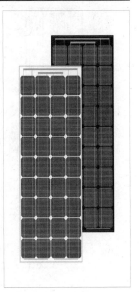

SPECTRUM SOLAR S125 MODULE

- Specially processed single crystal cells give the Spectrum Solar S125 module a conversion efficiency of 14%.
- High efficiency cells encapsulated between tempered glass and Tedlar.
- Modules available in either white or black Tedlar.
- Laminate framed in pre-drilled anodized aluminum combining structural strength with light-weight for ease in installation.
- Available with plug-in connectors.
- Designed to maintain charging voltage in high ambient temperatures and low or diffuse sunlight conditions.

APPLICATIONS

The S125 modules are used in off-grid and grid-tie system applications.

- Residential roof top systems
- Large commercial grid-tie systems
- Water pumping
- Telecommunications
- Outdoor lighting
- Rural electrification

QUALIFICATIONS

SPECTRUM SOLAR, INC. is ISO 9001 certified.

S125 has a 25-year limited warranty on power output.   (see warranty certificate for details)

QUALIFICATION TESTING

The S125 is UL 1703 listed and has passed rigorous JPL Block V environmental tests for weather resistance and electrical isolation, including:

- Thermal cycling test
- Thermal shock
- Thermal freezing and high humidity test
- Impact testing (535g steel ball dropped at1 meter)
- Electrical isolation (less than 50mA at 3000VDC)
- Corrosion resistance (5% salt fog at 35°C)
- Wind loading to 135 mph

MECHANICAL SPECIFICATIONS SPECTRUM S125

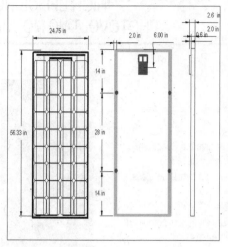

Length	56.33 in / 1429 mm
Width	24.75 in / 627 mm
Depth	2.0 in / 50.7 mm (aluminum frame) 2.6 in / 66 mm (w/ junction box)
Weight	25 lb / 11.4 kg
Front cover	Low-iron tempered glass
Encapsulant	Ethylene vinyl acetate
Solar cells	6 in x 6 in square cells 36 in series
Edge sealant	Butyl rubber
Frame	Anodized structural aluminum
Electrical isolation	3000 VDC, 10 microA (typical)
Fuse Rating	15 amps
Fire Rating	Class C

MODULE SELECTOR						
	Color	Typical power (W)	Voltage at peak power (V_{Max})	Current at peak power (I_{Max})	Short circuit current (I_{SC})	Open circuit voltage (V_{OC})
S125 WT	white	125	17.4	7.2	8	21.7
S125 BT	black	123	16.3	7.0	7.5	21.7

ELECTRICAL CHARACTERISTICS

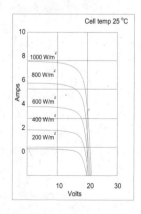

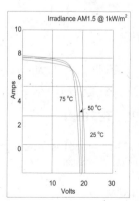

Chapter 10
SITE SELECTION
& MOUNTS

Site Selection

Selecting the site for your PV array is not complicated. Your array should receive full sunlight from at least 9:00 a.m. to at least 4:00 p.m. Early morning and late afternoon sunlight delivers less energy because the light spectrum passing through denser atmosphere is not the best wavelength for conventional solar cells. Look at the sun's path and note the hours when your proposed location is shaded.

If your home is surrounded by trees, don't start cutting them down. Spend some time sitting on the south side of your house and thinking. Consider the passive cooling and aesthetic aspects of the trees as opposed to harsh sun and stumps. It may be that judicious tree trimming or moving your solar array will eliminate any shading problems.

It is sometimes difficult to avoid shading from nearby utility poles and lines. Solar module bypass diodes (see Chapter 11) can protect cells and reduce losses from pole and wire shade lines that move across an array during the day.

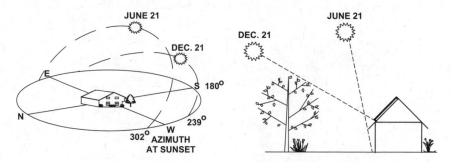

Figure 10.1. The sun at noon—summer and winter.

Solar angles (altitude and azimuth) define the shade pattern and the solar window, which is the area of the sky through which the sun appears to travel (Figure 10.1). Your solar window must be completely free of shading during the optimum solar radiation collection hours of 9:00 a.m. through 4:00 p.m. To estimate your solar window, stand or sit exactly where you plan to mount your collector and face true south. Extend your arm and point so that your finger and eye are horizontal to the horizon. As shown in Figure 10.2, place one fist on top of the other (see the table to determine the number of times). Sight over the top of your fist at true south and 30° azimuth to the east and west (with adjustments in fist height) to determine shading effects. Any object above your fists will cast a shadow on the collector; anything below will be below the lowest path of the winter sun.

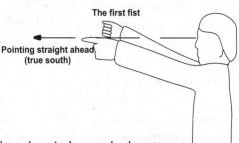

Figure 10.2. Determining the solar window or shade pattern.

Latitude	True South (12 o'clock Position)	11 o'clock Position (East) & 1 o'clock Position (West)
28°N	4½ fists (47° alt.)	3 fists (30° alt.)
32°N	3½ fists (34° alt.)	2½ fists (26° alt.)
36°N	3 fists (30° alt.)	2¼ fists (23° alt.)
40°N	2½ fists (27° alt.)	2 fists (20° alt.)
44°N	2¼ fists (23° alt.)	1½ fists (17° alt.)
48°N	2 fists (20° alt.)	½ fist (14° alt.)

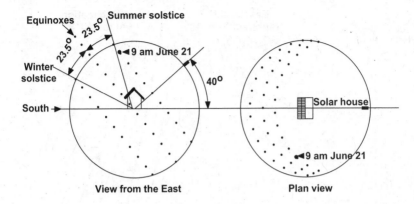

Figure 10.3. Seasonal changes in the sun's path and angle at latitude 40°.

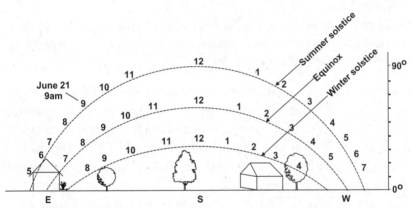

Figure 10.4. The sun's hourly location at latitude 40° facing south.

In the northern hemisphere your array should face true south. Magnetic south, or compass south, is not the same as true south in most areas. A compass needle responds to the distribution of magnetic forces emanating from the earth. These forces are mapped as "lines of equal magnetic declination." The angular difference between the direction of the needle and a true north-south line is called "compass declination." Use Figure 10.5 or ask a local surveyor for the compass declination of your area. The line for zero declination (where magnetic and true south are the same) falls roughly along the Mississippi River. On a compass north is 0° and south is 180°. This is a little confusing, but true south is east of magnetic south for locations west of the zero line. If you live in Seattle, which is west

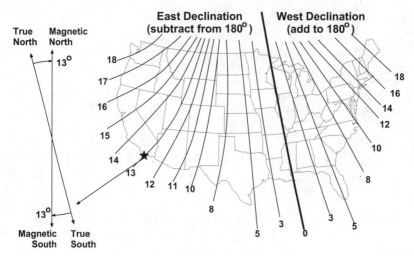

Figure 10.5. 2004 map showing the difference between magnetic north and true north. Los Angeles north compass reading is 13° east of true north, so true south is 13° east of magnetic south.

of the line of zero declination, true south is 18° east of magnetic south. You subtract 18° from 180° to find true south (162° on the compass). If you live in New York City, which is east of the line of zero declination, true south is 13° west of magnetic south. You add 13° to 180° to find true south. If you live in Minneapolis or New Orleans, magnetic south and true south are the same.

A simple way to find true south is to go outside at solar noon (1 p.m. during daylight savings time), and stick a pencil in the ground. The pencil's shadow forms a true north-south line. A line drawn at 90° to that shadow line gives the perfect east-west alignment for your PV array. Proper true south orientation is rewarded with maximum annual solar production.

Planning the layout of your PV system to minimize wire lengths and lower voltage drop is an important design factor, especially for low voltage DC wire runs. The basic rule is to always keep wire runs as short as possible. Batteryless residential PV systems may operate between 48 and 600 volts DC. Systems with batteries may be less than 50 volts DC. In Chapter 8, Ohm's Law is used to calculate voltage drop based on system voltage, wire length, and resistance. Thinner wires have more resistance, which increases voltage drop. The lower the voltage and longer the wire, the higher the voltage drop. The *National Electric Code* recommends keeping voltage drop below 3%. Longer wire lengths require thicker wires to reduce resistance and voltage drop. The distance between the PV battery bank and the inverter should be less than 10 feet or as short as possible. If your PV system includes batteries, the array should be situated close to the battery bank to minimize the length of DC wire runs. If the array is more than 100 feet from where the power will be used, DC wire size may be so large as to justify putting the battery bank and inverter at the array. In this case, higher voltage AC power can be wired to the home.

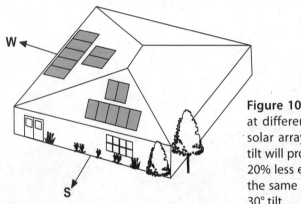

Figure 10.6. Solar production at different orientations. This solar array facing west at 30° tilt will produce approximately 20% less energy annually than the same array facing south at 30° tilt.

Sun Angle

The optimal tilt angle for a PV system is where the sun's rays at solar noon are perpendicular to the plane of the array or the face of the PV modules. This angle varies with latitude and season. For an off-grid PV system your optimum tilt angle at either equinox is the latitude of your location. If you adjust your array seasonally, you would tilt it down in the summer and up in the winter to optimize daily production. This is particularly beneficial for an off-grid PV system where seasonal adjustments can increase the PV array's daily output by 5% or more.

In states where grid-tied PV systems are annually net metered and the excess generation in one month's billing cycle is carried over to the next month, seasonal adjustments are not as critical, but the angle of PV modules will affect total year-round production. Use the National Renewable Energy Laboratory's "Performance Calculator for Grid-Connected PV Systems" (PVWATTS, http://rredc.nrel.gov/solar/codes_algs/PVWATTS/) to determine the optimal orientation and tilt angle for your array.

Most roofs do not face true south at the optimal tilt angle. A mounting system with tilt adjustment or legs to correct for roof angle will allow you to maximize your PV array output. In some cases production losses due to tilt angle or orientation are an aesthetic decision. Some homeowners opt to make a public "PV statement" and put a highly visible tilted rack on their roofs whereas others prefer to accept an annual production loss and follow the existing roof line with a low profile array.

Figure 10.7. This PV array in a Denver suburb is tilted south at the latitude degrees plus 15 degrees in an effort to get more winter solar input. Ten 45-watt solar modules charge eight 6-volt batteries for 24 volts at 900 ampere-hours to power a 2,500-watt inverter. This tilted mount produces 9.5% more annual energy (525 kWh/year vs. 497 kWh/year) than would a simple low profile stand-off array that followed the angle of the southeast roof. Loads can be switched from utility to solar and provide emergency power during utility blackouts. This is an example of a PV system in an area where net metering is not permitted. (Photo: Mark McCray)

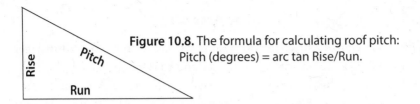

Figure 10.8. The formula for calculating roof pitch: Pitch (degrees) = arc tan Rise/Run.

Table 10.1. Converting Roof Pitch to Degrees
Roof pitch, or angle, is expressed as a ratio of rise to run (length).
For example, 3 inches of rise in 12 inches of run is expressed as 3:12.

Pitch (x":12")	Tilt Degrees	Pitch (x":12")	Tilt Degrees
½:12	2.39	5:12	22.62
1:12	4.76	6:12	26.57
1½:12	7.13	7:12	30.26
2:12	9.46	8:12	33.69
2½:12	11.77	9:12	36.87
3:12	14.04	10:12	39.81
3½:12	16.26	11:12	42.51
4:12	18.43	12:12	45.00

There are two easy ways to determine your roof angle. If it is possible to stand to the side of your house approximately level with the roof, you can sight the roof angle with a protractor held at arm's length. A more accurate method is to use a 24-inch bubble level and a sliding T-bevel. Standing on a ladder or the roof, hold the bubble level parallel to the ground, butting its end to the roof. Mark the angle between the roof and the top edge of the level with the sliding T-bevel, averaging out any surface unevenness. Use a protractor to measure the angle of the sliding T-bevel.

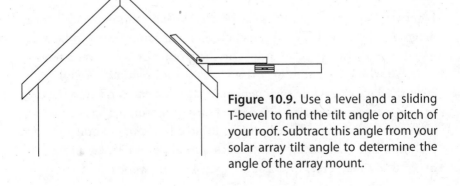

Figure 10.9. Use a level and a sliding T-bevel to find the tilt angle or pitch of your roof. Subtract this angle from your solar array tilt angle to determine the angle of the array mount.

Solar Access

Thirteenth-century English common law decreed that "to whomsoever the soil belongs, he also owns to the sky and to the depths." By the eighteenth century this had evolved into the Doctrine of Ancient Lights, which allowed property owners a prescriptive right or negative easement for the unobstructed passage of light onto their own land from adjoining properties. This is the basis for solar access laws that give U.S. property owners' solar rights. However, these laws vary from community to community. Before installing your array, ask your state energy office or solar contractor if your solar access is protected by law.

Mounts

The purpose of a mount is to hold the solar array securely in place and to optimize energy production by angling the array to the sun. A well-designed mount attractively fits the building and the environment, meets all structural engineering requirements, and may be expanded. Most home PV systems use a stand-off roof mount because it is inexpensive and easy to build and install. You should also investigate other mount systems to see if one might better serve your needs. Ground-mounted arrays have the advantage of easy access for cleaning and seasonal tilt angle adjustment. Pole-mounted arrays include everything from a single module fixed to a pole, to a larger array used as a carport or shade cover, to a tracking mount that automatically adjusts your array for both seasonal sun angle and the daily movement of the sun across the sky.

Portable PV Arrays

Lightweight portable modules can be used without a mount. A hiker can hang a module over his back and charge a small battery pack while walking. A camper can place a module on the ground or lay it over a tent.

Attaching a handle or a drawer pull to a module's metal frame creates a handy portable solar generator that you can carry into the field to charge tractor batteries, DC battery-driven motorized hay conveyors, or to recharge electric fence batteries located far from the house. When Joel lived in Arkansas, he would pack a module, a TV set, and a tasty beverage on his horse, find a sunny patch in the woods, and set up for

a little football watching. The 12-volt DC TV worked fine on power
directly from the module.

Boats have limited space for PV. You can buy hardware to mount
modules to the transom or rail. Flexible modules can be mounted
directly to the boat deck. Sailors who clear the decks for racing can make a
stowable array by fastening modules together with standard door hinges.
This folding, prewired array can be plugged into the battery with a trailer
quick disconnect (available at automotive parts stores) and stored below
deck when not in use. Rope rigging can also be used to string together
and tie down the array.

A PV array can be mounted on recreational vehicles with wing nuts
for easy removal. Instead of parking the RV in the hot sun, the vehicle
can sit in a shady spot while the array is set out in the sun. A folding
sawhorse, easy to store when not in use, can serve as a mount. There
are commercial RV PV mounts that lay flat when traveling and tilt up
when parked. The mount can be kept flat when the sun is high in the
sky (from March to September in the U.S.). During extended stays, the
array should be pointed south at latitude tilt plus or minus 15 degrees
to more than double daily energy production.

The PV array on your parked recreational vehicle can be used to
power 12-volt DC home emergency lighting. Park the RV close to the
house and connect the solar electric system to the house DC light circuits.
These house circuits cannot be connected to any circuit that is powered
at any time by the utility electrical grid. One recreational vehicle owner
with a kidney condition puts his dialysis machine into his PV-powered
RV during utility power outages. Portability works both ways.

Trailer-mounted solar electric systems provide power at temporary
work sites and in emergency situations. Farmers with stock watering wells
in several fields can move a trailer-mounted array from well to well. The
Sunrunner solar vehicle was transported on a trailer with batteries and an
inverter, and could be pulled up to a house to provide emergency power.
Almost anything on wheels can be turned into a mobile PV system.
Complete solar electric systems in skid-mounted containers or portable
enclosures can be moved to a location on a trailer or by helicopter. Once
in place, the PV array is either unloaded from and bolted to the container
or unfolded from the roof.

If you set up a temporary mount, make sure it is secure and won't fall or blow over. An unsecured mount is dangerous. One system owner, in a hurry to get his PV-powered well pump in operation, drove metal fence posts into the ground, strung wire between them, and tied his modules to the wire. A gust of wind knocked the modules to the ground. Luckily none broke. Another person made a temporary set-up to test his system, but he wasn't so lucky. He leaned his modules against a building he assumed was protected from the wind. It wasn't. All of his modules blew over and most were broken. If you must use a temporary set-up, lay the modules on the ground and make sure they cannot be walked on or driven over.

When off-grid modules are stored, they are not charging batteries. If you occasionally use a portable mount, don't let your batteries go dead. Batteries can be charged only if connected to PV modules that are in the sun. Whenever and wherever you set up a mount, consider carefully how you will ground the frame for lightning and shock protection.

Mounting Structure Design

Reasonably priced mounts are readily available. Using a commercial mounting system and following the manufacturer's instructions increases the chances that your installation will conform to residential building codes and may help you avoid costly structural engineering analysis. Most manufacturers supply the information your building and safety department will require for structural engineering documentation.

You can make your own mount. A good mount should be strong, secure, and maintain proper array orientation; it should not bend, flex, or move. Although its primary purpose is to support your array, the mount can also be used to support wire, conduit, junction boxes, and even inverters. However, most balance of system components are best kept off the roof and out of the sun, heat, and weather.

Before designing your mount, shop around and see what materials are locally available. Use aluminum or galvanized-steel and stainless-steel fasteners; 1½ × 1½ × ⅜-inch anodized aluminum angle is standard. Solar modules are usually fastened to mounts or racks with ¼-inch stainless-steel bolts. Star washers ensure a good electrical connection between module frames and mounting structure members for grounding, although most inspectors require a separate ground wire. Stainless-steel lag bolts, 5/16- or ⅜-inch, are often used to fasten roof mounts to rafter framings.

Mount framing and roof fasteners must resist wind, snow, and seismic loads. Use lateral bracing for added strength. If your roofing shingles are over ten years old, re-roof before installing a solar array mount. Roof-mounted arrays should be designed with the minimum required number of roof fasteners to reduce the possibility of leaks. Brackets or backing blocks may be needed in the attic to strengthen the sheathing or plywood at fastener penetrations. You may need to brace the roof to support the array load.

Make sure that fasteners and rails do not block drainage or allow snow or ice dams to form. While adjustable-leg roof mounts cost more than fixed-leg, they make it possible to mount an array on uneven surfaces or seasonally adjust the array tilt.

Assemble the ground mount first and mark the ground for the location of the footings. Bolt a few modules in place to align and strengthen the mount during handling. Dig the holes, pour the concrete footings, and set the mount in place with the anchor bolts on it. Press the anchor bolts into the wet concrete, and level and support the mount while the concrete is wet.

It may be convenient to preassemble and prewire your roof mount on the ground, but moving a fully assembled array onto a roof can be difficult and dangerous. If you assemble the mount on the roof, set up a safe work area on the roof for fastening and wiring modules.

Figure 10.10.
This low-cost owner-built ground mount shows simplicity in design and ease in construction. (Photo & design: Bruce Wheeler)

Roof Mounts

Roof mounts are easy to assemble or fabricate and easy to install. Solar water heater panel mounts can be used though they are often over-engineered for the lighter-weight PV modules. The growth of on-grid PV and the increase in the number of urban installations has led to some innovations in roof mounts.

Commercial roof mounting systems, complete with racks and installation hardware for most types of roofing materials, are available from several manufacturers. These mounting systems come with detailed installation instructions and engineering reports that will allay your inspector's concerns. Roof mounts have several advantages.

1. No additional land or building costs
2. Close to point of energy consumption to reduce wiring costs
3. Less subject to vandalism, theft, accidental damage
4. Usually have an unobstructed sun window
5. Most homes have sufficient south-facing roof space

Rack or tilt-up mounts can be installed over the existing roof surface. Sometimes called high-profile fixed stand-off mounts, they can be used on flat or pitched roofs. Rack mounts should be aluminum for long life.

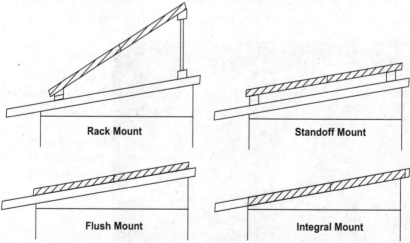

Figure 10.11. Four basic array mounting schemes. *Rack mounts* allow tilt adjustment. *Stand-off mounts* allow air circulation behind the modules. *Flush mounts* will cause some power loss due to module heating. *Integral mounts* replace roofing material with PV modules.

These mounts are low in cost and easy to install; they are able to withstand wind and snow loading if properly designed; and they permit passive module cooling since the back of the array is exposed to breezes.

Mounts and roof fasteners should be strong enough to withstand pulling and tugging. If you will not risk hanging all your weight from your mount, it is not strong enough. Although stand-off mounts have a low profile that does not expose the back of the array to the wind, the roof fasteners still must be designed to withstand wind lift pull-out. Conservative estimates of wind uplift (wind loading) are 25 lb per sq ft for 85-mph winds and about 55 lb per sq ft for 120-mph winds. The general rule is one roof fastener per 40 pounds of array weight. It may be necessary to install extra in-roof framing in order to put ¼- to ⅜-inch lag bolts directly into the rafters or framing members. Properly seal fastener mounting holes to prevent leaks.

Stand-off arrays follow the roof plane but may not be at the optimum tilt angle. Low-profile arrays also experience less air cooling and the modules operate slightly less efficiently than modules on high tilt arrays. Roof racks can be hinged to provide seasonal tilt, but remember that foot traffic can damage all roofing materials. Winter energy production is significantly less from low-pitched stand-off arrays, but the difference in annual production may be minimal because of the better summer sun angle, longer days, and less cloud cover. In Los Angeles a 10° tilt array produces 5% less energy per year than the optimum 30° tilt. In New York City a 10° tilt array produces 5.5% less than the optimum 40° tilt.

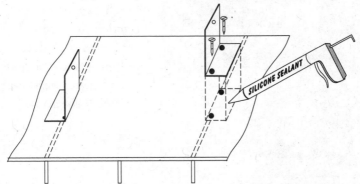

Figure 10.12. Stand-off mount feet. Fasten PV mounts directly to the roof framing members through the roofing material to ensure a strong mount. Be sure to use sealants at all roof penetrations to prevent leaks.

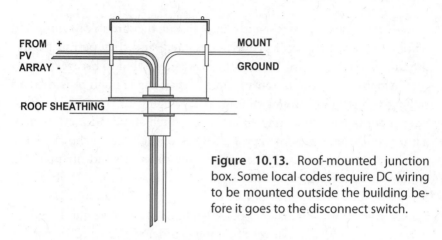

FROM +   MOUNT
PV
ARRAY -   GROUND

ROOF SHEATHING

Figure 10.13. Roof-mounted junction box. Some local codes require DC wiring to be mounted outside the building before it goes to the disconnect switch.

In Sacramento, California, PV homeowners with stand-off arrays have discovered a side benefit: Their air conditioning bills are lower, possibly because their arrays shade and protect their south roofs. Lower power consumption could offset the lower power production of the stand-off array.

Integral mounting allows modules to double as roofing material. Utilizing framed modules, this differs from a true building-integrated installation that uses either photovoltaic roofing membranes or shingles.

For this type of installation the modules and mount must be structurally strong. However, because there is much expansion and contraction in an integral solar array, it is difficult to make a leak-proof integral mount, so use adequate weatherproofing under the array.

Figure 10.14.
Fastening PV mounts.
(Photo: Photron, Inc.)

Framed modules and solar shingles can be directly mounted to the roof sheathing with wiring inside the attic space. Direct-mounted modules have an attractive low profile, but get very warm even with soffit and ridge vent cooling. Direct-mount modules should be mounted on top of, not instead of, waterproofing roof material. Flashing similar to the type used with skylights is necessary.

Tracking roof-mounted arrays are possible. However, the added cost and greater roof area needed to keep modules from shading each other can make them impractical. Before considering a roof-mounted tracker, be sure your roof can handle the increased dead load (equipment weight) and live loads (wind, snow, and workers).

Installing a Roof Mount

The first step in installing a roof-mounted PV array is to draw a layout of your array. The panels can be positioned in either portrait or landscape orientation. The roof plan must include the location and dimensions of any protrusions such as flues, vent stacks, and skylights. Make sure to leave enough room for you and your installers to move safely around the array. The area required for a crystal-cell PV array is approximately 110 square feet per kW DC. An amorphous silicon solar array requires approximately 250 square feet per kW DC. The final dimensions of the array are determined by the mounting structure hardware. Some mounts require an additional 1 inch between each module for a mid-clamp and 3 inches for end-clamps. The most important consideration is to center the fasteners over the rafters.

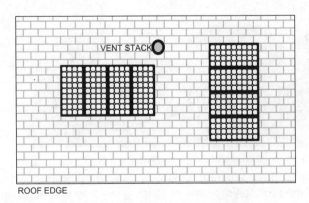

Figure 10.15. Solar array roof plan.

The method you use to fasten the mount to the roof will depend on local wind, snow, and seismic loads. Commercial mounts come with a detailed guide for code-compliant installation. Design specifications are calculated to wind loads, determined by your local building and safety department. The mount manufacturer, a structural engineer, or your local building and safety department can help you determine the type, number, and spacing of fasteners to use.

Figure 10.16.
These solar modules form part of the roof covering of this house. To prevent leaks, the roof under the modules was waterproofed and seams between modules were plugged with glazing tape and caulked. (Installer: Peter Talmage. Photo: Robert Sardinsky)

Figure 10.17. A simple solution to a difficult mounting problem. This 1,188-watt array on three separate mounts has a low profile and proper spacing between the sub-arrays to prevent shading in winter when the sun is low in the sky. (Haleakala Resources, Hawaii)

Ground Mounts

Ground mounts allow for ease in installation and adjustment, but ground-mounted glass-encapsulated solar modules can be more easily damaged by a rock flung from a lawn mower. Theft or vandalism of ground-mounted modules is another possible problem. Most people include their solar electric equipment in their homeowner's insurance. Coverage for fire, theft, vandalism, and accidental damage is not expensive compared to the cost of replacing modules.

Ground mounts can also attract the curious. At a solar energy fair, a ground-mounted array was used to supply power to the musicians. An ideal demonstration of PV's portability and power-producing capability, to some children the solar array also provided a great slide. Fortunately, the modules were strong enough to take the load.

Ground mounts do have their place. Many solar arrays that power telecommunications systems are situated on windy mountaintops and use low-profile ground mounts fastened to concrete footings. If you live away from the hustle and bustle, a ground mount may be the quickest and easiest way to install your array. Seasonal adjustment of ground mounts is possible. Taking a few minutes to adjust your array when you do your semi-annual maintenance in mid-March and mid-September can increase your yearly production by 10% or more.

Figure 10.18. Typical roof or ground mount. Adjustable rear legs allow for seasonal adjustment. (Direct Power & Water)

Figure 10.19. This 980-watt flat roof array in New Mexico has telescoping legs that can be easily adjusted seasonally to increase production. Seasonal adjustment adds about 10% to annual production, but few grid-tied homeowners actually make the adjustments each spring and fall. See the owner's comments on page 122 regarding the unique way he financed his PV system. (Designer/installer: Allan Sindelar, Positive Energy, Inc.)

Pole Mounts

A pole mount can be a good choice when the pole is already being used for some other purpose. For example, an amateur radio operator could mount solar modules on the side of an antenna pole. Pole-mounted solar-powered emergency telephones are a common sight on highways around the world.

Figure 10.20 shows four 10-module manually adjustable pole mounts. The arrays sit atop the poles and have locking nuts to adjust to various tilt angles.

Pole mounts have minimum impact on the environment, but they must be well-anchored to withstand wind loads and frost heave. Do not pole-mount a solar array where ice or rocks can fall and damage the modules or where someone might stumble into it at night.

Figure 10.20. If your home is in the shade or your roof is not large enough or not oriented south, use a pole mount. (Direct Power & Water)

Figure 10.21. Whether ground- or roof-mounted, locate your solar array to minimize snow build-up and shading from plants. This array was later raised four feet above ground level. (Photo: Robert Sardinsky, Rising Sun Enterprises)

Tracking Mounts

You can manually track the sun if you are at the array site daily. You tilt the array to the east in the morning, south at noon, and west in the afternoon. This will definitely increase the output of a small array. Don't forget to return the array to the true south position when you decide to stop daily adjustments.

There are two types of tracking mounts, active and passive. If you require more power in summer anywhere in the temperate zones, a tracker is ideal. On a clear summer day, single-axis tracker-mounted modules at latitude minus 15° tilt will deliver up to 55% more energy than a fixed array at latitude tilt. This means more power for water pumping or fan ventilation. In cloudy winter weather, a tracker can increase production about 10%. If your site has winds over 70 mph a tracker may not be suitable.

Active trackers use a DC motor to move the array. They can be accurate and are stable in high winds. A single-axis tracker automatically rotates 120° or more from east to west and can be manually tilted as the seasons change. A dual-axis tracker also rotates up to 75° up and down. A dual-axis tracker may consume 5 watt-hours per day from the PV array but tracks the sun to within 0.5° accuracy for increased energy production.

Concentrating solar arrays and most arrays with reflectors require dual-axis tracking to keep the concentration point of focus on the cells. However, only specially made modules can take the additional heat. Concentrating or intensifying the sun's energy on standard solar modules voids the warranty. Standard modules mounted on bright metal and white roofs, however, can collect reflected light without being damaged.

The Zomeworks passive solar tracking mount begins the day facing west. As the sun rises in the east, it heats the unshaded west-side canister, forcing liquid Freon® into the shaded east-side canister. As liquid moves through a copper tube to the east-side canister, the tracker rotates so that it faces east. The heating of the liquid is controlled by aluminum shadow plates. When one canister is exposed to more sun than the other, its vapor pressure increases, forcing liquid to the cooler, shaded side. The shifting weight of the liquid causes the rack to rotate until the canisters are equally shaded. The rack follows the sun at approximately 15° per hour, continually seeking equilibrium as liquid moves from one side of

---

*Freon is registered trademark of E. I. duPont de Nemours and Company, Wilmington, DE.

Figure 10.22. Two 12-module Zomeworks passive solar trackers.
(Offline Independent Energy Systems)

the tracker to the other. The rack completes its daily cycle facing west until it is "awakened" by the rising sun the following morning.

Single-axis tracking averages 30% year-round more PV energy production than a fixed latitude mount. The increase ranges from 10 to 20% in winter to 40 to 50% in summer. For an eight-module tracker, that's like having 1.6 extra modules in winter and 3.2 more modules in summer. In addition, Zomeworks trackers can be tilted easily up and down from 15° to 45°. In most cases, a tracker more than pays for itself.

A successful and cost-effective large tracking array is at the Sacramento's former nuclear power plant (see Figure 2.1, page 25). The two-megawatt PV power field uses horizontal single-axis tracking. The mount looks like a large handrail parallel to the ground, and the solar modules move east to west. This type of mount is simple, low in cost, and would be easy to install on a south-facing hillside.

Concentrators & Reflectors

Almost every concentrating PV array scheme requires tracking adjustments that add to the cost and maintenance of a solar array. Hybrid electric and thermal arrays appear attractive and seem simple, but they are complex when compared to a conventional fixed array mount. Seasonal adjustment of the array requires some form of flexible plumbing or ducting of thermal production.

A word of caution on reflecting. Commercial-grade solar modules are designed to be used without reflectors or concentrators. Substantial reflectance can cause overheating. Be sure to get written permission from the manufacturer that your module warranty will not be invalidated if you use reflectors or concentrators.

Summary

Whether you design your mounting structure yourself or purchase a mount, make sure your mount has these features:

- Maintains proper array orientation
- Uses standard tools for installation
- Requires minimal cutting & drilling
- Low cost
- Supports modules and other necessary equipment
- Resists wind loads
- Easy to earth-ground
- Easy to install
- Adjustable to an uneven base
- Suitable fasteners for existing roofing materials
- Minimum roof penetrations
- Does not block drainage
- Does not dam snow, ice, leaves, or debris
- Long lasting
- No electrolysis
- Resists seismic loads
- Lateral bracing
- Manufacturer's support
- Structural engineering documentation

It is important to plan ahead when designing your solar array mount. Leave room for your system to grow. Most important, make sure your mount is strong. If the mount cannot safely support your weight, would you trust it to support your modules?

Chapter 11
CHARGE CONTROLLERS, METERS & DIODES

Charge Controllers

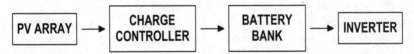

Charge controllers, also known as controllers or regulators, prevent current produced by the solar array from overcharging and damaging the battery bank. Charge control is necessary because a properly designed PV system can regularly produce excess power because of seasonal variations in solar input, variations in daily weather, and changes in power consumption. This regulation is essential because a solar array designed to meet winter power needs can produce as much as four times more energy in summer. A well-designed charge controller will allow as much array current as possible to flow safely into the battery bank, but not to excess.

Another benefit of using a charge controller is that it allows you to leave your PV system unattended for months. When you are away for extended periods your batteries should be connected to the array because they need a float charge to replace normal self-discharge energy loss.

Some building codes require that you use sealed batteries, which are particularly vulnerable to overcharging. Sealed batteries can be damaged by as little as one day of overcharging.

There are four types of charge controllers: shunt, series, pulse, and maximum power point tracking. Shunt controllers are solid-state devices with a transistor in parallel to the array that redirect excess array current to an earth ground, a power dissipating heat sink, or another load. Series controllers usually have a relay or switching transistor in series between the solar array and the battery that switches array current on and off. Pulse controllers also connect in series between the array and battery, rapidly switching or pulsing array current on and off.

Shunt controllers are simpler in design and usually cost less. A basic shunt controller is a switch with setpoints matched to the battery bank's fully charged and charge resumption voltage values. (*Practical Photovoltaics* by Richard Komp contains plans for building a shunt controller.)

Series controllers are also on-off switches. Some series controllers have additional modes for trickle-charging. A two-step or dual-mode

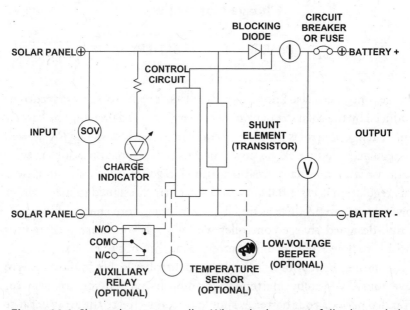

Figure 11.1. Shunt charge controller. When the battery is fully charged, the shunt transistor conducts across the solar array positive and negative inputs and no power reaches the battery. (Specialty Concepts, Inc.)

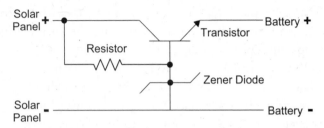

Figure 11.2. Series charge controller.

controller turns off most of the current produced by the PV array when the battery bank reaches around 80% state of charge. Then it allows a trickle, float, or finishing charge to pass to the battery. Two-step finishing charge rates may be fixed amperage or adjustable from 1% to 5% of the battery's ampere-hour capacity (1 to 5 amps charge for each 100 ampere-hours).

Pulse or pulse width modulation (PWM) controllers also connect in series between the array and the battery. Pulse controllers can vary the current pulse rate, pulse duration, or both. Some pulse controllers pulse charge current 1 to 5 cycles per second while others switch at over 20,000 cycles per second. PWM controller manufacturers claim that once a battery is charged to 80% of capacity, high frequency, full current pulses bring the battery up to full charge 10% faster than the trickle-charge of dual-mode controllers.

Maximum power point tracking (MPPT) controllers also connect in series between the solar array and battery bank. MPPT controllers use a microprocessor-based algorithm to repeatedly find the highest solar array voltage and current output. MPPT controllers are most effective in cold weather when solar array voltage is high. Some MPPT controller manufacturers say their devices increase production up to 25%, but the typical annual increase is 15% or less. This production increase is enough to justify the two- to four-times higher price of MPPT charge controllers.

Some MPPT controllers either buck or boost DC voltage in a manner similar to AC transformers. If the array voltage is low, the MPPT controller will boost the voltage to charge the battery bank. When the voltage is high, the MPPT controller will buck (reduce) the voltage to the proper value. The buck feature allows you to operate your solar array at a higher voltage and reduce the size of your array-to-battery homerun wire. As an example, a 120-VDC array can be used to charge a 12-, 24-, or 48-volt battery bank, while still adjusting the current times voltage (IV) to optimum power point.

MPPT controllers are not to be confused with solar array trackers that physically move the array to follow the sun. It should also be noted that most grid-tie inverters have MPPT electronics, but inverters do not regulate the charge from a solar array to the battery bank.

Which type of controller is best? Two-step, tapered-current finishing chargers were preferred until PWM controllers became popular. Now MPPT controllers are the most popular. Single-step shunt controllers have been in use for decades and work well. Designers are continuously changing solar charging algorithms. The best controller for your system should have charge termination and resumption setpoints that match your battery's type, specifications, age, temperature, and usage patterns.

The number of controllers in use and the number of years a particular model has been manufactured are measures of reliability. Older designs produced by established companies tend to be more reliable. Controllers that have been used in thousands of applications for three or more years usually have the design problems worked out.

Reliability is more important than novelty or the manufacturer's warranty period. Five-year warranties are becoming standard. Unfortunately only a few companies still provide Mean Time Between Failure (MTBF) data or the percentage of failed or returned units, but hopefully that will change as the industry adopts recognized quality standards. Ask other PV system owners and experienced installers for their recommendations.

Controller Sizing

The *National Electrical Code* requires that PV current-carrying devices like wire and charge controllers be sized to carry 125% of the array's short-circuit current (I_{SC}). In addition, another 125% current-carrying capacity is required to allow for current produced when sun conditions are greater than the industry standard of 1,000 watts per square meter. For example, a 12-module array of 80-watt modules rated 5.01 I_{SC} wired in series-parallel for nominal 24 volts with an array short-circuit current of 30.06 amperes [(12 modules ÷ 2 per string) × 5.01 = 30.06] should have a controller capable of handling (30.06 × 1.25 × 1.25) = 46.96 amperes.

Over the years, the size of the average PV system has increased. Today many homes with battery banks have 2,000 to 10,000 watt arrays or larger and multiple 2- to 5-kW inverters operating at 24- or 48-volts DC input. A 4,000-watt array with 32 Spectrum S125 125-watt modules ($17.4\ V_{pk}$, $21.7\ V_{OC}$, $7.2\ I_{pk}$, $8.0\ A\ I_{SC}$) wired for 48 volts will have 4 modules in series and 8 of the series strings wired in parallel:

32 modules ÷ 4 per voltage string = 8 strings

8 strings × 8.0 A I_{SC} × 1.25 × 1.25 = 100 amperes

The array can be wired into two sub-arrays, each feeding into a 60-ampere, 48-volt controller to handle this current. The system could also be split into three sub-arrays feeding into three 40-amp controllers. Three 40-ampere controllers may cost more than two 60-ampere units. However, splitting the array into three sub-arrays means that the cable from each sub-array has to conduct less current and can be a smaller gauge, which could reduce overall costs. Each charge-controlled sub-array requires its own combiner box.

In addition, splitting the array into sub-arrays and using more than one controller provides redundancy that can improve reliability. If one controller fails, its sub-array can be wired directly to the battery bank so that no energy is lost while the broken controller is in the repair shop. Be sure your wiring, safety switches, and fuses can handle the temporarily unregulated PV input, and carefully monitor system voltage.

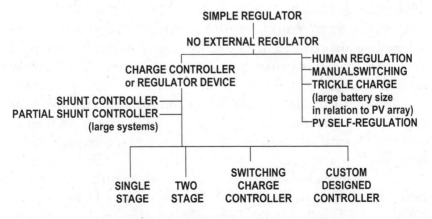

Figure 11.3. Charge controller decision flowchart.

Manual Regulation

Some campers, hikers, wilderness workers, people traveling light, or individuals on a very limited budget who are on-site with their PV all the time can charge batteries without a controller. Charging current can be controlled by manually tilting the PV array toward or away from the sun during the day while monitoring battery voltage. At night the PV module or panel can be disconnected from the battery. However, manual regulation is subject to human error. For peace of mind, get a controller and go automatic.

Self-Regulation

The principle of self-regulated battery charging is well established. Small, current-limiting, AC-powered battery trickle chargers are sold at most automotive supply stores. These "power-blocks" plug into an electric outlet, connect to the battery with clips, and put out about 0.5 amperes at around 13 volts so as not to overcharge a 50-ampere-hour vehicle battery.

In the past, designers have made self-regulating PV systems with peak PV current output of 5% or less than the battery's ampere-hour capacity and charge voltage below the battery float voltage. The NEC now requires a controller on PV systems with charging currents greater than 3% of battery amp-hour capacity.

Some 30-cell, low-voltage, 5- to 50-watt solar modules are called self-regulating. These modules do not have built-in regulators. The principle of 30-cell self-regulation is that as battery voltage increases, the voltage potential between the module and battery decreases; less current flows from module to battery until module and battery voltage are the same and charging ceases.

Most small trickle-charger modules are little more than toys made to give consumers the feeling of "going solar." Some may not even replace a battery's normal self-discharge losses, while others overcharge sealed batteries.

Overcharging and loss of electrolyte can expose battery plates, permanently damaging the battery, if a self-regulated system is not used for periods as brief as two months. For reliable solar battery charging under all conditions, 36-cell modules and a controller are recommended.

Which Controller for Your System?

Select a controller that meets your present and future needs. Controllers with similar features are interchangeable, but don't use windcharger or automotive regulators as they generally have energy-robbing solenoids, relays, and other high current consumption components.

A simple, reasonably priced PV controller is a good investment. A controller is just a switch. Complicated switches cost more and are less reliable than simple switches. The controller is usually one of the lowest cost components of a PV system, whereas failed batteries are one of the most expensive replacement parts.

Controllers can be damaged by lightning, current surges, and electrical shorts. Good controllers have built-in silicon oxide varistor (SOV) surge arrestors for protection. No controller can survive a direct lightning strike, but you can reduce the chance of destruction if you mount your controller indoors and ground it to earth.

Follow the manufacturer's installation instructions carefully. Some controllers have no polarity protection and are damaged if improperly connected to the array or battery bank. Almost any controller's printed circuit board will be destroyed if a loose array or battery wire or stray tool touches it. Most controllers will also be damaged if array or battery wires touch the controller's metal case or cabinet.

Most controllers are connected in the same sequence. The first wire to connect is the earth ground, which is typically the battery bank negative. Then the battery positive is connected to energize the controller electronics. Next connect the array negative and finally connect the array positive.

Controller Features

Controllers perform several essential functions. Few PV systems need all the features and options that are available. Time-tested, field-proven, uncomplicated controllers with a few desired options are best. You do not need to have a controller custom-designed for your specific PV system. Custom controllers generally have higher failure rates, and it takes longer to get them repaired—especially if the manufacturer is no longer in business.

Charge regulation is not an option. It is an essential requirement to prevent overcharging batteries.

Multiple voltages or adjustable voltages allow the same controller to be used for 12-, 24-, or 48-volt systems. Multi-voltage is useful if you start with a 12-volt DC PV system and plan to add a 24- or 48-volt inverter later. However, most people upgrade their controller when they upgrade their system.

Temperature compensation automatically adjusts charging on and off voltage setpoints. This is an important feature for systems where battery compartment temperatures drop below 59°F/15°C or rise above 95°F/35°C. Temperature compensation is typically -0.005 volts per degree centigrade per 2-volt battery cell for each degree above or below 77°F/25°C. A controller that normally switches on at 13 volts will allow array current to pass to a 6-cell battery at 13.75 volts at 32°F/0°C or 12.7 volts at 95°F/35°C.

The most accurate temperature compensation has a hard-wired temperature probe in contact with the battery. The probe should be fastened to the battery case firmly with tape and insulated from the surrounding air with rigid insulation. It is important that the entire battery bank temperature be uniform (see Chapter 12, "Battery Storage," for details).

Controllers that have internal temperature sensors instead of an external probe should be installed close to the battery. A wall-mounted charge controller with an internal temperature sensor is ineffective if the battery bank is sitting on a cold floor.

Reverse current leakage protection prevents nighttime discharge of batteries. Almost all controllers have a blocking diode or automatic switch that disconnects the array from the battery. Switches may consume less power than a blocking diode, but switch contacts can fail to open or close. Diodes can burn open or short out. However, reverse current problems rarely occur, and switches and diodes both work well. Failed reverse current devices can be identified if you have accurate system meters.

Low-voltage load disconnect switches automatically turn off power to attached devices and are used for load management, not regulation. This feature is designed to "think" for people who are not in the habit of turning lights and other electrical equipment off when not in use, and will help prevent overdischarge battery damage. Essential loads like refrigerators and communications equipment should not be connected to an automatic disconnect switch. Some load disconnects are adjustable and can be manually switched off.

A low-voltage audible alarm or light emitting diode (LED) visual indicator is useful. Reasonably priced battery **state of charge** indicators are available that graphically show when batteries are fully charged or at various states of discharge. Sound or light indicators are not a substitute for an accurate digital voltmeter. It may not be possible to accurately know the battery capacity of an active PV system because power is being produced and consumed at different rates at the same time. However, an indicator and voltage readings will help you develop a "feel" for your battery bank's normal conditions. Any deviation will then become obvious.

Overcurrent protection fuses and circuit breakers prevent equipment damage, fires, and electric shock. A fused controller can be used as a load center. Some controllers have PV array fuses or breakers. NEC requires this overcurrent protection.

System monitors, such as analog meters, digital meters, indicator lights, and warning alarms, are essential because they provide important information about your system. You need to know your system voltage and charging current.

Load controls can be used to start and stop standby generators, water pumps, and timed devices like lights. Timers that automatically turn lights and other equipment on and off at set times are convenient but can malfunction, resulting in damaged batteries. Automatic generator starters have been known to turn on at the wrong time or when the engine oil level is low. The convenience of automatic load control is not worth the risk without low voltage disconnect and other safety measures in place.

Load diversion of array power to a secondary load after the battery bank has been fully charged seems like a good idea. This option appeals to people who do not want to waste any PV power. Secondary loads may include an auxiliary resistance water heater element or a second set of batteries. Load diversion is needed with some wind generator systems that produce excess power, but owners of PV systems can usually find a use for any "extra" power.

Adjustability of controller functions is becoming a standard feature. Field-adjustable setpoints for charge termination and resumption, float voltage, load diversion, and low-voltage load disconnect are useful. Generic factory setpoints are usually unsuitable for nickel cadmium, nickel iron, or old or unusual batteries. Some sealed batteries require special setpoints. Be careful adjusting controls: Out-of-range setpoints will damage equipment.

A **central wiring terminal block** turns the controller into a power center that provides terminals to connect the array, battery, inverters, and loads. Terminals must be large enough to secure all cables. If your controller is your wiring center, be sure it complies with NEC disconnect requirements. Use a separate terminal block with large lugs if your controller's terminals are too small for your array and battery cables.

Reverse polarity protection prevents damage to the controller during installation. Installers like this feature. Even if your controller has polarity protection, take care not to reverse your array or battery wires. Do not touch wires together or to controller electronics. Label all wires and cables to facilitate future re-wiring.

Lightning protection may be a limited built-in feature or an add-on option. Direct lightning strike protection is impossible, but your controller should be able to handle transient currents from nearby strikes.

Battery equalization (controlled overcharge) may be a manual switch or an automatic option. Battery equalization timed for every 30 days or at selected intervals is useful for unattended systems but it is no substitute for regular battery maintenance. Never overcharge sealed batteries.

Installing a Controller

The controller is connected between the array and battery bank, and as close as practical to the batteries. Controller location should shorten, not increase, the length of cable (homerun) from the array to the battery. Mount your controller on a clean dry wall, a panel, or in an enclosure. Secure all wires, cables, and conduit that run to and from it.

Most controllers have their own metal or plastic box or enclosure that can be flush- or wall-mounted. Some controllers are electronic circuit boards and require a user-provided enclosure. Do not mount plastic or metal controller enclosures in direct sunlight even if they are designed for outdoor use. Also, mount controllers away from heat sources like engine generators, heaters, or roofs.

Controller meters, switches, and controls should be accessible and easy to see. NEC requires 3 feet of clearance in front of panels and controls. Some controllers produce heat and are hot to the touch during operation. Be sure to follow the manufacturer's instructions and provide proper ventilation. Do not hang wires, clothing, or tools on the controller or on system wires.

Some small controllers have been designed to go directly into a solar module junction box (j-box). This is not a good idea because temperatures can reach 180°F/82.2°C in j-boxes on roof-mounted PV arrays, and heat is bad for electronics.

Some sealed controllers are designed to be mounted in boat or vehicle battery boxes. They must not produce sparks that could ignite battery gases and should have non-corrosive terminals and screws because a battery compartment is a very harsh environment.

Controller terminal connector (lug) size is important. Array cables are often too big to connect directly to controller terminals. Many controllers are designed for 10-gauge wire, although more controllers are now made with terminals that can take 4-gauge or larger wire.

It is convenient to connect array cables directly to large controller terminals, but the *National Electrical Code* requires disconnects on the input and output of the controller. Most commercial disconnect switches have large terminals so you can use smaller gauge wire from the switches to smaller controller terminals. Be sure all wires are properly sized to carry the current.

How a Controller Works

A solid-state shunt controller allows full array current to flow into the battery until battery voltage reaches the charge termination setpoint. Then the shunt transistor turns on and shorts out the solar array by connecting array positive and negative, which stops battery charging. The shunt transistor turns off and charging resumes when battery voltage drops to the charge resumption setpoint. When battery capacity is low, charging is continuous. When the battery is full, the controller will pulse current to keep the battery full. At night the blocking diode prevents reverse current from the battery to the array.

A series controller operates in a similar manner. At sunrise when array voltage rises, the charging relay energizes and closes, connecting the solar array directly to the batteries. The constant-current full-charge mode is the first step of a two-step charging sequence. When the battery reaches the full charge termination threshold, the charging relay opens and the float controller takes over to supply a float current. In step two, as the battery approaches float voltage, the current tapers off to the battery maintenance current setpoint.

If a load is applied to a charger in float mode, the controller will continue to supply up to its maximum float current to maintain the battery's charge. Should voltage drop below the charge resumption setpoint, the controller will return to the first step and allow full array current through until the battery is recharged. Most controllers have LED indicators that show when the array is charging the batteries. Some have meters.

The series controller also disconnects the array to prevent reverse current loss. After hours of charging, and when the batteries are fully charged, the controller's charging relay may close and reopen to sense if current is still available from the array. The relay remains open to prevent nighttime reverse current losses.

Low voltage disconnection of some loads is useful. Should battery bank voltage drop below the setpoint, selected loads are disconnected to prevent deep discharging that could ruin the batteries. When charging brings the batteries back up to the load connect setpoint, the automatically disconnected loads turn on again. It is possible to automatically disconnect loads that are larger than the controller's disconnect relay rating by using a secondary external relay. This is most useful when operating an inverter-powered load and the inverter is wired directly to the battery bank. Be sure to use a relay with contacts rated for the inverter's output. Use a time-delay relay to allow for inverter initial surge currents and to avoid nuisance shutdowns.

Controller Troubleshooting

If you understand potential problems, you can design them out of your system. Proper installation will help prevent most problems.

In general:
- Check system wiring to ensure proper polarity and that all connections are sound and with minimum voltage drop due to corrosion or loose connections.
- Check that modules and batteries are in the correct series–parallel configuration for proper system voltage and current. Read instructions and specifications for array output, load ratings, and system sizing to ensure that ratings are not exceeded.
- Check that the array is not partially shaded or dirty.
- Check all system fuses and circuit breakers.
- Determine if the controller is operating at the correct setpoints by measuring the voltage at the battery voltage terminals when relay switching occurs. Allow for specification tolerances and the effect of temperature compensation (-5mv/°C/battery cell) if your controller has this option.

Is the battery undercharged?
- Batteries may be too cold and require higher voltage to achieve full charge. Insulate the battery bank and use a controller with temperature compensation.
- Not enough array for load. Check system sizing and add solar modules as needed.
- Charge rate too high causing battery voltage to rise too fast before charging is complete. Add batteries to the system or bypass the controller with a few solar modules to increase trickle-charge current.
- Battery capacity and ability to accept charge has been reduced by age or abuse. Replace the battery bank.
- Excessive voltage drop to battery caused by high current, small wire, and long wire runs. Replace wire with properly sized conductors.

Does the load disconnect improperly?
- Loads such as inverters can generate electronic noise. Wire inverters directly to the battery bank. Add filters to attenuate unwanted frequencies.

- Load has high current surge causing battery voltage to drop briefly. Check load specifications for load disconnect. Add battery storage to compensate for current surge draw-down. Load surges should not exceed 25% of battery capacity. Example: A 200-ampere surge load requires at least an 800-ampere-hour battery bank.
- Lightning damage to controller. Return to manufacturer for repair.

Do the controller relays buzz?

- Incorrect battery voltage. Check series-parallel wiring of batteries.
- Improper battery connection. Broken wires from battery. Check for tight, clean connections. Check wiring.
- Batteries dead. Check state-of-charge of batteries with hydrometer. If low, connect array directly to batteries until charged, then reconnect controller.

Does the array fuse blow? Does the circuit breaker blow or trip?

- Perform array short-circuit test with battery disconnected. Disconnect battery from the controller to perform the test. If the array exceeds rating of controller, split the array and add another controller or get a properly sized controller.

Is the battery overcharging or experiencing excessive water loss?

- Battery is too hot, gassing voltage is lower than normal. Insulate battery bank or replace controller with unit with temperature compensation.

Is the controller not switching to nighttime mode?

- Lightning strike or other high voltage source damaged controller. Return controller for repair.

Does the load fuse or circuit breaker blow?

- Load exceeds rating of controller. Check surge rating of load. Check for shorts in load circuit.

Meters

Meters, indicator lights, and alarms are essential parts of a PV system, not accessories. Digital meters are recommended. Analog, or needle, meters are better than indicator "idiot lights," and lights are better than nothing. Accurate analog meters are expensive and temperamental; they are most accurate at their mid-range. If your system produces 10 amperes, use a 15- or 20-ampere analog meter, not one that reads to 50 amperes.

Low-cost digital meters have changed how people think about measurement. We now measure time to the hundredth of a second and PV system voltage to the hundredth of a volt. While such accuracy is seldom needed for time, voltage accuracy is important. The useful range of a 12-volt battery is only about 0.8 volts from fully charged to empty. Meter accuracy of 1% is best. Meters with 5% accuracy cost less initially, but can end up costing more in the long run because a 5% meter may show 12.5 volts when the system's actual voltage is 11.875 (dead battery) or 13.125 volts (near float charge). One-tenth (0.1) or better current and voltage resolution is recommended.

Meters should be easy enough for a child to read and understand. Readings should not have to be translated, calculated, or compared to a chart. In a dual battery system, meter each battery bank separately.

There are several state-of-charge meters and indicators available. There are amp-hour meters that count electric consumption up (accumulative) or down (subtractive) from a fully charged battery bank. Multi-channel amp-hour meters can be used to track specific loads.

Multiplying amp times volt readings will accurately equal watts. However, amp-hours times system volts only approximates watt-hours because PV system voltage changes throughout the day. AC kilowatt-hour meters, like the meters used by utility companies for measuring inverter output, AC panel and subpanel loads, are available for under $100.

Meters that are "on" all of the time may use too much power. Analog ammeters continuously in series with current flow waste power. Meters that are off when not in use are best. Large-faced meters and meters that directly measure high current or voltage are expensive. Save money by using shunted meters to read high current or voltage. Resistive shunts (calibrated wires or metal bars) connected in series in the system wiring are available. A resistive shunt allows the use of properly matched, lower cost meters connected in parallel to the shunt. A shunt will turn a low-cost milli-ammeter into a high current reading ammeter.

Battery Equalizers

Battery equalizers allow you to safely tap 12 volts from a 24- or 48-volt battery bank. Tapping a small 12-volt load from a larger battery bank without an equalizer causes an imbalance between cells or batteries that will shorten the life of the tapped batteries. Frequently equalizing the entire battery bank helps balance out mismatched cells, as does frequently switching the tap from one group of cells to another, but a battery equalizer does the job automatically.

Economically priced 2,000-watt inverters generally have 24- or 48-volt DC input. Changing everything from 12-volt DC may not be practical; it may be too costly, and some people may want to continue powering favorite 12-volt equipment through an equalizer. In addition, it is wise that some lights and equipment be DC just in case the inverter should require repair.

Diodes

Blocking Diodes

Diodes are solid-state electric "gates" that permit current to flow in one direction only. PV system diodes must be capable of carrying at least 1.25 times the solar array's open-circuit voltage (V_{OC}) and at least 1.56 times (UL $1.25 \times$ NEC 1.25) the short-circuit current (I_{SC}) that will flow through them.

Blocking diodes prevent current from flowing back to, and dissipating through, the array at night or when array voltage is lower than battery voltage. Blocking diodes are also used between array source circuits to prevent current from flowing between source circuit strings. Most charge controllers have blocking diodes or relays to prevent reverse current.

Zener diodes (0.4-volt drop) lose less power than standard diodes (0.5- to 0.7-volt drop), but a 36-cell module's operating voltage is high enough above 12-volt system battery voltage to overcome most of the diode loss.

Bypass Diodes

Bypass diodes generally refer to diodes built into the PV module to allow current to flow around cells or groups of cells and protect shaded cells from conducting damaging current. Shading can happen when something—a leaf, bird droppings, a shadow—covers one or more cells. A bypass diode also allows current to flow around a damaged cell. If the undamaged cells are conducting, the shaded cells or broken cell must be bypassed or they will dissipate power as heat. The problem cells may even get hot enough to damage the rest of the module or cause a fire. Shading spread out over many cells is not as severe a problem as the shading of one or two cells. However, as little as 10% shading of a module or array can reduce power production by 50%.

Shading can damage modules without bypass diodes and rob power from the entire array. Bypass diodes across a module's plus and minus terminals will protect all 36 cells. For better protection, most modules have built-in bypass diodes every 18 cells. Solar arrays used in space and some terrestrial modules have bypass diodes in parallel with every cell. Underwriters Laboratories requires modules to have bypass diodes or some other form of shading or hot-spot protection.

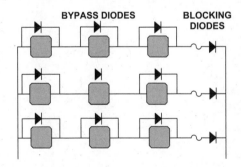

Figure 11.4. Bypass diodes allow current to flow around cells. Blocking diodes prevent reverse current flow.

Chapter 12
BATTERY STORAGE

The role that battery storage plays in PV systems has changed since the mid-1990s. Back then lead-acid batteries were an integral part of almost every PV system worldwide. Although the number of on-grid, battery-less PV installations has risen more dramatically, the number of off-grid PV installations with batteries continues to rise at a steady pace. The cumulative installed residential PV battery capacity in the U.S. as of December 2005 was over 2.4 million kWh, an increase of over 24% as compared to 2004.*

If your PV system is off-grid, it needs a battery bank to provide power at night and on cloudy days. If your PV system is grid-connected, a battery bank is not necessary unless you want back-up emergency power. Battery backup can add 10% to the system price and at least 10¢ to 50¢ per kWh to the system life-cycle cost. The added cost varies with the type and quality of battery used and can go up to over 26¢ to $1.30 per kWh if you factor in charging efficiency.

---

*2005 cumulative installed off-grid domestic PV: 100,000 kW; average one 12-V/100-AH battery (1.2 kWh) per 50 watts PV; $((10 \times 10^7)/50) \times 1.2 = 2.4 \times 10^6$.

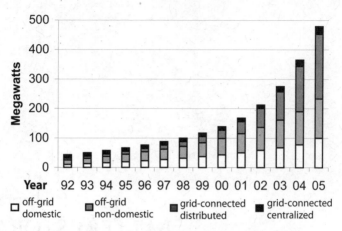

Figure 12.1. Cumulative installed PV in the U.S.

A battery is an electrical storage device. Batteries come in many shapes and sizes, but all have one thing in common: They store direct current energy for later use. In "A Short Course in Electricity" (Chapter 8), volts are compared to water pressure, amps to water current or flow, and ohms to the resistance in a pipe that slows the flow of water. Continuing the water analogy, a battery is like a bucket of electricity. Batteries do not have electricity any more than a bucket has water. Energy must be put into the battery. It could be said that a PV array fills the battery with energy. Some energy is lost when it is put into a battery: 10% of the ampere-hours splash out of the battery bucket in the form of heat as it is being filled. Nonetheless, batteries are the best way to store residential PV electricity.

Battery Construction

Types of Batteries

There are two basic kinds of batteries: primary and secondary. Primary batteries are designed to be used once and discarded. Americans purchase and throw away about 3 billion primary batteries each year to power toys, watches, radios, and tools. Rechargeable batteries are an alternative to single-use disposable batteries. Rechargeable, or secondary, batteries are designed to be charged, discharged, and recharged again and again. PV systems use secondary deep-cycle batteries.

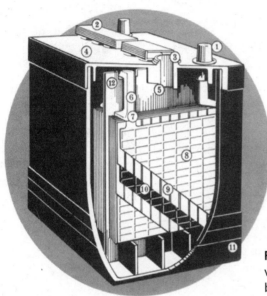

1. Terminal post
2. Gang vent plug
3. Vent
4. One-piece cover
5. Electrolyte level mark
6. Inter-cell connector
7. Plate strap
8. Negative plate
9. Separator
10. Positive plate
11. Container
12. Cell partition

Figure 12.2. Cut-away view of a deep-cycle battery. (ESB, Inc.)

There are different types of secondary batteries. Starting, Lighting, and Ignition (SLI) batteries in automobiles are shallow-cycle secondary batteries designed to provide powerful short spurts of electricity to turn a vehicle starter motor and engine. After the vehicle starts, its alternator or generator quickly recharges the battery. An SLI battery's useful life is shortened if it is deep discharged only a few times. SLI batteries can be discharged to only 10% of their capacity and have a habit of inconveniently failing in cold weather. SLI batteries are not recommended for PV systems.

Some flooded SLI batteries look like sealed no-maintenance deep-cycle batteries. Don't be fooled by their hard-to-remove tops. They are "can't maintain" shallow-cycle batteries unsuitable for home PV systems.

There are several types of deep-cycle batteries. The two major classifications are flooded- or wet-cell batteries and sealed batteries. Flooded batteries suitable for PV include golfcart, forklift, electric vehicle, deep-cycle marine/RV, and floor scrubber batteries; these are all "motive" batteries. Sealed batteries come in two major types—gel-cell and valve-regulated lead-acid batteries (VRLA).

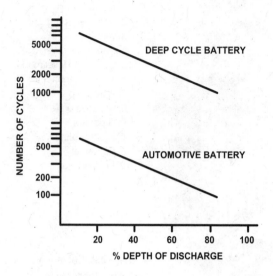

Figure 12.3.
Deep-cycle batteries will last many years more than automobile starting, light, and ignition (SLI) batteries.

Gel-cell or absorbed electrolyte true deep-cycle batteries can be used in PV systems. They cost about twice as much as flooded cell batteries and last about half as long when cycled regularly to 80% depth of discharge. Their primary advantage is that they require very little maintenance.

VRLA batteries have an oxygen recombination cycle that occurs at the negative plate to minimize hydrogen release during normal charging. VRLA batteries, as the name implies, have a pressure release valve to prevent excess built-up gas pressure. The volume of hydrogen gas emitted by VRLA batteries is very small under normal float charge. Equalizing sealed batteries usually voids the manufacturer's warranty and should not be done.

VRLA batteries are often used in uninterruptible power supplies (UPS) in offices, data centers, and rooms without special ventilation systems. They are the type of battery most familiar to many building inspectors and are the easiest to install in a code-compliant residential PV system. These batteries do emit small amounts of potentially explosive hydrogen gas under normal cycling; if overcharged they will emit more hydrogen and require the same ventilation as flooded batteries.

The NEC refers to battery installations in Articles 480 and 690.

How Batteries Work

A battery converts electrical energy into chemical energy and back. A flooded lead-acid battery has a series of cells made of lead plates immersed in an electrolyte solution. The chemical composition of the plates and the acid change when a battery is charged. The change reverses when the battery is discharged.

In a fully charged battery, the active material in the positive plates is lead peroxide (PbO_2) and the negative plates are sponge lead (Pb). Active plate material is very soft and is pressed into a stronger grid containing an alloy of lead and antimony or calcium. The electrolyte is a dilute sulfuric acid (H_2SO_4). When the battery is charged, all the acid is in the electrolyte and its specific gravity is high.

As stored electricity is used and the battery discharges, some acid separates from the electrolyte in the surface pores of the plates. This acid combines with the active plate material and becomes lead sulfate ($PbSO_4$), leaving behind water (H_2O). The specific gravity of water is lower than that of sulfuric acid. When the battery is discharged, its specific gravity is lower.

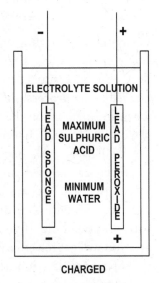

CHARGED

Figure 12.4.
A fully charged battery.

When the battery is being charged, the reverse chemical reaction takes place. Acid is driven out of the plates increasing the specific gravity of the electrolyte. If charging continues beyond the point where the plates can accept energy from the charging current, the electrolyte is broken down into hydrogen and oxygen. This process is called electrolysis and it is the reason water is added to batteries.

The chemical formula is:

$$\text{Discharge} \longrightarrow$$
$$PbO_2 + Pb + 2\,H_2SO_4 = 2\,PbSO_4 + 2\,H_2O$$
$$\longleftarrow \text{Charge}$$

Charged	Discharging	Discharged	Charging
• maximum sulfuric acid	• decreasing sulfuric acid	• minimum sulfuric acid	• increasing sulfuric acid
• minimum water	• increasing water	• maximum water	• decreasing water
• maximum lead sponge	• decreasing lead sponge	• minimum lead sponge	• increasing lead sponge
• maximum lead peroxide	• decreasing lead peroxide	• minimum lead peroxide	• increasing lead peroxide
• minimum lead sulfate	• increasing lead sulfate	• maximum lead sulfate	• decreasing lead sulfate

Charging, discharging, and recharging is not a 100% efficient process. Some energy is lost and some sulfate remains in the plates to form a deposit. This gradual deposit or "hardening of the arteries" is part of the normal aging process for a battery. When batteries are overcharged or when an equalization charge is applied, outgassing of hydrogen and oxygen occurs. Proper charging and discharging will extend battery life.

A battery cycle begins with a fully charged battery, continues as stored energy is used, and is complete when the battery is recharged. Deep-cycle batteries are designed to be discharged up to 80% of their capacity. The number of cycles and depth of discharge determine a battery's useful life. If you dip deeply and often into your battery storage, you will get fewer years of service than if you skim a little power off the top and replace it quickly.

Lead-acid batteries are made of cells nominally rated 2 volts. A 12-volt battery has six 2-volt cells. A battery bank is an assemblage of cells connected in series or series-parallel to match equipment operating voltage.

Battery Size

In 1953, U.S. automobile manufacturers began to standardize lighting and ignition systems to 12 volts. Earlier cars used 6, 12, 24, 30, and other voltages. By the early 1960s, 12 volts became the standard for most vehicle and marine equipment. PV manufacturers naturally developed 12-volt solar modules to charge 12-volt batteries. During the same period, small affordable 100- to 1,000-watt, 12-volt DC to 120-volt AC inverters became available. Larger capacity inverters that cost less per watt and operated more efficiently at 24 and 48 volts were also developed. However, 12 volts remains the most common small PV system voltage.

A 12-volt battery has 6 cells, a 24-volt battery has 12 cells, and so on. Batteries come in all sizes. Some deep-cycle batteries used in solar lanterns weigh under 4 pounds. Some batteries have 2-volt cells, each weighing hundreds of pounds. One kilowatt-hour of lead-acid battery storage weighs approximately 60 pounds, is slightly less than one cubic foot in volume, has a footprint approximately 11 by 8 inches, and is 12 inches tall. A PV home power system can have a ton or more of batteries.

Size and weight are important issues when transporting and handling batteries. Six-volt golfcart batteries and 2-volt industrial cells are used in many home PV systems. Golfcart batteries are generally available, reasonably priced (under 50¢ per ampere-hour or 10¢ per kilowatt-hour), have good charge/discharge characteristics, and can be carried by one person. Moving large batteries and cells can require extra labor, a forklift, skids, heavy-duty dollies, and block and tackle—all at added cost. Even so, it is important to limit golfcart batteries to no more than four parallel strings. Using many small batteries is inefficient due to wire and connector losses and cell mismatch. Use fewer, larger batteries. For example, instead of 32 golfcart batteries (96 cells) in four parallel 8-battery series strings for 48 volts, use one series string of twenty-four 1.35-kWh forklift cells.

Some 12-volt PV systems use 12-volt RV/marine batteries. Two deep-cycle 12-volt RV batteries have the same storage capacity at two 6-volt golfcart batteries of the same weight, but cost twice as much and last half as long. People sometimes buy 12-volt batteries because they do not understand that 6-volt batteries and 2-volt cells can be wired in series for 12, 24, 48, or higher voltage.

Battery Capacity

Just as bigger buckets can hold more water, bigger and heavier batteries generally have more electrical storage capacity. A 100-pound lead-acid battery generally has twice the storage capacity of a 50-pound battery.

Battery capacity is determined by size, weight, and depth of discharge (DOD), which is the percentage of energy that can be used and replaced. A 100-pound deep-cycle battery that can be discharged to 80% of its capacity may be smaller and weigh less than a 150-pound medium-cycle battery with a 50% depth of discharge rating, but it stores more usable energy (80 vs. 75 ampere-hours).

Your PV system battery bank size is determined by several factors: average daily amount of electricity used, the number of days of autonomy required for your climate, battery depth-of-discharge rating, and battery operating temperature. Battery bank design also takes into account the maximum electric load at any one time, what kinds of loads are being powered, what portion of the load is used at night, and, of course, your budget. Finally, your battery bank must be sized to meet winter energy requirements or whenever less solar radiation is available.

Depth of discharge is important. A golfcart battery regularly discharged to 11.98 volts or 1.145 specific gravity, which is 80% of its capacity, will last 400 or more cycles. Cycled this deeply every day, it will last around 12 months, which is about how long it would last in a golfcart that is used daily. Deep-cycle batteries discharged to 40% of capacity and sized for 5 days of autonomy should last 6 or more years in the southwestern U.S. Locations with less solar input require 8, 12, or even more days of autonomy for batteries to last 6 years.

Some medium-cycle (50% DOD) recreational vehicle and boat batteries are mislabeled deep cycle. They will not last as long as true deep-cycle batteries. If you must use an RV/marine battery, do not discharge it deeply. A 100-ampere-hour RV battery can be repeatedly discharged to 25% of capacity (25 ampere-hours) hundreds of times. RV batteries can be quickly recharged with up to 20 amperes (capacity divided by 5, or C/5) if kept below 100°F/37.7°C while charging.

Other factors affect battery capacity. Golfcart batteries can be discharged up to 500 amperes provided all terminals and cables are properly sized and in good condition. A high discharge current can occur when a large DC load or AC inverter load is applied. A high rate of discharge will cause capacity to fall off quickly and require equalization more often.

Discharging below terminal voltage gives very little additional service and significantly shortens battery life. A low-voltage cut-off switch on loads can be used to prevent deep discharge battery damage.

Operating temperature is very important. Batteries last longer if kept cool. However, a battery below 60°F/15.5°C has reduced capacity. Temperatures above 77°F/25°C increase capacity only slightly but they significantly reduce battery life. Batteries operating regularly above 85°F/29.4°C lose half their operating life.

Lead-calcium batteries have a lower self-discharge rate than lead-antimony batteries. The self-discharge rate of lead-calcium batteries remains relatively low throughout their life, whereas local chemical action in lead-antimony batteries increases about halfway through their useful life. Lead-antimony batteries cost about half as much per ampere-hour and have half the life-cycle kilowatt-hour cost as lead-calcium batteries.

Sulfation

Sulfation, the natural formation of sulfate crystals when a battery is less than fully charged, also affects capacity and useful life. Properly charged batteries do not get a chance to sulfate excessively. Batteries that are partially discharged for long periods of time develop permanent sulfate deposits and lose capacity.

Extreme sulfation must be avoided, but normal sulfation will occur when a battery is discharged. Sulfuric acid in the electrolyte acts on the positive and negative plates to form a new chemical compound called lead sulfate. When the battery is charged, the lead sulfate is expelled from the plates and returns to the electrolyte.

Prolonged periods of partial or complete discharge will cause permanent sulfate crystals to form and shorten the life of a battery. Very deep discharging and slow recharging can cause sulfate crystals to expand

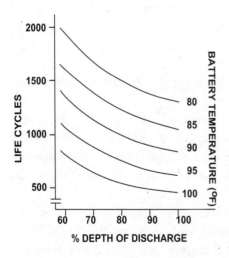

Figure 12.5.
The relationships of temperature and depth of discharge to battery life.

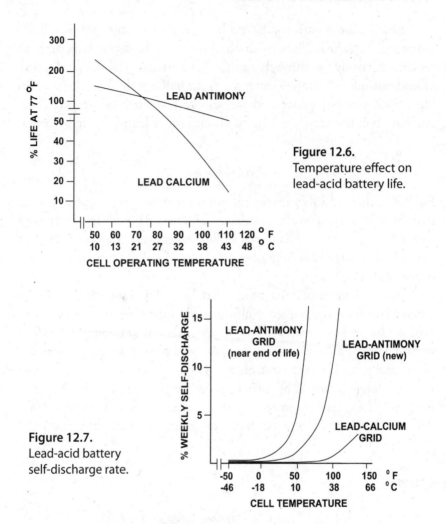

Figure 12.6.
Temperature effect on lead-acid battery life.

Figure 12.7.
Lead-acid battery self-discharge rate.

the negative sponge lead plates, separating the lead from the grid and permanently damaging the battery.

Sulfation can be a problem because PV systems often have very low recharge rates. Even if the PV array is large in relation to the battery bank, the batteries still may not get a full charge. This can happen when the charge controller senses the high array voltage instead of the actual battery voltage and shuts off the charge too soon.

The only way to prevent sulfation is to keep your batteries fully charged all the time, which is impossible with an active PV system. Minimize sulfation with a properly sized battery bank that is not deep discharged and quickly recharged every day. If your PV production matches your consumption on a daily basis and your battery bank is sized for 40% depth of discharge, and your charge controller is properly adjusted, sulfation should not be a problem. When you do discharge your batteries deeply, bring them quickly back to full charge and equalize them.

Some inverters with built-in battery chargers and battery desulfators can partially reclaim sulfated batteries. The charger that takes its current from the peaks of the AC waves causes battery plates to be charged by 60 small pulses of current per second. Desulfators may operate at even higher frequencies. The pulsing current causes the sulfation to break up and fall to the bottom of the battery as small particles. While pulse reclamation of badly sulfated batteries is not guaranteed, there is growing agreement that pulse charging is better than non-pulse charging.

Battery Efficiency

Battery efficiency can be expressed in ampere-hours or watt-hours. The ampere-hour recharge rate is 110% of the discharge rate or about 91% efficient. Voltage is higher during charging to "push" the current into the battery so voltage efficiency is about 85%. Combine the two (91% × 85%) for watt-hour or total energy efficiency of about 77%.

In other words, new batteries deliver 100 ampere-hours for every 110 ampere-hours of charge. Every 100 watt-hours of energy put into batteries will yield 77 watt-hours out. As batteries near the end of their useful life, efficiency drops to 50% or less.

Buying Batteries

Battery distributors are listed in the phone book. Find the dealers that have the freshest stock and offer the best service. Since golf courses use lots of batteries for their golfcarts, you could ask someone in the maintenance department where they buy batteries. If there are no golf courses or battery distributors nearby, ask at local automobile or truck service stations and auto parts stores.

Batteries are a commodity and prices are based on the amount and type of materials used, labor costs, shipping costs, and profit. Similar batteries have similar prices. If one battery is more expensive than a similar one, ask why. Sometimes suppliers sell one brand or type at a lower price because their sales volume is higher. Cheap batteries, however, can be a bad bargain. They may be poorly made, improperly shipped, and badly stored. On the other hand, a high price may just mean a high mark-up.

Commercial battery prices are rising after not having changed much from 1990 to 2005. Six-volt 220-ampere-hour golfcart batteries cost around $100. Sealed 12-volt batteries cost twice as much for the same capacity as flooded batteries. Good quality industrial-grade batteries cost more than golfcart batteries but usually have a longer warranty and should last longer.

Batteries are heavy. Buy locally and you can save hundreds of dollars in shipping costs. If your batteries must be shipped, insure them and use a freight company experienced in handling batteries and other materials classified as hazardous by the Department of Transportation. If you are buying a pallet load of batteries, ask for a volume discount. Make sure delivery and off-loading are included in the price.

Buy true deep-cycle batteries and get the specifications in writing. Do business with an established, reputable, local battery distributor who sells a lot of batteries and has rapid stock turnover. If offered a deal on some dusty batteries in the back room, forget it. Do not buy batteries that have been in stock over six months. The date of manufacture is stamped on the terminals or the case.

Tell the salesperson that the batteries will be used in a PV system. You may be surprised at the interest and cooperation this will generate. If he or she has sold batteries to other PV people, ask to be put in touch with them so you can gain from their experience. If the salesperson doesn't know what you are talking about, shop elsewhere.

Buy what you need, not what the shop has on hand. The price for good batteries shouldn't vary much from one supplier to another.

Sometimes you can find used batteries at a discount. If you find used batteries for sale, ask who owned them and how they were used. Are maintenance logbooks and service schedules available? Telephone companies usually take good care of their emergency storage batteries, often recycling them well before the end of their useful life. Golf courses, as mentioned earlier, use a lot of batteries in their golfcarts, as do warehouses with electric forklifts, but these batteries may have been misused. Abused batteries are no bargain, even if free.

Whether new or used, appearance tells a lot. Are the batteries clean and have they been kept out of the weather? Do they show signs of physical abuse or of having been recently cleaned up to look better? Use a flashlight to inspect the insides of the batteries. Is the electrolyte clear and clean? Are the plates in good shape and not warped or brownish or white from sulfate? Are terminals secure and in good shape? Test the cells properly with a good digital voltmeter and hydrometer. Are specific gravity readings for all cells within range of each other? Are cells completely discharged?

New lead-acid batteries may be shipped "dry" or "wet," but they all end up "wet." A wet battery has electrolyte added at the factory. A dry battery has electrolyte added in the field before it is used. Dry batteries, developed for the armed forces in World War I, ship and store easily and generally are fresher. However, dry batteries kept in a humid environment will absorb moisture from the air, begin aging, and possibly be ruined.

Aging is a normal chemical breakdown called "local action." Batteries begin to age as soon as the electrolyte is poured in. Wet batteries begin to degrade just sitting in a warehouse. When dry batteries are kept in a dry environment, aging is delayed until the electrolyte is added.

It may sound strange, but treat your batteries as if they are alive. Both batteries and living things are electrochemical processors. Keep batteries clean, not too hot, not too cold, dry, and full. If you use your batteries properly, inspect and test them regularly, and don't overwork or abuse them, they will be your dependable servants.

Figure 12.8. Telephone company batteries.

Eventually even the best-maintained battery grows weak and must be replaced. Sometimes individual cells fail, even in new batteries. If you test and maintain your battery bank properly, you will be aware of weak cells or any drop in specific gravity and can plan for its replacement.

Test your batteries regularly with a voltmeter and hydrometer during the first few months of the warranty period to identify bad or weak cells. Extended warranties are available, but they may not be worth the extra cost. Golfcart batteries typically have one-year prorated limited warranties. Some industrial- or commercial-grade batteries have 5-, 10-, and even 20-year very conditional warranties. Be sure you understand the terms of the warranty and keep accurate records in your battery service log.

Batteries can last up to 20 years. One of Joel's battery banks was manufactured in 1968 for a telephone company, which recycled them in 1978. He bought them in 1979 for the salvage price of lead. Although their self-discharge rate had increased over time, they were still in use three years later when he gave them to a new owner.

BATTERY REPORT

Date: _____   Time: _____
Date Installed: _____ System Voltage: _____
Last equalizing charge:
Date: _____ Volts: _____ Hours: _____

Cell Number	Cell Voltage	Specific Gravity	S.G. Temp.	Water Added (oz)
1				
2				
3				
4				
5				
6				
7				
8				
9				
10				
11				
12				
...				

Figure 12.9. A sample battery report. Your battery log should be kept weekly for the first two months. Then test one randomly selected pilot cell every month; test every cell four times per year.

Plan for replacing your battery bank when you design your PV system. If you design your battery bank to last five years at 80% depth of discharge, set aside one-fifth of its replacement cost each year. At the end of five years, you will have the money to buy new batteries, plus interest. If you have not been discharging your batteries to 80% DOD, your savings will continue to grow.

Batteries have salvage value. Recycle them. Do not pollute.

Alternatives to Batteries

Direct-Drive PV

Some water pumps and fans are powered directly by solar electricity without batteries. Direct-drive PV systems are elegantly simple and can save you money. Attic, ceiling, and ventilating fans driven directly by PV benefit from increased solar radiation. The brighter the sun, the faster the fan turns, and the more air is moved.

A PV-powered pump and storage tank can deliver water 24 hours a day. Water pumped during sunlight hours is stored in an elevated tank (or pond) to be used as needed. A water storage tank costs less than an equivalent battery electrical storage, is more energy-efficient, requires less maintenance, and should last longer. Direct-drive solar irrigation makes sense. The brighter the sun, the more plants need water, and the more solar energy there is to operate the pump.

Direct-drive PV works well but almost all PV systems use batteries because our modern world is "on" 24 hours a day.

Other Forms of Storage

There is a lot of speculation about new batteries and new forms of energy storage. For decades there have been rumors about imminent breakthroughs in battery storage. Science and technology have experienced revolutionary change in the past 100 years, but batteries remain pretty much the same—and not for lack of investment or research, but because of the physical limits of electrical energy storage chemistry.

Nickel-cadmium batteries (Ni-Cad) are excellent electrical energy storage devices, but impractical for most PV systems because they cost three to five times more than lead-acid batteries. Experimental recircu-

Figure 12.10. This fan and small solar module blow excess heated air from an attached greenhouse into the house to provide supplemental space heating. Direct-drive solar fans are also used to vent attics.

Figure 12.11. Lowering a Lorentz PV-direct DC submersible pump into a drilled well. PV water pumping is cost effective, even when the well is only a few hundred feet from power. (Photo: Windy Dankoff and Solar Energy International)

lating electrolyte, nickel metal hydride, and lithium ion batteries show promise, but are prohibitively expensive or not yet available. Some new battery designs have environmental and safety issues to resolve. Flywheels, hydrogen, and other gas storage have limited applicability in residences.

If you want a PV system, don't wait for a breakthrough in electrical storage technology. Don't design your system around a concept that has not reached the marketplace. You will only delay the day you begin using solar energy or frustrate yourself into inaction. When better batteries become available, millions of people will buy them, including us.

Installation & Maintenance

Always think **SAFETY** when handling batteries. Batteries are heavy. Battery acid is dangerous.

- Have protective clothes, gloves, safety glasses, baking soda (powdered sodium bicarbonate or $NaHCO_3$), and a water bucket on hand before your batteries arrive.
- Allow only qualified adults near batteries.
- Never slide batteries across a rough surface. You could damage their cases.
- Do not tip or tilt batteries more than 25° from vertical.
- Only use tools with insulated handles.
- Lift batteries with straps or carriers. Never lift batteries by their terminals: You might destroy internal electrical connections.
- If you must mix acid and water to make electrolyte, pour the acid into the water. Never pour water into acid as a violent chemical reaction will occur.
- Use dry baking soda to neutralize electrolyte spilled on the batteries, shelves, or floor. Sprinkle it on the spill until the foaming stops, then sweep it up. A solution of one pound baking soda to one gallon water will neutralize any remaining acid in crevices or porous surfaces.
- Be careful when adding electrolyte or acid to batteries. Wear gloves and safety glasses and protect your clothing from the acid. Do not smoke.
- If you get acid on your skin or in your eyes, immediately flush with lots of clean water and seek medical help.

Reliable energy storage begins with proper battery handling and installation. Inspect your batteries for damage when they arrive, not after you have signed for them. Do not accept damaged batteries. Look for signs of electrolyte leaks or spills on cartons, batteries, pallets, or crates. If a spill has occurred, your supplier may allow you to remove some electrolyte from each cell to put into the low cell. A battery with more than ¼-inch of plate exposed due to spillage may be permanently damaged. Re-inspect batteries within 15 days of receipt for concealed damage. If there are problems, make a claim against the carrier.

Never add electrolyte to a cell unless told to do so by the manufacturer. Increased concentration of acid will eat up battery plates. Only put distilled or clean, low mineral content water and a clean hydrometer into your batteries. Do not add water when you receive your new batteries. Wait until they have been charged a few times because electrolyte levels rise when batteries are charged. If you add too much water, the cells can overflow and you will have a mess to clean up. If you buy dry batteries, make sure you have the correct electrolyte with the proper specific gravity.

Figure 12.12. To maintain optimum system performance, it is very important to keep batteries and terminal connections clean. In corrosive environments, such as coastal areas, spray terminals with retardants or coat them with grease. (Photo: Robert Sardinsky, Rising Sun Enterprises)

Never use battery additives. Every manufacturer voids its warranty when additives have been used.

Install your battery bank as soon as possible after you get it. If you are not ready to install your batteries, prepare a temporary holding area near their permanent location that is out of the weather and between 57 to 70°F/18 to 21°C. If batteries must be kept in an unheated room, be absolutely certain they are fully charged and insulated from the cold. In a discharged battery, most of the acid is in the plates leaving behind water, which freezes at 32°F/0°C. Discharged batteries will freeze and their cases will crack.

Fully charged unused batteries stored at 0°F/-17.7°C can hold a charge for almost a year. Fully charged batteries stored at 130°F/54.4°C can completely discharge in a month. In any case, avoid storing batteries for more than three months.

If you must store your batteries, give them an occasional freshening charge with a generator or utility power to compensate for self-discharge. Keep the battery temperature under 130°F/54.4°C when charging. Normal battery operating temperatures are between 60°F/15.5°C and 90°F/32.2°C.

You can temporarily connect your PV modules to the batteries to maintain a trickle charge. If you set up your PV modules for a short time to charge your batteries, secure the modules so they cannot be blown or knocked over. Secure cables and wires to avoid tripping over them. Trickle-charge batteries at a current rate 2% of total ampere-hour capacity or follow manufacturer's recommendations.

You may have heard that putting a battery on a concrete floor will "sap" or drain its energy. Because the concrete floor is colder than the surrounding air, the bottom of the battery becomes cooler than the top of the battery. Unequal temperatures within the cells adversely affect the chemical reaction. Place batteries on wood planks (sleepers) or shelves to allow air to circulate evenly around them to keep their temperature even.

For the same reason, do not put your battery bank against a cold garage or basement wall. Use isulation, but maintain air circulation space between the wall and the batteries. For fire safety, protect insulation and other flammable materials with at least ⅝-inch gypsum or plaster board.

Code-Compliant Installation

NEC Article 480 applies to storage batteries. Vented and sealed batteries are treated alike in code-compliant installations. Neither requires electrical insulation if no voltage is present between the container and the ground. Metal racks should be treated to resist the deteriorating action of the electrolyte. Trays or racks directly in contact with the cells must be non-conductive. The battery location must be ventilated. Even though valve-regulated or sealed batteries emit only small amounts of hydrogen under normal use they can liberate large quantities of hydrogen if overcharged, so they must have the same ventilation as their vented counterparts. The battery room must be clean and dry and have sufficient workspace for inspections and maintenance. The wiring must be resistant to fumes and corrosion. Plastic conduit is recommended. All live parts are to be guarded.

Connecting Your Battery Bank

Make an accurate drawing of your battery layout before moving the batteries to their pallets, shelves, box, or compartment. Indicate how the batteries will be situated, which sides face what direction, and which terminals are connected. Note the polarity. Double-check your wiring diagram. Avoid unnecessary lifting and handling. Save your back.

Use properly sized battery interconnects. Use at least #4 AWG THW, RHW, and USE extra-flexible cable for ease in handling. Stranded and PVC-insulated battery cables can be used if they are UL-listed. If you do use stranded cable, be sure to use connectors that are intended and labeled for that use. Ask your battery supplier for advice and use the wire size charts in chapters 8 and 15 to determine the proper size conductor. Planned short, orderly wire runs save money and reduce voltage drop caused by wire resistance. Ring connectors should be securely crimped, soldered, and taped. Bare connectors should not touch each other or any metal shelves, equipment, or walls. Use stainless-steel bolts, nuts, and washers to fasten cables to terminals.

Exercise extreme caution when working around batteries. DC battery current is capable of welding steel or melting a wrench. Remove all rings, watches, and other metal before handling batteries or working on electrical circuits. Move your batteries into place. Neutralize and clean up any spills. Fasten interconnects tightly but do not deform the battery terminal. Torque nuts and bolts to 70 inch-pounds. Some manufacturers recommend coating terminals with petroleum jelly or other coatings. We don't like coatings that collect dirt and dust, and prefer to wait until after a few inspections to determine if an anti-corrosive coating is needed. If you do coat your terminals, protect them from dust and dirt. Do not put grease or other coatings on battery cases and cables.

If wires or terminals corrode, turn off your system, disconnect the cables from the battery terminals, and gently clean the metal parts with #00-grade sandpaper or a battery terminal cleaning tool, which is available at auto parts stores. It may be necessary to soak interconnect wires or cables in a solution of baking soda and water to remove corrosion.

Do not connect your battery bank to any equipment until all cells or battery bank interconnects are securely fastened. Electronic equipment, controllers, inverters, and meters must be connected in proper sequence according to manufacturer instructions.

The Battery Box or Battery Room

Small battery banks can be kept in a vented, lidded plastic storage box. Cut holes for the cables and screened vents. Some inspectors require pans under batteries to catch spilled electrolyte. Shallow plastic storage boxes without lids can be used as spill containment trays.

The battery room, cabinet, or area should be properly vented, clearly marked, and kept at 60 to 70°F/15.5 to 21.1°C. Lock the door to keep children and curious adults out. Place batteries on sturdy pallets off the concrete or dirt floor or on strong shelves. Cover the battery bank to prevent dirt and dust from collecting on the tops of the batteries. Give yourself plenty of room to service or replace batteries. If you stack batteries on shelves, leave enough shelf height to perform hydrometer tests and to add water. Fasten batteries and shelves with straps and braces to prevent them from falling over.

Battery shelves can be enclosed in a cabinet to keep the batteries clean, out of reach, and insulated from temperature extremes. Plywood doors hinged to the shelf frame will suffice. Vent the cabinet to allow hydrogen developed during charging to safely exit.

The floor, pallets, and wooden boards under the batteries must be strong enough to support their weight. A vented plywood enclosure with removable top can be placed over floor-mounted batteries. While a clean dirt floor is acceptable, a sloped concrete floor with a drain makes clean-up easier.

Controllers, inverters, and switches may have electrical relays and contacts that can spark when their contacts open. Equipment must be isolated from the batteries to prevent sparks from igniting the hydrogen gas emitted by the batteries. Fasten wire, cable, and conduit neatly to the battery area's fire-resistant wall.

Do not put equipment in a cramped, dark, hard-to-access corner. Provide at least three feet of clearance in front of all equipment. There should be no open flames or standing water in the equipment room. Do not use the battery or equipment room for storage. Never keep flammable liquids in the room. Mount inverters and other equipment that require venting where free-flowing cool air is present. Mount a fire extinguisher in a handy location.

The ideal battery storage area—

- has good ventilation.
- is not near any open flames.
- has no possibility for electrical sparks.
- is easy to maintain and inspect.
- is tidy and easy to clean.
- is kept at 60 to 70°F/15.5 to 21.1°C.
- is out of reach of non-authorized personnel.
- has an up-to-date fire extinguisher handy.

Building a Battery Box

An insulated and vented battery box keeps your batteries clean, stabilizes their temperature, protects them from dropped tools, and allows easy venting. This simple battery box can be modified to a battery cabinet with front doors, but be sure to allow enough space above the batteries on shelves so that you can test the batteries with a hydrometer and add water.

Build your insulated battery box large enough so that you can double the number of batteries if you choose. The box should be slightly higher

Figure 12.13. This commercially available tool shed serves as a battery and inverter enclosure. The solar array shelters the genset from rain and snow. (Offline Independent Energy Systems)

in back (the side against wall) and the top should have the hinges at the back. You can vent the box by using a piece of plastic pipe at the highest side of the box, with the vent going up through the wall or ceiling to the outside. Put a screened cover on the vent.

Put your batteries in a row along the wall on two 2″×4″ planks laid flat and spaced 2 to 4 inches apart. The batteries can be set in a shallow tray cut from a low-cost plastic storage box for spill containment. Leave about an inch between and around the batteries and the sides of the box for air circulation. Allow room behind the batteries for the plywood and insulation. When you take your measurements for the battery layout, be sure to allow for the high backside and for foamboard insulation on the sides and top.

This plywood battery box is bottomless. It is not going to bear weight, so it can be simple and light. Make the ¼- to ½-inch plywood sides, front, back, and top large enough to slip 2- or 4-inch foamboard insulation around the battery bank and still allow at least 1 inch of air space so that the battery bank temperature remains even. The top insulation is a piece of non-combustible foamboard laid on top of the batteries.

Assemble the topless and bottomless box and put the hinges on the backside. Use standard loose-pin door hinges so you can easily remove the box's top if you have to. Slip the plywood sides over batteries. Then slip in the foamboard sides. The hinged plywood box top keeps the dirt and kids out. The high-side vent allows gas to pass to the outside.

Battery Venting

Battery rooms must be properly vented to prevent the build-up of potentially explosive hydrogen gases that are released during charging. The correct charge controller and charge settings will limit out-gassing during normal charging. However, topping off and equalization charging, requirements for long battery life, will produce hydrogen.

Concentrations of 4% hydrogen are explosive. The recommended maximum concentration for battery storage areas is 2%. Use the following formula to determine the size of your battery room:

$$Q = 0.0135 \times I \times N$$

Q = the necessary quantity of air ventilation in
  cubic feet per minute (CFM)
I = the charge rate in amperes
N = the number of series connected cells

For example, a PV system with eight 100-watt modules (rated 17.1 V, 5.88 A) and sixteen 6-volt batteries is wired for 24 volts DC. The required venting is $0.0135 \times (5.88 \times 4) \times 12 = 3.81$ cubic feet per minute (CFM) or 228.6 cubic feet per hour. Many rooms are 8 feet in height and have two air changes per hour. A room 8'×6'×6' has 288 cubic feet volume and may naturally vent 576 cubic feet per hour, almost 2.5 times the required 288.6 cubic feet per hour.

Provide vent openings at the bottom of battery room doors and at the highest point in the ceiling. Ceilings should have no exposed rafters or framing that can trap hydrogen, which is lighter than air. If you cannot vent your battery compartment straight up and out through a ceiling, use a vent with a fan to move the air out of the battery area. Never run

Figure 12.14. Some electrical inspectors require battery spill containment. Acid-resistant plastic trays or enclosures can be used. When the lockable lids on these plastic enclosures are closed, batteries are kept clean and accidental contact with terminals is prevented. Covers have small holes and provide adequate ventilation even during equalization. (John Wiles)

the battery vent pipe horizontally or downward as gas will collect in it. Energy-efficient homes might have as low as one-half an air change per hour and may require a fan to vent battery gasses.

If you cannot provide enough space or air changes for natural venting, make a battery compartment or cabinet and vent it to the outside. Put a screen over the vent to prevent mice, bugs, or birds from nesting in the battery box. Use the battery vent only for venting batteries—not as wire conduit.

VRLA or sealed batteries emit very little gas. Typically, 12-volt, 100-ampere-hour VRLA batteries produce 20 cc of gas per hour at 2.4 volts per cell. Twenty batteries in a $10' \times 10' \times 8'$ unvented room would produce 400 cc of hydrogen gas per hour. There are 28,320 cubic centimeters per cubic foot. Dividing 400 cc of hydrogen gas per hour by 28,320 cc/cubic foot = 0.0141 cubic feet of hydrogen gas emitted per hour. It would take 1,134 hours or 47 days (16 cubic feet ÷ 0.0141 = 1,134 hours) to reach the allowable maximum 2% gas accumulation in the 800-square-foot unvented room. Venting is still required for code compliance and safety.

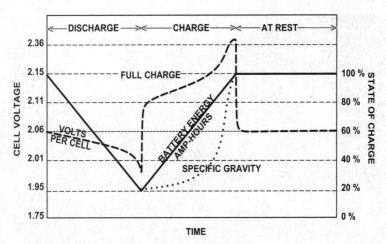

Figure 12.15. Typical voltage and specific gravity characteristics during a constant rate discharge and recharge. (Exide)

Do not tempt fate. Do not enter a battery room or open a battery compartment or cabinet with a cigarette in your mouth. If the lights are out, use a flashlight, never an open flame, match, or candle. Open the battery compartment carefully and let it air for a few minutes before you begin work. Be overcautious.

For peace of mind, lock the battery room or compartment and put up a sign that reads: DANGER! BATTERY ROOM! KEEP OUT!

Battery Testing

Use a three-ring binder or file folder to keep a complete history of your battery bank and PV system. Note dates of purchase and installation, prices, parts, initial voltage, hydrometer readings, comments, and anything else you feel is important. Your equipment log will provide valuable system performance information.

Battery testing begins the day you receive your batteries. You only need three tools: a thermometer, hydrometer, and voltmeter.

Chart the specific gravity and voltage to estimate your battery's state of charge (SOC), depth of discharge (DOD), and capacity to extrapolate how much stored energy is available. You can use a battery amp-hour meter as a fuel gauge, but the amp-hour meter does not substitute in any way for regular battery maintenance and testing.

Specific gravity (SG) is the measurement of electrolyte density. The more sulfuric acid in the electrolyte, the higher the SG. When the battery is fully charged, all of the acid is in the electrolyte and SG is high. When the battery is discharged, SG is low. In a fully charged deep-cycle battery, SG is typically between 1.265 to 1.280 or higher. Anything above 1.300 SG may mean that there is too much sulfuric acid. Different batteries have different SG so refer to the manufacturer's specifications.

A hydrometer is used to read SG. A good hydrometer costs around $20 and will last for years. Get one with a large calibrated float and easy-to-read scale. Don't buy a cheap plastic toy with floating balls.

Specific gravity is affected by temperature. Use a clean, accurate thermometer to measure the electrolyte's temperature. Thermometers built into hydrometers are usually hard to read and often inaccurate. Add or subtract 1 point (0.001) SG for every 3°F/1.67°C above or below 77°F/25°C.

To read SG, remove the battery cell cap and stick the clean, dry hydrometer nozzle deep into the cell. Do not jam it into the cell or between the plates. At each cell, squeeze to fill and empty the hydrometer a couple of times. Flushing the hydrometer with each cell's electrolyte gives a more accurate reading for that specific cell. Leave the hydrometer nozzle in the cell when taking the reading to avoid drips.

Carefully avoid getting dirt into the cell or on the hydrometer nozzle. Do not siphon electrolyte from one cell into another. Work clean and avoid contamination. When finished, rinse the hydrometer with clean water, dry it, and store it in a clean box.

Manufacturers recommend that batteries sit at rest for several hours being neither charged nor discharged before reading SG. This is not possible with an active PV system. Your system may be at rest at night, but waking at 3 a.m. to test SG is not necessary. Pick a convenient time to test your batteries. Note date, time, weather conditions, ambient temperature, system voltage, charging or discharge current, and any other conditions.

If you test your batteries under similar conditions each time and keep good records, you will develop an accurate history of your battery bank's state of charge.

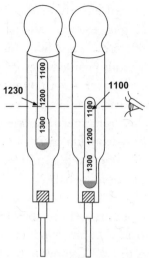

Table 12.1.
Typical Deep-Cycle Battery Capacity

Depth of Discharge	Specific Gravity at 80°F (27°C)	Volts per Cell
0%	1.2650	2.100
10%	1.2500	2.090
20%	1.2350	2.075
30%	1.2200	2.060
40%	1.2050	2.045
50%	1.1900	2.030
60%	1.1750	2.015
70%	1.1600	2.000
80%	1.1450	1.985
100%	1.1300	1.750

Figure 12.16.
The correct way to read a hydrometer.

Specific Gravity Test Guidelines

1. Plate "surface charge" during and immediately after charging affects readings. Test in the early morning when loads are light or charging has not begun.

2. If your electrolyte is ½-inch low, SG will read 15 points (0.015) above normal and vice versa. If there is no electrolyte level mark, fill to ⅜ inch over the tops of cell separators after readings and after equalization charging.

3. SG readings taken during the early stages of recharge will lag as the heavier acid at the bottom of the cell has not yet mixed with electrolyte at the top of the cell.

4. When adding water to a cell, expect several days' lag until water is diffused.

5. The SG of a fully charged cell will drop 0.001 per year due to a slight loss of electrolyte at each reading. Change the pilot (or sample) cell every five readings.

6. A weak cell will vary 20 points (0.020) or more than other cells in an equalized battery. A bad cell will vary 30 points (0.030) or more than other cells in the same battery.

The difference between a charged cell and a discharged cell at rest is less than 0.4 volts. Make your voltage tests with a good digital voltmeter that reads to 0.001 volts. There are excellent digital multi-testers that cost less than $100. First measure total battery bank voltage, making sure your test probes are clean and in good contact with the battery terminal. Then measure each cell or battery voltage. Record all readings.

There is a direct relationship between voltage and specific gravity of a cell at rest. Voltage equals specific gravity plus 0.84 ($V = SG + 0.84$). For example, a cell with SG 1.280 will read 2.12 volts.

The first two months you have your batteries, check a few different cells each week for voltage and specific gravity. After two months, when everything seems stable, test one cell every month. Four times a year test every cell and record the readings. A complete voltage and hydrometer test of 24 cells takes about an hour. Most of the time is spent removing and replacing caps. This is a good time to examine the caps to make sure that the vent holes are not blocked. It is also a good time to clean the battery room, shelves, racks, and cables and to tighten loose cable connectors.

Battery Charging

Battery charging and discharging are expressed as a function of time. For example, a 100-ampere-hour battery discharging at a constant rate of 2 amperes has a 50-hour discharge rate or capacity divided by 50 (C/50). This means the 100-ampere-hour capacity is discharged over a period of 50 hours. The same battery may have a maximum charge rate of C/5 or 20 amperes and a trickle-charge rate as low as C/100. C is typically the battery's capacity at the manufacturer's 20-hour rating.

PV battery charging through a charge controller is automatic. In the morning the controller senses when the array's voltage is higher than battery voltage and allows full current to flow. As battery voltage rises, multi-stage controllers adjust current flow to top off, trickle, or maintain a charge. At night array voltage stops and solar charging ends.

Lead-antimony batteries begin gassing when they are 90 to 95% charged. The last 5% of charging is the least efficient because most of the energy is used in electrolysis (gassing). However, this gassing is beneficial because it mixes up the acid and prevents the stratification that occurs in stationary batteries.

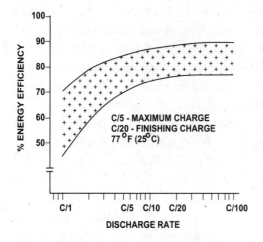

Figure 12.17. Typical efficiency range for lead-acid batteries as a function of discharge rate. (Note: C = capacity. Example: The maximum charge rate for a 200 ampere-hour battery is 200/5 = 40 amps.)

A full charge should not produce excessive gassing, but can occasionally raise battery temperature above 110°F/43.3°C and cell voltage above 2.4 volts. If you have a two-stage charge controller, its final period of charging or topping-off charge rate should be between 4 and 10 amperes per 100 ampere-hours of battery capacity.

A discharged battery can take a very high initial rate of charge, up to ten times its finishing rate. However, excessive charge rates corrode the grids of the positive plates into lead peroxide, which weakens them physically, increases their electrical resistance, and shortens battery life. Limit charging to 20 amperes per 100 ampere-hours or to manufacturer recommendations.

Some single-stage charge controllers prematurely cut off charge current to a battery bank that is too small for the PV array because the controller sees the high charging voltage instead of the actual battery voltage, which is lower. For example, PV array output is 16 volts at high current because the battery was deeply discharged to 11.75 volts. System voltage under bright sun conditions may reach the controller cutout set-point of 14.5 volts before the battery is 100% recharged. The array must be properly sized to the battery bank to prevent premature charge cutout. See Chapter 11 for additional information on charge controllers.

"Mossing," the formation of a sponge-like deposit on the negative plates and strap, is an indication of overcharging. Sediment at the bottom of the battery container also indicates overcharging or too high a charge rate. Overcharging literally tears batteries apart.

The float current requirement increases as batteries age. Float current for lead-calcium batteries is low, in the 4- to 11-milliampere (0.004 to 0.011 amperes) range per 100 ampere-hours, rising only slightly as batteries age. Average float voltage for lead-calcium cells is 2.17 to 2.20 volts.

Float current for lead-antimony batteries is 30 to 50 milliamperes (0.030 to 0.050 amperes) per 100 ampere-hours, increasing up to 10 times this rate toward the end of a battery's life. Lead-antimony float voltage is 2.15 to 2.20 volts per cell.

An increase of 15°F/8.3°C will double the float current drawn by either type of battery. Float current draw doubles for each 0.05 volt per cell increase and decreases by one-half for a similar decrease in voltage. Batteries age rapidly at 90°F/32.2°C.

New batteries require several charge cycles for their chemistry and capacity to fully develop. The initial capacities of golfcart and industrial-grade batteries are only about 75% of rated capacity. During their first 8 to 10 cycles batteries may have limited capacity and greater voltage fluctuations.

Equalization

Variable current at constant voltage is a common method of charging batteries. PV systems are variable current chargers with relatively low charge rates early in the morning and late in the afternoon. Low charge rates also occur when controllers switch to trickle or float charge. At midday on a sunny day charge rates are relatively high.

Flooded batteries that are variable-current charged require equalization charging. Overall battery bank voltage does not show voltage differences between individual cells. Only individual cell testing shows cell voltage and state-of-charge differences. Only equalization charging can bring all cells up to full charge.

Some charge controllers have automatic equalization, but it may be an unreliable option. Most PV system owners manually override or bypass the controller to equalize or "boil" their batteries. "Boiling" or bubbling is not caused by heat. Bubbling is electrolysis, electric current breaking down the electrolyte water into hydrogen and oxygen. A controlled equalization charge extends battery life and increases capacity. It agitates the electrolyte, breaks up stratification, and mixes in any water that has been added. Equalization also consumes water that must be added later.

The time to equalize is—

• at quarterly inspection or when the float voltage of any cell is 0.04 volts below the average for the battery,
• when temperature-corrected specific gravity is more than 10 points (0.010) below its full charge value, or
• after any unusually deep discharge period.

The chemical reaction that occurs during charging and discharging is temperature determinant. A 5°F/3.3°C or less variation in cell temperatures within a battery bank will reduce the frequency for equalization charging.

Lead-antimony battery equalization voltage is generally 2.48 volts per cell. Equalization current can be 20 amperes per 100 ampere-hours of capacity or higher depending on manufacturer specifications. At float voltage, individual cells should be equalized within 0.02 volts of each other.

Monitor battery temperature during equalization. Stop equalization if any cell exceeds 130°F/54.4°C. Wait for the temperature to fall to below 90°F/32.2°C and resume charging at a reduced current. Monitor the electrolyte level of any cell that boils more rapidly than others.

Monitor the cell with the lowest voltage to determine when voltage is up and equalization is complete. Charging for 24 hours may be possible if utility or generator power is available to equalize your batteries. Lead-antimony batteries that show no voltage change after 24 hours of equalization are completely charged. If the difference between the highest and lowest cell is 50 points (0.050) SG or more, the battery bank may be fully charged but near the end of its useful life.

Table 12.2. Typical Equalization Charging Time

Electrolyte Specific Gravity	Equalizing Charge Time
1.260 to 1.280	None
1.240 to 1.259	4 hours
1.220 to 1.239	8 hours
Below 1.220	12 hours

Schedule regular PV array equalizations. It should be done four times a year during sunny periods when consumption is low and your battery is fully charged. Good records will help you determine the best times to equalize. On a sunny day bypass the charge controller and charge your batteries for 4 to 6 hours. That evening test your battery voltage. If some cells are still low, equalize the next day. Add distilled water a few hours after equalization is complete, if needed. Vent your battery room or compartment during equalization.

All manufacturers warn against tapping part of a battery bank to power lower voltage equipment. Some people tap 24-or 48-volt battery banks to power 12-volt lights and other equipment. Tapping shortens battery life. The untapped cells overcharge and the tapped cells undercharge. If you must tap, use a DC-to-DC equalizing converter or voltage-dropping resistors on small loads.

If you will not be using your PV system for a month or longer, equalize and then float charge the batteries. A good charge controller will have a float charge setting. When you are away, float lead-antimony batteries at 2.15 volts per cell. Be sure all cells have enough electrolyte to last for the duration of your trip.

If cells are always 0.04 volts or more below float voltage then either your electric consumption is greater than your PV production or your charge controller is cutting out too soon.

Battery caps are available that contain a catalyst that recombines the hydrogen and oxygen released during charging. Tests in hot climates have shown that the need to add water to golfcart batteries can be lessened when these recombinant caps are used. Recombinant caps require servicing and they do wear out. Automatic battery watering devices are also available, but are not fail-safe. They are not a substitute for periodic testing and maintenance.

Equalizing, testing, and cleaning your batteries takes very little time. An investment of less than one hour four times a year will ensure long battery life.

Replacing Your Batteries

PV cells last much longer than batteries. Most PV system designers use a system design life of 20 to 30 years or longer; during that time the array changes very little, but the batteries age and must be replaced.

Battery type, environmental conditions, use patterns, and maintenance affect battery life. Long periods of operation while partially discharged and failure to equalize will drastically shorten battery life.

Buy all your batteries at the same time so that they age at the same rate. Buy only the number of batteries that your PV array or back-up generator/charger can charge. If you must add batteries later, match them as closely as possible to the original set. Separate old and new batteries with blocking diodes. Mismatched batteries or cells require equalization more often and will shorten the life of the new batteries.

People tend to undersize their battery banks. Frequent deep cycling of an undersized battery bank drastically shortens its life. Frequent charges from a back-up generator or utility power will not extend the life of an undersized battery. The more you run your generator, the worse the economics. Not only will an undersized battery bank age more rapidly, generator running time and fuel costs increase. Besides, running a generator to charge your battery bank defeats the purpose of having a clean energy system.

Batteries still have salvage value when they reach the end of their useful life. Most battery suppliers will give recycling credit for old batteries when you buy a new set.

Battery Safety

Low-voltage, low-wattage PV systems are relatively safe, but there is always the possibility that you can get shocked, burned, or killed if you aren't careful. The current in your battery bank can melt metal and start fires. A tool dropped across battery terminals will weld to the terminals and spark dangerously. Remove all rings, watches, bracelets, and metal before working with batteries. A battery bank can discharge thousands of amperes of current if a tool or some other metal object touches battery terminals. A golfcart battery has more than a kilowatt-hour of energy, enough to weld heavy steel plate or a ring on your finger.

Hydrogen can explode. Vent your battery storage compartment.

Sulfuric acid in battery electrolyte is dangerous. Battery acid will eat holes in clothes and carpets. Acid on your skin will cause a rash or burn and scar. Acid in your eyes can blind you.

Overcharging can cause the "misting" of an acid film on the batteries. Microscopic bits of metal in the dust that accumulates in the acid film provide an electrical path from the battery terminal to the case. You can measure the effect of dirt on battery cases with a DC voltmeter. Put one test lead to a battery terminal and the other lead to the dirty case. A voltage reading indicates that dirt is discharging stored energy to the case, wasting power, shortening battery life, and creating an unsafe condition.

Carefully rinse batteries with a solution of water and baking soda (sodium bicarbonate) to dilute acid from spills or misting. Mix one cup of baking soda to two gallons of water. You may see bubbling as the baking soda reacts with the acid. Once acid is neutralized, rinse the battery case with fresh water and dry with a paper towel or a disposable cloth. Do not let any of the baking soda solution or rinse water get into your batteries, on you or your clothes, or in your eyes.

Batteries are heavy. Avoid straining yourself or dropping batteries on your toes or fingers. Wear protective clothing, gloves, and safety glasses. Be careful.

Chapter 13
INVERTERS

An inverter changes direct current (DC) from solar modules or batteries to alternating current (AC). Inverters allow off-grid PV homeowners to use direct current from their battery banks to power conventional AC equipment. On-grid PV homeowners can power their homes and also feed their excess PV production through their inverters into their utility meters to offset the electricity they would otherwise have consumed from the utility grid.

Inverter History

In the 1970s almost all PV power systems were off-grid, operating at 12 volts DC. PV pioneers spent a lot of time hunting for DC appliances or modifying AC equipment to operate on DC—or they would do without. They crafted DC lamps from automotive taillights and listened to music on automobile radios and homemade 12-volt DC record turntables. They retrofitted fans, pumps, and washing machines with DC motors and searched through automotive catalogs and secondhand stores for 12-volt DC appliances and tools. Those with electronic skills tore into electrical equipment—ignoring electrical hazard warnings and "No Serviceable Parts Inside" labels—looking for transformers and rectifiers to bypass so they could wire directly to the DC electronics.

In the early 1980s scores of 12-VDC products were introduced for the recreational vehicle (RV) market. These tools and appliances could also be used in PV homes. Unfortunately, most were poorly made. During that same period PV homeowners began experimenting with inverters. The first inverters they used were simple 12-volt DC to 120-volt AC square wave output mechanical switches called vibrators designed for use in vehicles. A few talented people built their own inverters. These were the prototypes for a new generation of inverters that significantly changed PV use. Adding inverters to PV systems gave off-grid PV homeowners the option to choose any AC appliance or equipment rather than having to rely on 12-volt DC equipment.

The development of sine wave inverters included the development of inverters that could be synchronized to the utility grid. The first net-metering law went into effect in California in 1996. At that time a few small companies were making batteryless utility inter-tie inverters. Inverter manufacturers recognized that there was a demand for non-polluting electricity from renewable resources to offset peak utility loads. When the second round of solar incentives in the U.S. began in 1998, more inverter companies started making grid-tie inverters. Today the number of grid-connected residential PV systems far exceeds off-grid installations.

How Inverters Work

The basic electronics of an inverter are fairly simple. There are two transistors (semiconductor switches), one on either side of a transformer (coils of wire around a core). One transistor opens, which causes current to flow from negative through the transformer's primary winding to positive. This induces a current on the other side of the transformer in the secondary winding. Halfway through every 60 Hz (every 120th of a second), one transistor closes and the other opens. This on-off or push-pull configuration produces a square wave.

Square wave inverters can be very efficient. Many operate at 85% efficiency at their peak power rating. However, some devices cannot

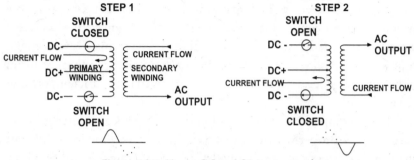

Figure 13.1. How a DC to AC inverter works.

use square wave power efficiently because square waves, some modified square waves, and out-of-phase sine waves cause unwanted heat in reactive loads, like motors, reducing their efficiency and operating life. Square wave inverters can power almost all radios, televisions, computers, and video recorders, although some inverters emit radio signals that interfere with AM radio reception and amateur radio stations. Some devices with electronic speed controls, laser printers and copying machines as well as rechargeable batteries, can be damaged by incompatible inverters.

As inverters evolved, transistors and transformers were added in different types, numbers, and configurations, but the basic conversion of DC to AC electricity remains the same.

The next step up from a square wave inverter is a stepped square wave inverter, often called a modified sine wave inverter. These inverters are square wave inverters that have three or more transistors that close between cycles and clear the current from the transformer. More devices operate without problems when this type of inverter is used because its AC output more closely approximates a sine wave. However, timers and clocks may run fast when powered by a stepped square or modified sine wave inverter and fluorescent lights will flicker if inverter output is mismatched to the ballast electronics.

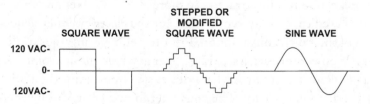

Figure 13.2. Wave forms.

Sine wave inverters, an improvement over square and stepped wave inverters, have been in use for decades. Until recently they were too expensive and inefficient to be practical for PV systems. Developments in power electronics and mass production have changed that, resulting in the current generation of sine wave inverters that are efficient, reliable, and affordable.

Some equipment, such as high-end stereos and computer printers, will only work properly with pure sine wave inverters. Modern, solid-state sine wave inverters can power any device that operates on utility power. They can power resistive loads like incandescent lights, coffeemakers, toasters, and hair dryers. Reactive loads, like home appliances and shop tools with motors, work well with power from sine wave inverters.

Most newer inverters are designed to handle the high starting current draw of induction motors. Water pumps, garbage disposals, dishwashers, refrigerators, air conditioners, and washing machine motors can draw up to six times their operating current at startup. A refrigerator compressor motor rated 300 watts may need an 800- to 1,500-watt inverter to start. An inverter's surge rating determines its motor starting capability.

The new digital sine wave inverters have fewer parts and often the lowest life-cycle cost. Most digital inverters are over 90% efficient across their operating range and often put out better quality electricity than grid power.

The PV-Grid Connection

You can connect a residential PV system, either with or without batteries, to the utility grid. No batteries mean no battery maintenance, but it also means no stored power for emergencies, selective autonomy, or load shifting to get reduced electric rates.

An on-grid PV system can be net metered, which makes the PV grid connection economically attractive to more PV users. Net metering is a method of crediting PV system owners for the electricity they generate in excess of their own consumption. Prior to net metering, most utilities bought PV power through a separate meter at wholesale or avoided-cost rates. The avoided-cost rate is what utilities save by avoiding the addition of new production capacity. The consumer, however, must buy energy at retail, which is often more than three times the avoided-cost rate.

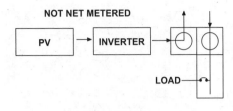

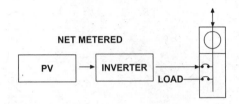

Figure 13.3.
Utility-connected PV systems:
not net metered and net metered.

With net metering, the PV system owner buys and "sells" electricity at the same retail rate.

PV net metering is the law in most U.S. states. If it is available in your state, you can install a PV system and get full value for your PV energy. This reduces your electric bill and offsets your PV system cost. When the sun shines, PV power not immediately used by the home is fed into the grid, which is the electricity equivalent of putting money in the bank. At night "deposited" electricity is withdrawn.

As recently as 1998 there were only a few thousand grid-connected PV homes in the world. Today megawatts of grid-connected residential PV are installed daily. Without net metering, a grid-connected PV system amortizes faster than a 30-year mortgage, but net metering can cut payback time to less than 15 years (see Chapter 6).

Connecting PV to the utility grid requires a synchronous, line-commutated inverter. Grid-interconnect inverters take unregulated DC power and track for the highest AC output with maximum power tracking electronics. Some grid-tie inverters change 48 VDC to 120 or 240 VAC. Others require 150- to 600-volts DC input. PV modules are usually 12- and 24-volt output. Using a 24-volt, 100-watt panel as an example, a 48-VDC input inverter can operate with as few as two modules for a 48-volt, 200-watt array. High voltage DC input inverters would require an 18- to 24-module array connected in series.

Installing a synchronous inverter and tying into the grid is often beyond the do-it-yourself capacity of the average homeowner. Some

Figure 13.4. Net-metered inverter without batteries. From left to right are the solar array DC disconnect switch, inverter, AC kWh meter, and AC disconnect switch. (Energy Outfitters, Ltd.)

homeowners with a background in electronics, engineering, or the building trades have installed their own systems. Most hire a licensed electrician familiar with PV, the NEC, and utility company safety requirements to do all or part of the installation.

As grid-connected PV becomes more affordable, electricians are becoming experienced with PV. If you cannot find a PV electrician in your area, contact the American Solar Energy Society (www.ases.org) or the Solar Energy Industries Association (www.seia.org) for assistance.

PV UPS

An uninterruptible power supply (UPS) consists of a battery for electrical storage, a battery charger, and an inverter. A UPS is used to protect electrical equipment from power interruptions and poor quality grid power. Since battery power is steady, without spikes, surges, or interruptions, a regulated battery-powered UPS can be more reliable than grid power and can provide emergency electricity for computers, lights, appliances, or a water pump.

Our grid-connected office was hit by a utility power surge that killed our computer's power supply. After costly repairs, we installed a PV-powered UPS and our office has been trouble-free. A better surge suppressor may have sufficed, but it does not protect against power outages like a UPS.

Our office PV system was installed before the introduction of batteryless grid-tie inverters. Since their introduction, uninterruptible power supplies have become a consideration separate from the PV system. Before batteryless inverters came on the market, grid-connected inverters required batteries. This made the PV system a UPS whether you wanted it or not. However, maintaining a charge on the battery bank with grid power at night uses power and the equipment and batteries add about 10% to the initial cost of a PV system. If you want UPS for emergency power it is usually more economical to install it separately from your PV system.

To set up a grid-connected UPS without PV, you need an inverter with a built-in battery charger and a battery. Connect the battery to the inverter and plug the inverter into the grid. Then plug your computer or, if the inverter can handle it, your whole home office into the inverter. You can put a UPS with sealed batteries in your office, workshop, home, or apartment under a desk, in a closet, or in a cabinet, depending on the size of the UPS.

Selecting an Inverter for Your Off-Grid PV System

When selecting an inverter for your off-grid PV system, the inverter peak continuous rating must be able power all simultaneous AC loads. The inverter must also handle load surges. Refer to your power requirements list (see Chapter 16, page 314).

Will you be using motors with high starting currents? Will complex electronic equipment be powered? In the kitchen, do you run the garbage disposal and dishwasher at the same time? What about the well pump? Note which AC loads will be on at the same time. Don't forget that your refrigerator and any other automatically cycling load can come on at any time while other loads are running and add to your simultaneous loads.

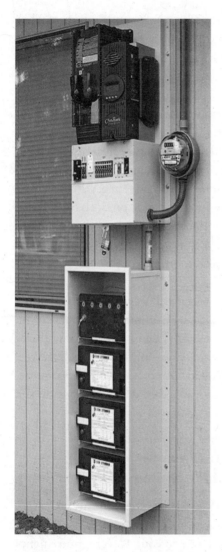

Figure 13.5.
Net-metered inverter with batteries. Right to left at top are the charge controller and inverter. Below are the DC and AC circuit breakers and AC kWh meter. Several equipment configurations are available. This outdoor inverter enclosure was designed for mild climates. (Outback Power Systems, Inc.)

Add up your AC loads that automatically cycle on and off. List their starting currents separately. Add demand loads like the dishwasher and washing machine. Then add convenience loads like blenders, toasters, and hair dryers in possible combinations of simultaneous operation. You can monitor combined loads to reduce inverter size and cost, as Bill and Jackie Perleberg do (see Chapter 3).

Take into consideration the inverter's capacity over time if it must handle continuous loads or loads that are on for several hours. An inverter rated 2,500 watts may only operate continuously at 2,200 watts at 77°F/25°C. Ambient and internal heat decrease an inverter's capacity and shorten its life so be sure to provide adequate ventilation.

There are good reasons to go up one size to the next higher wattage rating for your inverter than you first estimated. You may have construction or remodeling work planned. A bigger inverter will be useful and is definitely quieter and cleaner than a generator. As you settle in to your home, you may need or want more appliances. Too often people pinch pennies when they buy an inverter and then push a marginally sized device to its limit. Get an inverter with sufficient capacity for your loads and operate it within its factory ratings.

For off-grid PV systems, if a small inverter is all you can afford, or if you are installing your PV system in stages, you can use a couple of smaller inverters for specific loads. Some inverters are specially designed to be coupled or "cascaded" for increased capacity. Cascading inverters work together when power is needed to operate big loads or individually to power small loads. Cascade inverters can be more efficient than one single large inverter when they are operated at or near their peak efficiency. They also provide useful redundancy. If one cascade inverter is in the shop for repairs, the other can be powering at least part of your load. Cascade-type inverters also allow for expansion. You may be starting on a limited budget or initially require less AC power. You can install the first (master) inverter to get started, and add the second (slave) unit and its connecting control cable later.

Many off-grid inverters have automatic load sensing, which allows them to remain on using very little power, idling under no load. When a load is turned on, power is automatically available. When the load is turned off, the inverter returns to its no-load or idle state. Some inverters consume as little as one watt on idle.

Phantom loads—devices that appear to be off but actually are still drawing power—defeat the savings gained by automatic load sensing. Some examples of phantom loads are instant-on televisions and computer monitors, devices with built-in clocks and timers, satellite dishes, and ground fault detectors. The power blocks on battery chargers, answering machines, some phones, and many other electronics also draw power continuously. Small phantom loads trip the inverter automatic load sensor and cause it to consume a lot of power. Even though the phantom load itself is small, it causes the inverter to operate at an inefficient power point and waste power.

Loads that are too small to trigger an inverter from its energy-saving idle state are another concern. Be sure to ask your supplier if nightlights, clocks, and other necessary small loads can be inverter-powered and how they affect inverter load sensing and idle-state power consumption.

Warning!

Use safety switches to prevent two AC inputs from being on the same circuit at the same time. Never put generator or grid power into a circuit powered by an inverter. It is illegal, and dangerous to powerline workers, to feed generator or inverter power into the utility grid without permission. Keep children, animals, and the unqualified away from inverters.

Buying an Inverter

Make a list of the features you want in an inverter. For example:

•Utility interconnect	•Automatic generator start/stop
•Sine wave form	•Automatic load sensing
•Desired wattage	•Integral DC/AC disconnect switches
•High surge capability	•Expandability
•Battery charging	•Stacking capability
•Maximum power point tracking	•Integral displays & metering
•High efficiency	•Remote displays & metering
•FCC compliance	•Computer interface & communication port
•Desired DC input voltage	ication port
•Desired AC output voltage	•Low idle current
•Ease of installation	•Indoor/outdoor installation
•Programmable	•Warranty
•Multiple DC input connections	•Field service history

Be specific, be patient, shop around, and ask lots of questions. You usually cannot try an inverter before you buy it, so visit a few PV/inverter systems in action. The trips will be worth it. Most inverter sales are final; if not, a substantial restocking and handling fee may be charged on returns or exchanges of undersized or inappropriate inverters.

Buy as much inverter as you can afford. It is false economy to buy a small inverter that you will soon outgrow. It is unwise to buy a discontinued model inverter. Buy one that can be repaired, and buy it from a manufacturer that is likely to stay in business. Be sure technical support will be available when you need it. When shopping for an inverter, ask "What if . . . ?" If you don't like the answers you get, shop elsewhere. A reputable inverter sales outlet will satisfactorily answer every question.

If you hire an installer, consider purchasing a service contract that includes inverter replacement. Some manufacturers will ship replacement parts overnight so that circuit boards can be swapped out by a technically competent person. The manufacturer will give you a list of factory authorized repair people.

Inverter Problems

You may be told that the inverter is often the weakest link in a PV system. PV modules are the most reliable electric generator and inverters can fail, although quality inverters are very reliable. The right inverter can give you years of trouble-free service. The wrong choice or a poorly designed and built inverter can be a nightmare. New inverter models have more unresolved design problems than tried-and-true models. Much patience, tinkering, and repair is to be expected if you buy a new model inverter.

Most inverter problems can be avoided from the start with proper installation. You increase the probability for problems if your inverter has been handled roughly, improperly installed, overloaded, or if undersized DC input cables are used. Some inverters must be adjusted to your specific system's parameters after installation. This means you or your installer will need system instrumentation and test equipment to measure input and output voltage and current.

It is unwise to rely on one inverter for all of your off-grid power. Even if your inverter has a good reputation, is fully warranted, and repairs are free, it can fail. If it fails, you will be without AC power. Keep a few DC lights for emergency use. A self-sufficient water supply is a must. A generator can carry you through as you wait for inverter parts and repairs. Utility companies have emergency plans. When you are your own utility company, you should have an emergency plan too.

Electromagnetic fields (EMFs) are produced any time current flows through wire. EMFs are extremely low frequency waves that decrease rapidly proportional to the distance from the source. Inverters have a transformer that is basically current flowing through coiled wire. This current causes an EMF to form around the coil. Fluorescent light ballasts, motors, clock radios, power blocks, microwave ovens, kilowatt-hour meters, and electric service panels all emit EMFs.

There is no practical way to block EMFs. They pass through almost everything including walls and lead. EMFs have been anecdotally linked to birth defects, developmental defects, and cancer. Despite extensive research into the effects of EMFs, no study has been acknowledged in the U.S. as definitive. There is no U.S. safety standard for EMFs. Some say 8 milligauss or more is dangerous and 2.5 milligauss or less is safe. We tested for EMFs in our home using a milligauss meter. Our utility meter service panel has a strong EMF field, but that's not a problem because it is mounted on an outside wall with a closet between the living space and the meter. The panel's EMF falls to a safe level between the wall and the closet door. The bedroom clock radio was the second largest EMF source in our home. Moving the clock one foot farther away from the bed brought the EMF level under 2 milligauss. Our 4,000-watt inverter emitted a field that fell to a safe level 2 feet from the inverter.

If you are concerned about EMFs, ask your inverter supplier for recommendations. Install inverters away from bedrooms and at least three feet from where people or animals spend a lot of time. You will probably want to install your inverter away from occupied rooms in any case because it emits an audible noise. It may only be a slight buzz, but even that can be irritating. Some inverters also produce electronic noise, called radio frequency interference (RFI), which disrupts radio and other wireless communications. More inverters are being made Federal Communications Commission (FCC) compliant.

Square and sine waves have resonant or harmonic waves above and below the primary wave. Measured as total harmonic distortion (THD), harmonics adversely interact with devices being powered and other devices on the same circuit. THD greater than 5% may be unacceptable. THD less than 3% is good.

Inverter Wiring

A fundamental of PV system design is that the inverter input voltage determines the system's primary DC voltage. Many people mistakenly plan to use 12 volt DC because most PV modules are designed to charge 12-volt batteries and historically most PV systems have been 12 volts DC. Your PV system's primary DC voltage can be 12, 24, 36, 48, or higher depending on your inverter input. Higher system voltage reduces wire size and cost and other system costs allowing you to use readily available standard wiring, switches, breakers, and the whole range of hardware designed for grid-connected homes.

Using higher input voltage and an inverter will not allow you to reduce the size of all wires. The ampacity of the cables from your inverter to the battery bank must be sized for the lowest operating voltage at full inverter power. For example, if you use a 4,000-watt, 24-VDC inverter, 4,000 watts divided by 22 volts at 85% inverter efficiency and the NEC safety factor of 125% equals 267 amperes (4,000 W ÷ 22 V ÷ 0.85 × 1.25 = 267 A). For the proper ampacity at 86°F/30°C ambient temperature and 167°F/75°C-rated insulation in free air, two #3/0 AWG cables are required—one for positive and one for negative. #3/0 AWG is rated 310 amps at 86°F/30°C.

You also need high current DC protection devices (fuses or circuit breakers) mounted near the battery between the battery bank and the inverter. A 4,000-watt, 24-volt DC input inverter/battery charger requires a 275-ampere fuse.

The wire from the AC output of the 4,000-watt inverter/battery charger in the above example to the AC load center must be sized to handle a maximum steady-state output of 93 amperes and the NEC safety factor of 125% (93 × 1.25 = 116 A). At 86°F/30° and 167°F/75°C insulated #2 AWG cable can handle 115 amps. Refer to Chapter 15 for complete formulas and calculations for wire sizing.

If you install your own system, your inverter supplier should help you determine wire and cable sizes. If you hire someone to install your system, be sure that person has experience installing PV, inverters, batteries, and other high current DC power devices.

Consumer demand for easier to install PV systems has resulted in factory prewired, UL-approved combination inverter and power centers. These come in a variety of configurations that include charge controllers and DC and AC disconnects. The UL label gives building inspectors immediate confidence that the device is NEC-compliant. The floor-mounted power center illustrated in Chapter 4, figures 4.18 and 4.19, includes a code-compliant battery compartment. Other models are designed to be wall-mounted. Some inverter power panels can weigh over 300 pounds. Decide how you will fasten heavy equipment securely to the wall before you buy it.

Inverter technology has evolved rapidly with the growth of off-grid and utility-interconnected PV. Today's inverters are loaded with features, reliable, and affordable. Improvements in inverters designed for off-grid use have expanded the capabilities and convenience of off-grid PV systems. Good inverters put out better quality electricity than most utility companies and are often more reliable than grid power. Homeowners connecting PV to the grid are leading the way to a cleaner, safer, and more secure environment by replacing electricity produced by fossil fuel and nuclear power plants. They are helping to change the grid into a diversified, bi-directional network of distributed generation supplemented by clean solar power.

Chapter 14
WIRES, GROUNDS, CONNECTIONS & LABELS

Wire Basics

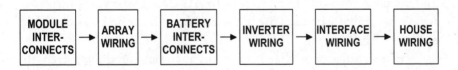

Wires connect the components of a PV system. Wire is the conductor through which your electricity flows. Wiring is not a minor component. Proper wiring is required for proper performance. Since it costs far less than the other components in a PV system, skimping on wiring is foolish. Only copper wire and cable are acceptable for PV installations.

The cross-sectional area, or size, of a wire determines its capacity to carry current. The larger the cross-sectional area, the more current the wire can carry. The larger the wire, the smaller the gauge number ascribed to it. A #6 AWG (American Wire Gauge) wire is a larger conductor than a #14 AWG wire. It makes no sense to throw power away by using undersized wires.

All things being equal, there is little difference between solid and stranded wire. Electricity travels through the wire conductor, not on its surface. Current does skim the surface of wire carrying thousands of volts, but that isn't the case in home wiring carrying 12 to 600 volts. The main difference between solid and stranded wire is ease in handling. Stranded wire is more flexible so it is easier to bend and run through conduit or wrap around a screw terminal. Large diameter solid wire can be difficult to bend without pliers.

Both solid and stranded wire must be securely supported at connections to avoid flexing and eventual breakage. Secure your PV array wiring so that the wind doesn't cause it to flap or move. All wires, solid or stranded, should be strain-relief fastened or placed in conduit (a metal or plastic pipe made specially for electrical use). Wire should never hang loose. If you use stranded wire, all of the terminals and connectors should be constructed and marked for use with stranded wire.

Solid wire is almost always used in home wiring. Romex®* (a brand name) has become so standard that electricians will call almost any home wiring "romex." Romex is typically 12/2 or 10/2 with ground, which means the outer insulator holds three wires. One wire is bare and is used as the ground wire. The other two wires are insulated separately and are used in the electrical circuit. In AC circuits the white insulated wire is neutral; the black is the hot wire. Never wire the DC side of PV systems with Romex.

Wire insulation is characterized by material, durability, strength, temperature rating, insulating properties, recommended use, flammability, and resistance to moisture and chemicals. Some insulation types must be used in conduit. Other wire types can be used overhead outdoors. Some wire can be used underground for direct burial. Some insulation can resist ultraviolet (UV) rays. Still other wire is oil-resistant. Use only UL-listed wire for your PV system.

Grounding & Ground Fault Protection

Grounding is a connection to the earth for conducting electrical current to and from the earth. A grounded conductor, or ground wire, carries current and is intentionally grounded. In PV systems the negative conduc-

*Romex is a registered trademark of Southwire Company, Carrollton, GA.

tor is usually grounded except in systems with ground fault protection. An equipment or chassis grounding conductor connects the exposed metal portions of equipment and enclosures to the grounding electrode (grounding rod). The grounding conductor can run in the same raceway or cable assembly as the DC conductor as it leaves the PV array. The equipment grounding conductor must be sized in accordance with NEC 690.45, and not smaller than specified in Table 15.11 (page 287). The equipment grounding conductor should never be smaller than #14 AWG. If the ground conductor is #6 AWG or larger it must be in conduit (NEC 250.120) unless it is located in a place where it is protected from physical damage, such as inside a wall.

The DC grounding electrode is common with, or bonded to, the AC grounding electrode. The system grounding conductor and the equipment grounding conductor are also tied to the same grounding electrode. A separate equipment grounding electrode should be installed at the site of a ground-mounted array and for any roof-mounted array that is on a building that contains any power equipment for the PV system.

Copper-clad steel ground rods, 8 to 10 feet long by ¾ inch in diameter, are standard. The conductivity of the earth-ground connection is determined by soil conditions, which in turn determine the total surface area of the ground rod. Ground rods should be driven into the ground at the point where power enters the building, with the ground wire fastened securely to the rod with a ground-rod clamp. Tag (label) the ground wire at the ground rod.

If it is impossible to drive an 8- to 10-foot rod into the ground (sometimes the whole world seems like solid rock), clamp the ground wire to a metal water supply pipe that has good earth-ground conductivity. In some dry, rocky regions it is necessary to dig a trench and bury a measured length of bare wire to provide an adequate earth ground. Your local utility company or an electrician can advise you.

The 2008 NEC requires a ground fault interruptor (GFI) and an array disconnect switch for PV arrays. A GFI disconnects a circuit when it senses a fault or short circuit between conductors, to other equipment, or to ground and limits the fault current to a safe level. There are a limited number of commercially available DC GFIs. A ground fault protection device should be located where it can be readily seen and easily reset. It should be as close as possible to the DC conductors that come from the

PV array and it should be labeled, both at the GFI and at the batteries, to warn of electric shock hazard if a ground fault occurs. Residential ground-mounted arrays are required to have ground fault protection. All commercial arrays must have a GFI and an array disconnect switch.

AC Ground

Ground your PV home with a standard 3-wire 120/240-volt AC service neutral wire grounded on the line side of the main contactor. Your house should have three-wire circuitry throughout for proper appliance grounding. The house wiring neutral conductor should be grounded at the house panel on the panel neutral bus or bus bar. (A bus or bus bar is a conductor that makes a common connection between two or more circuits.) The neutral bus is connected to the panel enclosure. Ground the negative side of a 2-wire DC system to the same AC ground. Ground all equipment, cabinets, and enclosures in the PV system to this same ground. A ground conductor must be used to tie all the ground rods together if separate ground rods are used at two or more locations (i.e., at the array and the house).

Disconnect switches must not open the grounded conductors. In an off-grid PV system the NEC considers the charge controller to be part of the PV output circuit. The grounding point of connection can be either before or after the charge controller, but making the connection on the battery side of the controller provides less lightning protection.

Lightning Protection

While the chance of a direct lightning strike is low in most areas, it is important to protect your system from the induced effects of nearby lightning. If you live in a lightning-prone region, check with local authorities and neighbors to learn how they protect barns, TV antennas, and other high-profile equipment and structures. In any case, nominal surge protection in the form of varistor surge arrestors across both AC and DC entrance cables is recommended. Lightning arrestors connected to the system's earth ground provide a shunt or path for transient currents and side effects of nearby lightning.

Color Coding

The NEC requires that DC and AC wires be properly color-coded. Grounded negative conductors should be white or marked with white paint or white electrician's tape at each termination. This means that the negative grounded conductor in your PV system will be white. Conductors used for grounding frames should be either bare, green, or green and yellow striped. Ground wires smaller than #6 AWG should not be marked with white unless the local building and safety department approves the marking.

Ungrounded DC conductors are colored black (negative) and red (positive). By convention the positive conductor on an ungrounded DC system is red or any color marked with red, except green or white. If only black wire is available, wrap it with colored electrician's tape to identify the positive (red), negative ungrounded (black), negative grounded (white), AC hot (black), AC neutral (white), and ground (green or bare).

Connections

The electrical connections in a PV system can be its weakest link. There are several code-compliant ways to make secure connections.

Weatherproof twist-on wire connectors that are UL-listed and installed according to the manufacturer's instructions are an option, but they do not provide the strength of a splice or a terminal block. Connections should be made inside a junction box. Put strain relief clamps on conductors entering the junction box. The connector manufacturer's instructions will tell you if the connector is weatherproof, its temperature rating, voltage rating, gripping strength, and the size conductor it is suited for. Most connectors are designed for only one conductor per lug.

Soldering is allowed by the NEC. UL-listed heavy-duty shrink tubing, either dry or with an internal meltable sealant, provides insulation for soldered connections. Spade or ring connectors can be mechanically fastened and soldered to wire ends to give a secure connection at the module terminals.

A split bolt can be used for heavy-duty connections, to securely clamp two conductors together. It can also be used to avoid "double lugging" or putting two conductors on a single terminal rated for only one conductor. The two conductors are brought to the split bolt where one conductor is terminated and the second conductor is carried through to the terminal. Split bolts must be adequately insulated with either rigid or flexible snap-on insulating covers. An alternative to a split bolt is an insulated multiple conductor connector.

In response to consumer demand many PV modules and other system components are equipped with plug-in male/female connectors called "quick connects." In situations where the connector is not compatible with other equipment, it may have to be cut off and replaced with a ring or spade connector, but confirm with the module supplier that this will not void the manufacturer's warranty.

The 2008 edition of the NEC requires a tool to open any easily accessible AC or DC connection carrying over 50 volts. Latching or locking quick connectors make connections secure.

Labels

All circuits, switches, breakers, and boxes must be clearly identified with labels. This is important because you may not remember every detail of your wiring or someone else may have to work on the system when you are not available. Keep good records: an equipment log with wiring and other diagrams; installation, operation, and maintenance manuals; a record of receipts, warranties, and dates put in service, maintenance checks, repairs, modifications, notes, and so on. Keep your log book near the main breaker box or control panel.

The NEC has specific requirements for the location and wording of markings and labels. All power sources must have a permanent plaque. This is an example of a label located at the system disconnect switch:

PHOTOVOLTAIC POWER SOURCE	
OPERATING CURRENT	34 AMPS
OPERATING VOLTAGES	48 VDC, 120 VAC
OPEN-CIRCUIT VOLTAGE	85.6 VDC
SHORT-CIRCUIT CURRENT	37.6 AMPS

The grid disconnect switch and battery disconnect switch must also have labels. PV modules should have their polarity marked and a label listing the overcurrent device rating, open-circuit voltage, operating voltage, maximum system voltage, short-circuit current, operating current, and maximum power.

A label near the ground fault indicator and the batteries should read:

WARNING
ELECTRICAL SHOCK HAZARD
If a ground fault is indicated—all normally grounded conductors may be ungrounded and energized at array open-circuit voltages.

Any ungrounded PV power source with exposed, uninsulated terminals should be marked at each junction box, combiner box, or device with the following label:

WARNING
ELECTRICAL SHOCK HAZARD
The DC conductors are ungrounded and may be energized with respect to ground due to leakage paths and/or ground faults.

Equipment containing more than one circuit supplying power to a bus bar or conductor should have the following label at the PV supply connection or at the circuit breaker:

WARNING
THIS PV SUPPLY CONNECTION MUST REMAIN CONNECTED OR INSTALLED AT THIS LOCATION, WHICH IS FARTHEST FROM THE FEEDER OR SERVICE CONNECTION.

Utility interconnect inverters should be labeled:

WARNING
ELECTRICAL SHOCK HAZARD

If a ground fault is indicated—all normally grounded conductors
may be ungrounded and energized.

The inverter output on a stand-alone or off-grid system should be labeled. For example:

WARNING

Single 120-volt supply
Do not connect multi-wire branch circuits.

Battery boxes or battery rooms should have a warning sign to encourage safety:

WARNING
ELECTRICAL SHOCK HAZARD

Dangerous voltage and currents

EXPLOSIVE GAS

No spark or flames
No smoking

CORROSIVE MATERIALS

Wear protective clothing when servicing to prevent acid burns

---

The back-feed circuit breaker in utility-interconnected PV systems should be labeled, as well as switches and circuit breakers that can still be energized when in the open position.

Ask your local building inspector for the preferred wording for all labels.

Putting a PV System Together

Starting from sunlight converted to current flow within the solar cell, the first actual wiring in a PV system is the solar cell interconnect. This usually flat wire is built into the solar module to carry current from one cell to another, in series, to increase voltage. Cell interconnects are internally connected to the module's plus and minus terminals.

Your PV system wiring starts at the solar module junction box or quick connect cables. There are either one or two junction boxes or two cables on the back of each PV module. Most modules have one junction box with both positive and negative terminals and additional unconnected terminals that can connect modules in series. If a module has two junction boxes, one contains the positive terminals and the other will contain the negative terminals.

Just as cell interconnects tie solar cells together for higher voltage or current, module interconnects tie modules together. If you install a 12-volt system with 12-volt modules, connect the modules in parallel for increased current. Connect 12-volt modules in groups of two in series for 24 volts or four in series for 48 volts and so on for higher voltage.

The module manufacturer's installation guide should have complete instructions for bringing wires into, and making connections with, the junction box. The wire that interconnects modules should be rated for

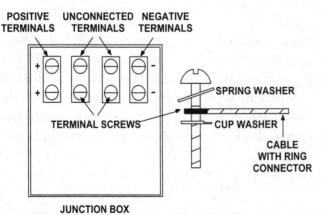

Figure 14.1. A typical PV junction box.

wet locations, weather- and UV-resistant single- or two-conductor stranded copper #10 AWG to #14 AWG in size, as recommended by the manufacturer or as determined by the amperage. Since roof-mounted equipment temperatures can exceed 167°F/75°C, USE #2 wire with 194°F/90°C-rated insulation is used for module interconnections.

The terminals on some modules and other equipment are not adequate for wires larger than #10 AWG. If you use larger wire, install a terminal block and add a small section of #10 AWG to fasten to the small terminal screw. Remember that when you connect small wires to undersized terminals, you are restricting the flow of electricity. This is the same as putting the squeeze on a garden hose. Too much resistance will cause power loss, heat up the wire and insulation, and possibly cause a fire. Always consult wire size charts.

Series module strings are grouped together as close as possible within the PV array with terminal blocks or combiner boxes. Commercially available weather-tight combiner boxes have fuses or circuit breakers and a negative terminal block and bus bars. Each solar module frame must be grounded with a conductor from the designated frame grounding point to the grounding bus bar in the combiner box.

Connecting modules, first in series for system voltage and then in parallel in the combiner box, has the advantage that if one module should fail the array will still work, less the failed module and its partner(s) in the series string.

Partial shading of a module in a series string can cause reverse voltage across the shaded module. Current is forced through the shaded cells causing the module to overheat. Bypass diodes in the module will prevent overheating by allowing current to flow around the shaded cells. Most modules are built with bypass diodes. Modules that do not have factory-installed bypass diodes should have a diode installed in the junction box or as recommended by the manufacturer's installation manual.

Array Wire

The next wiring element in a solar electric system is the array wire. This wire, which runs between the array and either the battery bank or inverter, can be very large if it is carrying high current at low voltage for a great distance. The array wire, sometimes call the homerun, may be the most expensive wire in your system.

At this point batteryless, on-grid, net-metered systems differ from either off-grid PV systems or on-grid systems with batteries. In a batteryless net-metered system, the array wire goes through a DC disconnect switch directly into the inverter.

In PV systems with batteries, the array wire goes through the GFI, the DC disconnect switch, the charge controller, and on to the batteries. The wire from the array to the charge controller does not need to be weather wire if it is in conduit. The allowable amperage of a conductor in conduit is reduced and must be adjusted proportional to the number of conductors in the conduit. If you bury the wire from the array to the charge controller or house, use wire that is designated "direct-burial."

Since the charge controller is usually located indoors near the battery bank, wire and protective conduit are often used. The wire from the charge controller to the battery bank carries the entire output of the solar array.

Voltage Drop

The final step in sizing the wires for your PV system is to calculate voltage drop. After you have drawn your system block diagram and determined the lengths of all the wire runs, voltage drop is calculated using the method in Chapter 8 (page 141). Keep all wire lengths as short as possible to minimize voltage drop and wire cost. The maximum allowable voltage drop for your PV system is 3%. Many systems designers limit voltage drop to 2% to optimize power production, and some inverter manufacturers recommend less than 1.5% DC voltage drop. If the correct ampacity wire results in a voltage drop that exceeds 3%, the wire and fuses must be resized. Minimizing voltage drop is important because a 1% increase in voltage drop means a cumulative 1% decrease in the total output of your PV system over its lifetime.

Remember: Electricity can kill, and bad wiring can cause fires. Hire a professional electrician to install your PV system or to check your work.

Chapter 15
CALCULATING WIRE, FUSE, BREAKER & CONDUIT SIZES

In the traditional method for calculating PV system wire, fuse, breaker and conduit sizes, wire size is initially determined by circuit amperage and ampacity. Starting at the array, go through the system—wire by wire, component by component—to check for *National Electrical Code* compliance, adjusting wire, fuse, and breaker sizes as needed. Calculate the voltage drop after you have confirmed that all components meet *NEC* requirements. If the voltage drop is greater than 3%, the entire process must be repeated until you arrive at an acceptable voltage drop.

A more straightforward approach is to calculate voltage drop first and apportion the voltage drop to each circuit based on total wire length. Wire sizing through this "path of least resistance" virtually guarantees *NEC* ampacity compliance and avoids circular and repetitious calculations.

A Step-by-Step Guide to Wire Sizing

Step 1: Draw an accurate wiring diagram of the system. Include all equipment, components, conductors, conduit, fuses, and breakers. Indicate the length of all wire runs, which wires will be in conduit, design temperature, and ambient temperature of all conductors.

Step 2: List the watts, volts, and amps of all equipment.

a. Solar modules	Watts STC	
	V_{max}	
	I_{max}	
	V_{oc}	
	I_{sc}	
	Nominal module volts	
	Table 15.8 temperature correction factor	
	V_{oc} corrected (see V_{oc} note)	

V_{oc} note: For crystalline and multi-crystalline silicon modules, multiply V_{oc} by Table 15.8 correction factor (page 286). V_{oc} × correction factor × quantity modules in series = V_{oc} (corrected). Use this value for all calculations using V_{oc}. For other type modules refer to manufacturer's instructions for corrected V_{oc}.

b. Series source circuit	Quantity of modules in series	
	Quantity of series source circuits	
	Watts STC = (module watts × quantity modules in series)	
	V_{max} = (module V_{max} × quantity modules in series)	
	I_{max} = (same as module I_{max})	
	V_{oc} (corrected) = (V_{oc} × correction factor × quantity modules in series)	
	I_{sc} = (same as module I_{sc})	
	Nominal system volts = (nominal module volts × quantity modules in series)	
c. Combiner box (output)	Watts STC = (series string watts × quantity series source circuit strings)	
	V_{max} = (same as series string V_{max})	
	I_{max} = (series string I_{max} × quantity series strings)	
	V_{oc} = (same as corrected series string V_{oc})	
	I_{sc} = (series string I_{sc} × qty series strings)	
	Nominal system volts	

d. Charge controller (skip for battery-less systems)	Current rating	
	MPPT (yes or no)	
	PV V_{oc} (corrected)	
	Nominal battery volts	
	Maximum watts = (current rating × nominal battery volts or V_{max} if MPPT)	
e. Inverter	Continuous power rating. VA (watts)	
	Lowest DC input volts. V	
	Output volts. VAC	
	Continuous AC current output. A	
	Maximum AC current output. A	
	Efficiency %	

Do not use inverter peak efficiency. Use average inverter efficiency (usually 80 to 90%) or inverter efficiency at 75% load available on the California Energy Commission's eligible inverters list.

Step 3: Note the series and parallel circuits. Select the parallel circuit with the highest current and/or the longest conductor length. Voltage and voltage drop from parallel circuits (such as module series strings) are not additive.

List the design temperatures and conditions for the conductors and components. The design temperature is the highest temperature the wire and insulation will experience. Wire in solar module junction boxes can reach 167°F/75°C or over 86°F/30°C above the highest ambient temperature. Wire in conduit on a roof can reach 140°F/60°C. Wire in conduit in other outdoor locations can reach 113°F/45°C. Wire in conduit indoors is usually designed for 86°F/30°C. Module interconnects in free air (not in conduit) must still be rated for the PV module's highest design temperature. Use the recommended temperature ratings for ambient temperature (pages 266, 272, 273) or the actual highest temperature measured under the specified conditions (indoor, outdoor, in conduit, in free air, etc.). The highest measured temperature on a roof is very different from the temperature at ground level.

Where there is a difference between the manufacturer's temperature rating for the lugs or terminals and the conductor, use the lower temperature rating when making the calculations. The lugs on almost all off-the-shelf equipment are rated 167°F/75°C.

Design Temperature Chart

Wire	Design Temperature	Insulation Rating	Lug Rating	Number of Wires	Conduit Yes / No
Solar module interconnects	167°F/75°C roof	usually 194°F/90°C	194°F/90°C		
Conductor from array to junction box	167°F/75°C roof	usually 194°F/90°C	usually 194°F/90°C		
Conductor from module junction box to combiner box	140°F/60°C roof	194°F/90°C	194°F/90°C		
Conductor from combiner box to batteries	86°F/30°C indoor	167°F/75°c	167°F/75°C		
Conductor from battery to inverter	86°F/30°C indoor	167°F/75°C	167°F/75°C		
Conductor from inverter to service panel	86°F/30°C indoor	167°F/75°C	167°F/75°C/		

Step 4: List the wire lengths for each circuit in feet. If there are multiple parallel circuits, select the circuit with the highest current and longest wire length to calculate the sum of the wire lengths.

Important! This applies to all systems: Do not round off numbers. All apportioning and voltage drop calculations must be carried out to eight decimal points because the proportional differences between the conductors are very small. For example, 1 foot of #4 AWG wire has only 0.000308 ohms resistance. The resistance of either 1 foot or 10 feet of #4 wire rounded off to two decimal points would equal zero resistance! That is the wrong number.

Systems with Batteries

	quantity	each length in feet	total length feet
a. Module interconnects			

		length in feet	× 2 = round-trip feet
a. Module interconnects			
b. Series string to junction box			
c. Junction box to combiner box			
d. Combiner to charge controller			
e. Controller to battery bank			
f. Battery bank to inverter			
Total DC wire length (do not include module interconnects or battery to inverter cable)			
g. AC inverter to service panel			
h. Sum of DC and AC wire lengths			

Calculate the proportional length of DC conductors (each conductor's round-trip length divided by the sum of the DC and AC round-trip wire lengths)

a. Module interconnects (not included in this calculation)

	round-trip length	÷ total length (h) = proportional length
b. Series string to junction box		
c. Junction box to combiner box		
d. Combiner to charge controller		
e. Controller to battery bank		
f. Battery bank to inverter (not included in this calculation)		
g. AC inverter to service panel		

Batteryless Systems

	quantity	each length in feet	total length feet
a. Module interconnects			
		length in feet	× 2 = round trip feet
a. Module interconnects			
b. Series string to junction box			
c. Junction box to combiner box			
d. Combiner to inverter			
b + c + d = Total DC wire length (do not include module interconnects)			
e. AC inverter to service panel			
f. Sum of DC and AC wire lengths			
Calculate the proportional length of DC conductors (each conductor's round-trip length divided by the sum of the DC and AC round-trip wire lengths)			
a. Module interconnects (not included in this calculation)			
	round-trip length	÷ total length (f) = proportional length	
b. Series string to junction box			
c. Junction box to combiner box			
d. Combiner to inverter			
e. AC inverter to service panel			

Step 5:

    a. Calculate the voltage drop for the module interconnects. Use #10 AWG USE-2 or the cables provided or recommended by the manufacturer. Terminal connectors can be used to accommodate larger wires, if needed. The formula for voltage drop is ohms per 1,000 feet from Table 15.2 (page 281) divided by 1,000 times the round-trip feet times the series string I_{max} equals volts lost to resistance. Use the value from Step 4(a) for round-trip feet and the value from Step 2 (b) for I_{max}.

(____ ohms per 1,000 ft ÷ 1,000) × (____ ft$_{round\ trip}$ × ____I_{max})

= ____volts lost to resistance

____ volts lost to resistance ÷ ____ system volts = ____ voltage drop

____ voltage drop × 100 = ____% voltage drop

b. Calculate the voltage drop for the battery to inverter cable. Use #4/0 (or larger as recommended by inverter manufacturer). Use type USE cable in free air or THHN/THWN in conduit.

The battery to inverter cable requirements are determined by inverter specifications. Use the inverter input requirements (from Step 2e) calculated as:

[(inverter rated watts ÷ lowest DC input volts ÷ inverter efficiency) × 1.25] = Calculated input current in amps

(____ inverter rated watts ÷ ____ lowest DC input volts

÷ ____ inverter efficiency) × 1.25

= ____ amps calculated input current

Use this calculated inverter input current for the next calculation and to size the fuse between the battery and inverter.

(____ ohms per 1,000 ft ÷ 1,000) × (____ ft$_{round\ trip}$

× ____ calculated input current) = ____ volts lost to resistance

____ volts lost to resistance ÷ ____ system volts = ____ voltage drop

____ voltage drop × 100 = ____% voltage drop

Step 6: Select the acceptable voltage drop for your system and express it as a decimal. The maximum total acceptable voltage drop for a PV system is 0.03 (3%). Many system designers use 2% total voltage drop. From the inverter to the service panel the voltage drop should always be less than 1.5%. Remember that each 1% voltage drop means 1% of the total output of your PV system is lost for its entire lifetime.

Selected acceptable voltage drop = ____ (____%)

Use the decimal value for all calculations, not the percentage.

Step 7: Subtract the voltage drop for the module interconnects and battery cable (steps 5a and 5b) from the acceptable voltage drop in Step 6.

_____ acceptable voltage drop – _____ module interconnect

voltage drop – _____ battery cable voltage drop

= _____ net remaining voltage drop

Step 8: Multiply the proportion of DC wire and the proportion of AC wire calculated in either Step 4 by the net remaining voltage drop from Step 7. This will give you the allowable voltage drop for the DC and AC conductors. Do this for all the conductors in Step 4, except the module interconnects and the battery to inverter cable (calculated in steps 5a and 5b).

Circuit name _____: (list each circuit)
_____ proportion DC wire × _____ net remaining voltage drop
= _____ acceptable DC voltage drop

Inverter to service panel:
_____ proportion AC wire × _____ net remaining voltage drop
= _____ acceptable AC voltage drop

Step 9: Calculate conductor resistance. Perform this calculation for each conductor listed in Step 8 using the defined acceptable voltage drop values calculated in Step 8.

For systems with maximum power point tracking (MPPT) charge controllers or batteryless systems use V_{max}. For all other systems use the nominal system voltage.

(defined acceptable DC voltage drop × $V_{max \ or \ nominal}$ × 1,000)

÷ ($ft_{round \ trip}$ × I_{max}) = ohms per 1,000 ft

AC conductor requirements are determined by system output voltage and the inverter's continuous AC output amperage:

(defined acceptable AC voltage drop × V_{output} × 1,000)

÷ ($ft_{round \ trip}$ × $A_{AC \ continuous \ output}$) = ohms per 1,000 ft

Circuit name_____: (list all circuits in system separately)

(____ DC voltage drop × ____ $V_{max \ or \ nominal}$ × 1,000)

÷ (____ $ft_{round \ trip}$ × ____ I_{max}) = ____ ohms per 1,000 ft

Inverter to service panel:

(____ AC voltage drop × _____ V_{output} × 1,000)

÷ (____ $ft_{round\ trip}$ × ___ $A_{AC\ continuous\ output}$)

= ____ ohms per 1,000 ft

Determine the proper wire size for each conductor based on the resistance in Table 15.2 (page 281), in the "ohms per 1,000 feet" column. Most installers prefer stranded wire because it is easier to bend. Most available wire is uncoated. If the calculated resistance falls between two values, always use the larger wire size—that is, the wire size with the lower resistance.

Step 10: List each wire size and actual resistance from Table 15.2 (page 281).

Series string to junction box:
_____ wire size, _____ ohms per 1,000 ft

Junction box to combiner box:
_____ wire size, _____ ohms per 1,000 ft

Combiner box to charge controller:
_____ wire size, _____ ohms per 1,000 ft

Charge controller to battery bank:
_____ wire size, _____ ohms per 1,000 ft

Inverter to service panel: ____ wire size, ____ ohms per 1,000 ft

Calculate the actual voltage drop for your PV system using the ohms per 1,000 feet from above and the formulas from Step 5.

Module interconnects: _____ actual voltage drop (from Step 5)

Series string to combiner box: _____ actual voltage drop

Combiner box to charge controller: _____ actual voltage drop

Charge controller to battery bank: _____ actual voltage drop

Battery to inverter: _____ actual voltage drop (from Step 5)

Inverter to service panel: _____ actual voltage drop

Total actual voltage drop: _____ (sum of all conductors)

The total should be equal to or less than the voltage drop you selected in Step 6.

Specifying Conductors, Overcurrent Protection Devices, Grounds & Conduit

The next set of calculations will confirm all of the conductor sizes, determine the size of the conduits and grounds, and determine the size and placement of fuses, circuit breakers, and disconnects to ensure that your PV system conforms to the NEC. Guidelines and rules for specifying conductors, grounds, conduit, and overcurrent protection devices are:

- Conductors are rated by ampacity, insulation temperature rating, installation temperature extremes, whether they are in free air or in conduit, and how many wires are in the conduit.

- The NEC requires an additional safety margin of 125% on fuses and AC breakers.

- Use solar module I_{SC} rating × 1.25 × 1.25 for DC conductors. [NEC 690.8 (A)(1), (B)(1)].

- Use Table 15.4 (page 283) or Table 15.5 (page 284) and the temperature correction factors from Table 15.6 (page 285) for temperature multipliers for conductors.

- Use adjustment factors from Table 15.10 (page 287) for more than three conductors in conduit.

- Use V_{OC} times the temperature correction factor from Table 15.8 (page 286) for low ambient design temperature.

- Use Table 15.9 (page 286) for high ambient design temperature.

- Use no temperature correction factor for 86°F/30°C indoor design temperature.

- Use 167°F/75°C design temperature for module interconnects or your site's record high ambient temperature plus 62.6°F/17°C.

- Use 140°/60°C or your high ambient temperature plus 62.6°/17°C for roof-mounted conduit.

- Use 113°F/45°C for wire in conduit in sun but not on the roof.

- Use 167°F/75°C for conductors and lugs unless otherwise specified. Use 167°F/75°C conductor insulation temperature rating for ampacity unless otherwise specified.

- Avoid multiple conductor factors by running no more than 3 current-carrying conductors in a conduit if possible.

- Do not run DC and AC conductors in the same conduit.

- Do not run communications or data wires in the same conduit as DC and AC conductors.

- Use terminal blocks to combine multiple conductors instead of feeding multiple conductors into one lug.

- The sum of the ampacity ratings of overcurrent devices coming into a bus bar or conductor should not exceed 110% of the bus bar or conductor rating.

- Use 167°F/75°C conductor ampacity and temperature deratings when 194°F/90°C conductors are connected to 167°F/75°C lugs (almost all off-the-shelf-equipment has 167°F/75°C lugs).

- Use AC conductor ampacity times 1.25.

- Use inverter AC current × 1.25 for continuous AC current rating.

- Use Table 15.12 (page 288) and Table 15.11 (page 287) for sizing grounding conductors.

- PV adjusted V_{OC} must be less than 600 volts.

- Use a fuse or circuit breaker rated greater than V_{OC} × 1.25 in series with current-carrying conductors.

- The fuse or circuit breaker must be rated less than the conductor's ampacity in order to protect the conductor.

Conductors

Step 1: Select the type of wire that will be used for each conductor. USE-2 is recommended for module interconnects in free air because the insulation is UV-stable, designed for wet locations, and the USE-2 meets roof design high temperature requirements. Refer to Table 15.1 (page 280). Conductors used for PV systems include USE-2, XHHW, XHHW-2, THHN and THHN-2, and THWN and THWN-2.

Step 2: List conductors, their design temperatures, insulation types, and temperature ratings. For any conductor with a design temperature other than ambient (86°F/30°C), use the temperature correction factors from Table 15.6 (page 285). Use the column that corresponds to your conductors' insulation temperature rating.

Step 3: List the short circuit current for each conductor from the array to the battery bank. Multiply the DC short-circuit current (I_{SC}) by 1.25 × 1.25 = 1.56 for PV array to battery circuits.

_____ I_{SC} × 1.56 = _____ A

The ampacity of the conductor from the battery to inverter is:

(inverter rated watts ÷ lowest DC input volts÷ inverter efficiency) × 1.25 = ampacity

(Note: The second "× 1.25" factor for irradiance isn't necessary.)

Use this ampacity in Step 4 for overcurrent devices. The ampacity of the conductor from the inverter to the service panel is determined by continuous AC output of the inverter times 1.25.

Step 4: Refer to Table 15.4 (page 283) for conductors in conduit and Table 15.5 (page 284) for conductors in free air.

Multiply conductor amperage by the design temperature correction factor from Step 2.

amperage × temperature correction factor = amperage corrected for design temperature

If you have more than three current-carrying conductors (not counting the ground conductor) in a conduit, the allowable ampacity of the conductors (from Table 15.4, page 283) after temperature adjustment must be adjusted for conduit fill using the percent allowable ampacity from Table 15.10 (page 287).

Example: There are 6 current-carrying #6 AWG XHHW-2. Ambient design temperature is 140°F/60°C with a temperature correction factor of 0.71. The #6 AWG XHHW-2 has an allowable ampacity of 75 A. Temperature correction is 75 A × 0.71 = 53.25 A. Referring to Table 15.10 (page 287), the percent allowable ampacity for six conductors in conduit is 70% (0.70). Calculate the maximum allowable ampacity of the #6 AWG: 53.25 A × 0.70 = 37.25 A.

Determine the allowable ampacity for each conductor using either the insulation temperature or the temperature rating of the lugs, whichever is lower. Compare this value to the value obtained in Step 3. In every case the value from Step 4 should be equal to or greater than the value in Step 3. If it is not, use the next larger size of wire, that is, the one with the higher ampacity rating.

Overcurrent Devices (Fuses & Circuit Breakers)

Step 1: To size a combiner box fuse or breaker multiply the series string short-circuit current by 1.56. Use a fuse or breaker with a higher rating than this value.

_____ I_{SC} × 1.56 = _____ A

Fuse or breaker rating = _____ A

Multiply the I_{SC} by 1.25 only if you are using a 100% DC-rated circuit breaker.

Step 2: To size the GFCI (ground fault circuit interruptor) between the array and balance of system (BOS), multiply the combined series string short-circuit current by 1.56. Use a GFCI with a higher amperage rating

_____ I_{SC} × 1.56 = _____ A

GFCI rating = _____ A

If the GFCI is a 100% DC-rated circuit breaker, multiply the I_{SC} by 1.25.

Step 3: To size each DC disconnect (into and out of the charge controller) use the amperage in Step 2.

Step 4: To size the fuse or breaker between the battery and the inverter, use the ampacity of the cable from the previous conductor section, Step 3. Use the fuse or breaker size recommended or supplied by the inverter manufacturer or use a fuse with a higher rating than the value calculated in the previous conductor section on sizing the conductor between the batteries and the inverter.

fuse rating = _____ A

Step 5: To size the AC disconnect switch (or system disconnect switch) and fuse, use the inverter AC output voltage and the inverter continuous output current times 1.25.

_____ V inverter AC output voltage
_____ A inverter continuous output current × 1.25
= _____ AC disconnect switch specifications

Step 6: To size the circuit breaker in the service panel, use the continuous output current of the inverter. Use a breaker with an equal or higher rating than this value. AC breakers are rated to 100% capacity (specified on the breaker or the packaging) and do not require the 125% adjustment.

Grounding Conductors

Step 1: All equipment must be earth-grounded with the exception of solar module circuits, which must be protected by a GFCI. The metal frame of all solar arrays must be earth grounded. The grounding conductor for PV source and output circuits must be sized for 125% of the DC short-circuit current. Refer to Table 15.11 (page 287) for DC and Table 15.12 (page 288) for AC. For DC circuits, use the amperage rating of the overcurrent device that is immediately before the equipment. For AC circuits, use the size of the AC conductor. Use the amperage rating to determine the minimum size of the grounding wire. Any grounding conductor smaller than #6 AWG must be in conduit unless it is already protected from physical damage.

Step 2: The size of the service entrance grounding conductor (the ground wire from your AC service panel to the grounding electrode or rod) is based upon the size of the service entrance conductor (wires from the utility service, generator, or inverter AC out). Use Table 15.12 (page 288) to determine the minimum size and type of the grounding electrode conductor.

Conduit

Step 1: List all of the wires that will be in a conduit including the ground wire if insulated. Uninsulated grounding conductors must not go into the conduit.

Step 2: Refer to Table 15.3 (page 282). Determine the area in square inches, based on the type of insulation, for each wire. Add the cross-sectional areas together. Find this sum in the appropriate column of Table 15.7 (page 285). Ask your local building inspection office which type of conduit is allowed. Some jurisdictions do not allow non-metallic conduit.

Important Note

The design guidelines in this book do not supersede *National Electrical Code* and/or local building code requirements. These requirements change, so be sure that you are complying with the most current national and local codes.

Tables for Calculating Wire, Fuse, Breaker & Conduit Sizes

Table 15.1

Conductor Applications and Insulations					
Trade Name	Type	Maximum Operating Temperature	Application Provisions	Insulation	Outer Covering[1]
Moisture resistant thermoset	RHW	75°C/167°F	Dry and wet locations	Flame-retardant, moisture-resistant thermoset	Moisture-resistant, flame-retardant, non-metallic covering
Moisture resistant thermoset	RHW-2	90°C/194°F	Dry and wet locations	Flame-retardant, moisture-resistant thermoset	Moisture-resistant, flame-retardant, non-metallic covering
Heat-resistant thermoplastic	THHN	90°C/194°F	Dry and damp locations	Flame-retardant, heat resistant thermoplastic	Nylon jacket or equivalent
Moisture and heat-resistant thermoplastic	THW[2]	75°C/167°F 90°C/194°F	Dry and wet locations Special applications within electric discharge lightning equipment. Limited to 1000 open-circuit volts or less (size 14-8 only)	Flame-retardant, heat resistant thermoplastic	None
Moisture and heat-resistant thermoplastic	THWN[2]	75°C/167°F	Dry and wet locations	Flame-retardant, moisture- and heat-resistant thermoplastic	Nylon jacket or equivalent
Underground feeder and branch-circuit cable - single conductor	UF	60°C/140°F 75°C/167°F[3]	See NEC Article 340	Moisture resistant Moisture- and heat- resistant	Integral with insulation
For Type UF cable employing more than one conductor see NEC Articles 339,340					
Underground service-entrance cable - single conductor	USE[2]	75°C/167°F	See NEC Article 338	Heat- and moisture- resistant	Moisture-resistant non-metallic covering (NEC 338.2)
For type USE cable employing more than one conductor see NEC Article 338					
Moisture-resistant thermoset	XHHW[2]	90°C/194°F 75°C/167°F	Dry and damp locations Wet locations	Flame- retardant, moisture-resistant thermoset	None
Moisture-resistant thermoset	XHHW-2	90°C/194°F	Dry and wet locations	Flame- retardant, moisture-resistant thermoset	None

[1] Some insulations do not require an outer covering.
[2] Listed wire types designated with the suffix "2" such as RHW-2, shall be permitted to be used at a continuous 90°C/194°F operating temperature, wet or dry.
[3] For ampacity limitation see NEC 340.80

Table 15.2

						DC Resistance at 75°C (167°F)		AC Resistance	
Size (AWG or kcmil)	Area Circular mils	Stranding Quantity	Diameter inches	Diameter inches	Area inches²	Uncoated ohms/1000 ft.	Coated ohms/1000 ft.	Uncoated in Conduit ohms/1000 ft.	Size (AWG or kcmil)
18	1620	1	-	0.040	0.001	7.77	8.08	-	18
18	1620	7	0.015	0.046	0.002	7.95	8.45	-	18
16	2580	1	-	0.051	0.002	4.89	5.08	-	16
16	2580	7	0.019	0.058	0.003	4.99	5.29	-	16
14	4110	1	-	0.064	0.003	3.07	3.19	-	14
14	4110	7	0.024	0.073	0.004	3.14	3.26	3.10	14
12	6350	1	-	0.081	0.005	1.93	2.01	-	12
12	6350	7	0.030	0.092	0.006	1.98	2.05	2.00	12
10	10380	1	-	0.102	0.008	1.21	1.26	-	10
10	10380	7	0.038	0.116	0.011	1.24	1.29	1.20	10
8	16510	1	-	0.128	0.013	0.764	0.786	-	8
8	16510	7	0.049	0.146	0.017	0.778	0.809	0.78	8
6	26240	7	0.061	0.184	0.027	0.491	0.510	0.49	6
4	41740	7	0.077	0.232	0.042	0.308	0.321	0.31	4
3	52620	7	0.087	0.260	0.053	0.245	0.254	0.25	3
2	66360	7	0.097	0.292	0.067	0.194	0.201	0.20	2
1	83690	19	0.066	0.332	0.087	0.154	0.160	0.16	1
1/0	105600	19	0.074	0.372	0.109	0.122	0.127	0.13	1/0
2/0	133100	19	0.084	0.418	0.137	0.097	0.101	0.10	2/0
3/0	167800	19	0.094	0.470	0.173	0.0766	0.0797	0.082	3/0
4/0	211600	19	0.106	0.528	0.219	0.0608	0.0626	0.067	4/0
250	-	19	0.082	0.575	0.260	0.0515	0.0535	0.057	250
300	-	37	0.090	0.630	0.312	0.0429	0.0446	0.049	300
350	-	37	0.097	0.381	0.364	0.0367	0.0382	0.043	350
400	-	37	0.104	0.728	0.416	0.0321	0.0331	0.038	400
500	-	37	0.116	0.813	0.519	0.0258	0.0265	0.032	500
600	-	61	0.099	0.893	0.626	0.0214	0.0223	0.028	600
700	-	61	0.107	0.964	0.730	0.0184	0.0189	-	700
750	-	61	0.111	0.998	0.782	0.0171	0.0176	0.024	750
800	-	61	0.114	1.030	0.834	0.0161	0.0166	-	800
900	-	61	0.122	1.094	0.940	0.0143	0.0147	-	900
1000	-	61	0.128	1.152	1.042	0.0129	0.0132	0..019	1000
1250	-	91	0.117	1.289	1.305	0.0103	0.0106	-	1250
1500	-	91	0.128	1.412	1.566	0.00858	0.00883	-	1500
1750	-	127	0.117	1.526	1.829	0.00735	0.00756	-	1750
2000	-	127	0.126	1.632	2.092	0.00643	0.00662	-	2000

These resistance values ar only good for the parameters given

Formula for temperature change : $R_2=R_1[1= aT_2-75)$, $a_{Cu} = 0.00323$

Table 15.3

	TW, THW, THHW, THW-2, RHH*, RHW*, RHW-2*		THHN, THWN, THWN-2		XHHW, XHHW-2, XHH		
AWG	Diameter (inches)	Area (inches²)	Diameter (inches)	Area (inches²)	Diameter (inches)	Area (inches²)	AWG
14			0.111	0.0097	0.133	0.0139	14
12			0.130	0.0133	0.152	0.0181	12
10	0.206	0.0330	0.164	0.0211	0.176	0.0243	10
8	0.266	0.0556	0.216	0.0366	0.236	0.0437	8
6	0.304	0.0726	0.254	0.0507	0.027	0.0590	6
4	0.352	0.0973	0.324	0.0824	0.322	0.0814	4
3	0.380	0.1134	0.352	0.0973	0.350	0.0962	3
2	0.412	0.1333	0.384	0.1158	0.382	0.1146	2
1	0.492	0.1901	0.446	0.1562	0.442	0.1534	1
1/0	0.532	0.2230	0.486	0.1855	0.482	0.1825	1/0
2/0	0.578	0.2640	0.532	0.2223	0.528	0.2190	2/0
3/0	0.630	0.3117	0.584	0.2679	0.580	0.2642	3/0
4/0	0.688	0.3718	0.642	0.3237	0.638	0.3197	4/0
250	0.765	0.4596	0.711	0.3970	0.705	0.3904	250
300	0.820	0.5281	0.766	0.4608	0.760	0.4536	300
350	0.871	0.5958	0.817	0.5242	0.811	0.5166	350
400	0.918	0.6619	0.864	0.5863	0.858	0.5782	400
500	1.003	0.7901	0.949	0.7073	0.943	0.6984	500
600	1.113	0.9729	1.051	0.8676	1.053	0.8709	600
700	1.184	1.1010	1.122	0.9887	1.124	0.9923	700
750	1.218	1.1652	1.156	1.0496	1.158	1.0532	750
800	1.250	1.2272	1.188	1.1085	1.190	1.1122	800
900	1.314	1.3561	1.252	1.2311	1.254	1.2351	900
1,000	1.372	1.4784	1.310	1.3478	1.312	1.3519	1,000
1,250	1.539	1.8602			1.479	1.7180	1,250
1,500	1.662	2.1695			1.602	2.0157	1,500
1,750	1.776	2.4773			1.716	2.3127	1,750
2,000	1.882	2.7818			1.822	2.6073	2,000

*Types RHH, RHW, and RHW-2 without outer covering.

Table 15.4

	Allowable Ampacity of Insulated Copper in Raceway or Conduit Based on Ambient Temperature of 30°C/86°F			
	Temperature Rating of Conductor			
	60°C/140°F	75°C/167°F	90°C/194°F	
Size (AWG or kcmil)	Types TW, UF	Types RHW,THHW,THW, THWN, XHHW, USE, ZW	Types TBS, S, SIS, FEP, MI, RHH, RHW-2, THHN, THHW, THW-2, THWN-2, USE-2, XHH, XHHW, XHHW-2, ZW-2	Size (AWG or kcmil)
18	-	-	14	18
16	-	-	18	16
14	20	20	25	14
12	25	25	30	12
10	30	35	40	10
8	40	50	55	8
6	55	65	75	6
4	70	85	95	4
3	85	100	110	3
2	95	115	130	2
1	110	130	150	1
1/0	125	150	170	1/0
2/0	145	175	195	2/0
3/0	165	200	255	3/0
4/0	195	230	260	4/0
250	215	255	290	250
300	240	285	320	300
350	260	310	350	350
400	280	335	380	400
500	320	380	430	500
600	355	420	475	600
700	385	460	520	700
750	400	475	535	750
800	410	490	555	800
900	435	520	585	900
1000	455	545	615	1000
1250	495	590	665	1250
1500	520	625	705	1500
1750	545	650	735	1750
2000	560	665	750	2000

Table 15.5

Allowable Ampacity of Single Insulated Copper in Free Air Based on Ambient Temperature of 30°C/86°F			
Temperature Rating of Conductor			
60°C/140°F	75°C/167°F	90°C/194°F	
Size (AWG or kcmil)	Types TW, UF	Types RHW,THHW,THW, THWN, XHHW, ZW	Types TBS, SA, SIS, FEP, FEPB, MI, RHH, RHW-2, THHN, THHW, THW-2, THWN-2, USE-2, XHH, XHHW, XHHW-2, ZW-2	Size (AWG or kcmil)
18	-	-	18	18
16	-	-	24	16
14	25	30	35	14
12	30	35	40	12
10	40	50	55	10
8	60	70	80	8
6	80	95	105	6
4	105	125	140	4
3	120	145	165	3
2	140	170	190	2
1	165	195	220	1
1/0	195	230	260	1/0
2/0	225	265	300	2/0
3/0	260	310	350	3/0
4/0	300	360	405	4/0
250	340	405	455	250
300	375	445	505	300
350	420	505	570	350
400	455	545	615	400
500	515	620	700	500
600	575	690	780	600
700	630	755	855	700
750	655	785	885	750
800	680	815	920	800
900	730	870	985	900
1000	780	935	1055	1000
1250	890	1065	1200	1250
1500	900	1175	1325	1500
1750	1070	1280	1445	1750
2000	1155	1385	1560	2000

Table 15.6
Use with Tables 15.4 & 15.5

Correction Factors for Allowable Ampacity Tables				
	Temperature Rating of Conductor			
	60°C/140°F	75°C/167°F	90°C/194°F	
Size (AWG or kcmil)	Types TW, UF	Types RHW,THHW,THW, THWN, XHHW, USE, ZW	Types TBS, S, SIS, FEP, MI, RHH, RHW-2, THHN, THHW, THW-2, THWN-2, USE-2, XHH, XHHW, XHHW-2, ZW-2	Size (AWG or kcmil)
Ambient Temperature °C	For ambient temperatures other than 30°C/86°F, multiply the allowable ampacity in the chart by the appropriate factor below.			Ambient Temperature °F
21-25	1.08	1.05	1.04	70-77
26-30	1.00	1.00	1.00	78-86
31-35	0.91	0.94	0.96	87-95
36-40	0.82	0.88	0.91	96-104
41-45	0.71	0.82	0.87	105-113
46-50	0.58	0.75	0.82	114-122
51-55	0.41	0.67	0.76	123-131
56-60	-	0.58	0.71	132-140
61-70	-	0.33	0.58	141-158
71-80	-	-	0.41	159-176

Table 15.7

Dimensions and Percent Area of Electrical Metallic Tubing (EMT)				
Trade Size	Nominal Internal Diameter (inches)	2 Wires 31% (inches^2)	Over 2 Wires 40% (inches^2)	1 Wire 53% (inches^2)
1/2	0.622	0.094	0.122	0.161
3/4	0.824	0.165	0.213	0.283
1	1.049	0.268	0.346	0.458
1 1/4	1.380	0.464	0.598	0.793
1 1/2	1.610	0.631	0.814	1.079
2	2.067	1.04	1.342	1.778
2 1/2	2.731	1.816	2.343	3.105
3	3.356	2.742	3.538	4.688
3 1/2	3.834	3.579	4.618	6.119
4	4.334	4.573	5.901	7.819

Table 15.8

Voltage Correction Factors for Crystalline and Multicrystalline Silicon Modules		
Ambient Temperature (°C)	Correction Factors for Low Ambient Temperatures (multiply Voc by the correction factor)	Ambient Temperature (°F)
24 to 20	1.02	76 to 68
19 to 15	1.04	67 to 59
14 to 10	1.06	58 to 50
9 to 5	1.08	59 to 41
4 to 0	1.10	40 to 32
-1 to -5	1.12	31 to 23
-6 to -10	1.14	22 to 14
-11 to -15	1.16	13 to 5
-16 to -20	1.18	4 to -4
-21 to -25	1.20	-5 to -13
-26 to -30	1.21	-14 to -22
-31 to -35	1.23	-23 to -31
-36 to -40	1.25	-32 to -40
Note: Use the manufacturer's data for non-crystalline solar modules.		

Table 15.9

Voltage Correction Factors for High Ambient Temperatures					
Ambient Temperature (°C)	Temperature Rating of Conductor				Ambient Temperature (°F)
	60°C / 140°F	75°C / 167°F	90°C / 194°F	105°C / 221°F	
30	1.00	1.00	1.00	1.00	86.00
31-35	0.91	0.94	0.96	0.97	87-95
36-40	0.82	0.88	0.91	0.93	96-104
41-45	0.71	0.82	0.87	0.89	105-113
46-50	0.58	0.75	0.82	0.86	114-122
51-55	0.41	0.67	0.76	0.82	123-131
56-60	-	0.58	0.71	0.77	132-140
61-70	-	0.33	0.58	0.68	141-158
71-80	-	-	0.41	0.58	159-176

Table 15.10

Adjustment Factors for More Than Three Current-Carrying Conductors in a Raceway or Cable	
Number of Current-Carrying Conductors	Percent of Allowable Ampacity (after ambient temperature adjustment)
4–6	80
7–9	70
10–20	50
21–30	45
31–40	40
41 and above	35

Table 15.11

Minimum Size Equipment Grounding Conductors for Grounding Raceway and Equipment		
Rating or Setting of Automatic Overcurrent Device ahead of Equipment (amperes)	Size (AWG)	
	Copper	Copper-clad Aluminum or Aluminum
15	14	12
20	12	10
30	10	8
40	10	8
60	10	8
100	8	6
200	6	4
300	4	2
400	3	1
500	2	1/0
600	1	2/0
800	1/0	3/0
1,000	2/0	4/0
1,200	3/0	250
1,600	4/0	350
2,000	250	400
2,500	350	600
3,000	400	600
4,000	500	800
5,000	700	1,200
6,000	800	1,200

Table 15.12

Grounding Electrode Conductor for AC Systems			
Size of Largest Ungrounded Service-Entrance Conductor or Equivalent Parallel Conductors (AWG)		Size of Grounding Electrode Conductor (AWG)	
Copper	Aluminum or Copper-Clad Aluminum	Copper	Aluminum or Copper-Clad Aluminum
≤2	≤1/0	8	6
1 or 1/0	2/0 or 3/0	6	44
2/0 or 3/0	4/0 or 250	4	2
>3/0 to 350	>250 to 500	2	1/0
>350 to 600	>500 to 900	1/0	3/0
>600 to 1100	>900 to 1750	2/0	4/0
>1100	>1750	3/0	250

National Electrical Code Tables

NEC Table 310.13 — Conductor Applications and Insulation

NEC Chapter 9, Table 8 — Copper Conductor Properties

NEC Chapter 9, Table 5 — Dimensions of Insulated Conductors

NEC Table 310.16 — Allowable Ampacity of Insulated Copper in Raceway or Conduit

NEC Table 310.17 — Allowable Ampacity of Single-Insulated Copper in Free Air

Temperature Correction Factors for Tables 310.16 & 310.17

NEC Chapter 9, Table 4 — Dimensions and Percent Area of Electrical Metallic Tubing (EMT)

NEC 690.7 — Voltage Correction Factors for Crystalline & Multicrystalline Silicon Modules

NEC Table 690.31(c) — Voltage Correction Factors for High Ambient Temperatures

NEC Table 310-15 — Adjustment Factors for More than Three Current-Carrying Conductors in a Raceway or Cable

NEC 250.122 — Minimum Size Equipment Grounding Conductors for Grounding Raceway and Equipment

NEC 250.66 — Grounding Electrode Conductor for AC Systems

Sample Calculation #1—Off-Grid System

Step 1: See Granby, Colorado, PV system wiring diagram, Figure 16.8 (page 336).

Step 2: List the watts, volts, and amps of all equipment.

a. Solar modules	Watts STC	125 W
	V_{max}	17.4 W
	I_{max}	7.2 A
	V_{oc}	21.7 V
	I_{sc}	8 A
	Nominal module volts	12 V
	Table 15.8 temperature correction factor	1.25
	V_{oc} corrected (see V_{oc} note)	27.13 V

V_{oc} note: For crystalline and multi-crystalline silicon modules, multiply V_{oc} by Table 15.8 correction factor (page 286). V_{oc} × correction factor × quantity modules in series = V_{oc} (corrected). Use this value for all calculations using V_{oc}. For other type modules refer to manufacturer's instructions for corrected V_{oc}.

b. Series source circuit	Quantity of modules in series	4
	Quantity of series source circuits	3
	Watts STC = (module watts × quantity modules in series)	500 W
	V_{max} = (module V_{max} × quantity modules in series)	69.6 V
	I_{max} = (same as module I_{max})	7.2 A
	V_{oc} (corrected) = (V_{oc} × correction factor × quantity modules in series)	108.5 V
	I_{sc} = (same as module I_{sc})	8 A
	Nominal system volts = (nominal module volts × quantity modules in series)	48 V

c. Combiner box (output)	Watts STC = (series string watts × quantity series source circuit strings)	1500 W
	V_{max} = (same as series string V_{max})	69.6 V
	I_{max} = (series string I_{max} × quantity series strings)	21.6 A
	V_{OC} = (same as corrected series string V_{OC})	108.5 V
	I_{SC} = (series string I_{SC} × quantity series strings)	24 A
	Nominal system volts	48 V
d. Charge controller (skip for battery-less systems)	Current rating	40 A
	MPPT (yes or no)	No
	PV V_{OC} (corrected)	108.5 V
	Nominal battery volts	48 V
	Maximum watts = (current rating × nominal battery volts or V_{max} if MPPT)	1920 W
e. Inverter	Continuous power rating. VA (watts)	4000 W
	Lowest DC input volts. V	44 V
	Output volts. VAC	120 V
	Continuous AC current output. A	33 A
	Maximum AC current output. A	78 A
	Efficiency %	85%

Do not use inverter peak efficiency. Use average inverter efficiency (usually 80 to 90%) or inverter efficiency at 75% load available on the California Energy Commission's eligible inverters list.

Step 3: List circuits.

Wire	Design Temperature	Insulation Rating	Lug Rating	Number of Wires	Conduit Yes / No
Solar module interconnects	167°F/75°C roof	194°F/90°C	194°F/90°C	n.a.	no
Array junction box to combiner box	167°F/75°C roof	194°F/90°C	194°F/90°C	6	yes
Combiner box to charge controller	140°F/60°C roof	167°F/75°C	167°F/75°C	2	yes
Charge controller to batteries	86°F/30°C indoor	167°F/75°C	167°F/75°C	2	yes
Batteries to inverter	86°F/30°C indoor	167°F/75°C	167°F/75°C	2	no
Inverter to service panel	86°F/30°C indoor	167°F/75°C	167°F/75°C/	2	yes

Step 4. Systems with Batteries

	quantity	each length in feet	total length feet
a. Module interconnects	3	2	6

		length in feet	× 2 = round trip feet
a. Module interconnects		6	12
b. Series string to junction box		—	—
c. Junction box to combiner box		50	100
d. Combiner to charge controller		15	30
e. Controller to battery bank		15	30
f. Battery bank to inverter		5	10
Total DC wire length (do not include module interconnects or battery to inverter cable)		80	160
g. AC inverter to service panel		20	40
h. Sum of DC and AC wire lengths		100	200

Calculate the proportional length of DC conductors (each conductor's round-trip length divided by the sum of the DC and AC round-trip wire lengths)

a. Module interconnects (not included in this calculation)

	round-trip length	÷ total length (h) = proportional length
b. Series string to junction box	—	—
c. Junction box to combiner box	100	0.50000000
d. Combiner to charge controller	30	0.15000000
e. Controller to battery bank	30	0.15000000
f. Battery bank to inverter (not included in this calculation)		
g. AC inverter to service panel	40	0.20000000

Step 5: a. Calculate voltage drop for #10 AWG module interconnects.

[**1.24** ohms per 1,000 ft] ÷ 1,000] × (**12** ft$_{round\ trip}$ × **7.2** I$_{max}$)
= **0.10713600** volts lost to resistance

0.10713600 volts lost to resistance ÷ **48** system volts
= **0.00223200** voltage drop

0.00223200 voltage drop × 100 = **0.2232**% voltage drop

b. Calculate the voltage drop for the #4/0 AWG battery to inverter cable.

(**4,000** inverter rated watts ÷ **44** lowest DC input volts ÷ **0.85**
inverter efficiency) × 1.25 = **133.69** amps calculated input current

The formula for voltage drop is:

[**0.0608** ohms per 1,000 ft ÷ 1,000] × (**10** ft$_{round\ trip}$ × **133.69**
calculated input current) = **0.08128342** volts lost to resistance

0.08128342 volts lost to resistance ÷ **48** system volts
= **0.00169340** voltage drop

0.00169340 voltage drop × 100 = **0.16934**% voltage drop

Step 6: Specify total acceptable voltage drop: **0.03** = **3.0**%

Step 7: Subtract voltage drop for module interconnects and battery to inverter cable:

0.03 voltage drop – **0.00223200** module interconnect voltage
drop – **0.00169340** battery cable voltage drop = **0.0260746**
net remaining voltage drop

Step 8: Calculate the voltage drop for each circuit.

Series string to combiner box:
0.5000000 proportion DC wire × **0.02607460** net remaining voltage
drop = **0.01303730** acceptable DC voltage drop

Combiner box to charge controller:
0.15000000 proportion DC wire × **0.02607460** net remaining
voltage drop = **0.00391119** acceptable DC voltage drop

Charge controller to battery bank:
0.15000000 proportion DC wire × **0.02607460** net remaining
voltage drop = **0.00391119** acceptable DC voltage drop

Inverter to service panel:
0.20000000 proportion AC wire × **0.02607460** net remaining voltage drop = **0.00521492** acceptable AC voltage drop

Step 9: Calculate conductor resistance.

Series string to combiner box:
(**0.01303730** DC voltage drop × **48** $V_{nominal}$ × 1,000)
÷ (**100** ft$_{round\ trip}$ × **7.2** I_{max}) = **0.86915318** ohms per 1,000 ft

Combiner box to charge controller:
(**0.00391119** DC voltage drop × **48** $V_{nominal}$ × 1,000)
÷ (**30** ft$_{round\ trip}$ × **21.6** I_{max}) = **0.28917732** ohms per 1,000 ft

Charge controller to battery bank:
(**0.00391119** DC voltage drop × **48** $V_{nominal}$ × 1,000)
÷(**30** ft$_{round\ trip}$ × **21.6** I_{max}) = **0.28917732** ohms per 1,000 ft

Inverter to service panel:
(**0.00521492** AC voltage drop × **120** V_{out} × 1,000)
÷(**40** ft$_{round\ trip}$ × **33** $A_{AC\ output}$) = **0.47408355** ohms per 1,000 ft

Step 10: List each wire size and its actual resistance from Table 15.2 (page 281).

Series string to combiner box:
8 wire size, **0.7780** ohms per 1,000 ft

Combiner box to charge controller:
3 wire size, **0.245** ohms per 1,000 ft

Charge controller to battery bank:
3 wire size, **0.245** ohms per 1,000 ft

Inverter to service panel:
4 wire size, **0.308** ohms per 1,000 ft

Use the ohms per 1,000 feet from above and the formula from Step 5 to calculate the actual voltage drop for each circuit.

Module interconnects:
0.00223200 actual voltage drop (from Step 5)

Series string to combiner box: **0.011670** actual voltage drop

Combiner box to charge controller:
0.003307500 actual voltage drop

Charge controller to battery bank:
0.003307500 actual voltage drop

Battery to inverter:
0.00169340 actual voltage drop (from Step 5)

Inverter to service panel: **0.00338800** actual voltage drop

Total actual voltage drop:
0.02559840 (sum of all conductors) = **2.559%**

Code Compliance

Module interconnects:

Required ampacity = 8 A I_{SC} × 1.56 = 12.48 A, #10 AWG wire, type USE-2 rated 194°F/90°C has an ampacity of 40 A times the ambient temperature factor of 0.41 for its 167°F/75°C location = 16.4 A. This is acceptable because it is greater than 12.48 A.

Series string to combiner box:

Required ampacity = 8 A I_{SC} × 1.56 = 12.48 A, #8 AWG wire, type XHHW-2 rated 194°F/90°C, the lugs are rated 194°F/90°C. The ampacity in the 194°F/90°C column is 55 A. The ambient design temperature of 168°F/75°C requires a temperature correction factor of 0.41 (50 × 0.41 = 22.55 A). This is acceptable because it is greater than 12.5 A.

Combiner box to charge controller to battery bank:

Required ampacity = 8 A I_{SC} × 3 string circuits × 1.56 = 37.44 A, #3 AWG wire, type XHHW-2 rated 194°F, the lugs are rated 167°F/75°C. The ampacity in the 167°F/75°C column is 100 A. The ambient design temperature of 140°F/75°C requires a temperature correction factor of 0.71 (100 × 0.71 = 71 A). This is acceptable because it is greater than 37.44 A.

Battery to inverter:

Required ampacity = [(inverter rated watts ÷ lowest DC input volts ÷ inverter efficiency) × 1.25] = (4,000/44/0.85) × 1.25 = 133.69 A.

4/0 AWG wire, type XHHW-2 rated 194°F/90°C, the lugs are rated 167°F/75°C. The ampacity in the 167°F/75°C column is 230 A. The ambient design temperature of 86°F/30°C requires a temperature correction factor of 1.0. (230 A × 1.0) = 230 A. This is acceptable because it is greater than 133.69 A.

Inverter to service panel:
Required ampacity = inverter continuous output amps × 1.25 = 33 A × 1.25 = 41.25 A, #4 AWG wire, type THHN/THWN, rated 167°F/75°C, the lugs are rated 167°F/75°C. The ampacity in the 167°F/75°C column is 85 A. The ambient design temperature of 86°F/30°C requires a temperature correction factor of 1.0 (85 × 1.0 = 85 A). This is acceptable because it is greater than 41.25 A.

Overcurrent Devices (Fuses and Breakers)

Combiner box:
8 I_{sc} × 1.56 = 12.5 A, fuse rating = 15 A = 3 each 15-A fuses

GFCI between the combiner box and the charge controller:
3 × 8 = 24 I_{sc} × 1.56 = 37.5, GFCI rating = 40 A

DC disconnect:
3 × 8 = 24 I_{sc} × 1.56 = 37.5 A
Buy two 125-VDC, 40-A DC circuit breakers or fused disconnect switches.

Fuse between the battery and the inverter:
(4,000 inverter-rated watts ÷ 44 lowest DC input volts ÷ 0.85 inverter efficiency) × 1.25 = 133.69 A, fuse rating = 150 A.

Breaker in the service panel:
120-V inverter AC output voltage at 33 A
Buy 120/240-V, 40-A breaker for service panel.

Grounding Conductors

Equipment ground wires:

Module and mount frame:
Minimum #14 (determined by the energized circuits' overcurrent device = 15 A) (in protected area) or #10 (same size as circuit conductor) or #6 (which does not require conduit).

GFCI box:
Minimum #10 (overcurrent device = 40 A) or same size as circuit conductor or #6

Metal battery rack:
Minimum # 6 (overcurrent device =150 A) or same size as circuit conductor

Inverter: Minimum #10 (overcurrent device = 40 A) or #6

Service panel:
#4 AWG conductor from inverter
Use #8 AWG or #6.

Circuit grounds:

Inverter to service panel:
#4 AWG current-carrying conductor
Use #8 in conduit.

Conduit

Series string to combiner box:
2 #8 AWG XHHW-2, #10 insulated ground THHN, area = 2 × 0.0437 in.^{2}
+ 0.0211 in.^{2} = 0.1085 in.^{2}
Use ½-inch conduit.

Combiner box to batteries:
2 #3 AWG XHHW-2, #10 insulated ground THHN, area = 2 × 0.0962 in.^{2}
+ 0.0211 in.^{2} = 0.2135 in.^{2}
Use 1-inch conduit.

Batteries to inverter:
2 #4/0 AWG THHN/THWN, #8 insulated ground THHN, area = 2 × 0.3237 in.^{2} + 0.0366 in.^{2} = 0.6844 in.^{2}
Use 1½-inch conduit.

Inverter to service panel:
2 #4 AWG THHN/THWN, #8 insulated ground THHN, area = 2 × 0.0824 in.^{2} +0.0366 in.^{2} = 0.2014 in.^{2}
Use ¾-inch conduit.

Sample Calculation #2—Batteryless On-Grid System

Step 1: See Los Angeles system wiring diagram, Figure 17.2 (page 348).

Step 2: List the watts, volts, and amps of all equipment.

a. Solar modules	Watts STC	**125 W**
	V_{max}	**17.4 V**
	I_{max}	**7.2 A**
	V_{oc}	**21.7 V**
	I_{sc}	**8 A**
	Nominal module volts	**12 V**
	Table 15.8 temperature correction factor	**1.10**
	V_{oc} corrected (see V_{oc} note)	**23.87 V**

V_{oc} note: For crystalline & multi-crystalline silicon modules, multiply V_{oc} by Table 15.8 correction factor (page 286). V_{oc} × correction factor × quantity modules in series = V_{oc} (corrected). Use this value for all calculations using V_{oc}. For other module types refer to manufacturer's instructions for corrected V_{oc}.

b. Series source circuit	Quantity of modules in series	**24**
	Quantity of series source circuits	**1**
	Watts STC = (module watts × quantity modules in series)	**3000 W**
	V_{max} = (module V_{max} × quantity modules in series)	**417.6 V**
	I_{max} = (same as module I_{max})	**7.2 A**
	V_{oc} (corrected) = (V_{oc} × correction factor × quantity modules in series)	**572.88 V**
	I_{sc} = (same as module I_{sc})	**8 A**
	Nominal system volts = (nominal module volts × quantity modules in series)	**288 V**

c. Combiner box output (not used in this example)		
d. Charge controller (not used in this example)		

e. Inverter	Continuous power rating. VA (watts)	**2500 W**
	Lowest DC input volts. V	**207 V**
	Output volts. VAC	**240 V**
	Continuous AC current output. A	**10.4 A**
	Maximum AC current output. A	**12 A**
	Efficiency %	**94%**

Do not use inverter peak efficiency. Use average inverter efficiency (usually 80 to 90%) or inverter efficiency at 75% load available on the California Energy Commission's eligible inverters list.

Step 3: List circuits.

Wire	Design Temperature	Insulation Rating	Lug Rating	Number of Wires	Conduit Yes / No
Solar module interconnects	167°F/75°C	194°F/90°C	194°F/90°C	n.a.	no
Array to DC disconnect	167°F/75°C	194°F/90°C	194°F/90°C	2	yes
DC disconnect to inverter	140°F/60°C	167°F/75°C	167°F/75°C	2	yes
Inverter to service panel	86°F/30°C	167°F/75°C	167°F/75°C	2	yes

Step 4: Batteryless system

	quantity	each length in feet	total length feet
a. Module interconnects	23	2	46
		length in feet	× 2 = round trip feet
a. Module interconnects		1 circuit	46
b. Series string to inverter		100	200
c. Junction box to combiner box (not included in this calculation)			
d. Combiner to inverter (not included in this calculation)			
b + c + d = Total DC wire length (do not include module interconnects or battery to inverter cable)		100	200
e. AC inverter to service panel		10	20
f. Sum of DC and AC wire lengths		110	200
Calculate the proportional length of DC conductors (each conductor's round-trip length divided by the sum of the DC and AC round-trip wire lengths)			
a. Module interconnects (not included in this calculation)			
	round-trip length	÷ total length (f) = proportional length	
b. Series string to inverter	200	0.90909091	
c. Junction box to combiner box (not included in this calculation)			
d. Combiner box to inverter (not included in this calculation)			
e. AC inverter to service panel	20	0.09090909	

Step 5: a. Calculate the voltage drop for the module interconnects.

[(**1.24** ohms per 1,000 ft) ÷ 1,000] × (**46** ft$_{round\ trip}$ × **7.2** I$_{max}$)

= **0.41068800** volts lost to resistance

0.41068800 volts lost to resistance ÷ **417.6** system volts
= **0.00098345** voltage drop

voltage drop × 100 = **0.099834**% voltage drop

Step 6: Specify total acceptable voltage drop: **0.02** = **2.0**%

Step 7: Subtract the voltage drop for the module interconnects:

0.02 voltage drop – **0.0009834** module interconnect voltage drop
= **0.0190166** net remaining voltage drop

Step 8: Calculate the voltage drop for each circuit.

Series string to inverter:
0.90909091 proportion DC wire × **0.01901660** net remaining voltage drop = **0.01728777** acceptable DC voltage drop

Inverter to service panel:
0.09090909 proportion AC wire × **0.01901660** net remaining voltage drop = **0.00172878** acceptable AC voltage drop

Step 9: Calculate conductor resistance.

Series string to inverter:
(**0.01728777** DC voltage drop × **417.6** V$_{nominal}$ × 1,000)
÷ (**200** ft$_{round\ trip}$ × **7.2** I$_{max}$) = **5.01345455** ohms per 1,000 ft

Inverter to service panel:
(**0.00172878** AC voltage drop × **240** V$_{out}$ × 1,000)
÷ (**20** ft $_{round\ trip}$ × **10.4** A$_{AC\ output}$)
= **1.99474319** ohms per 1,000 ft

Step 10: List each wire size and its actual resistance from Table 15.2 (page 281).

Series string to inverter: **12** wire size, **1.98** ohms per 1,000 ft

Inverter to service panel: **12** wire size, **1.98** ohms per 1,000 ft

Use the ohms per 1,000 feet from above and the formula from Step 5 to calculate the actual voltage drop for each circuit.

Module interconnects: **0.00098345** actual voltage drop (from Step 5)

Series string to inverter: **0.00682759** actual voltage drop

Inverter to service panel: **0.001716** actual voltage drop

Total actual voltage drop:
0.00952703 (sum of all conductors) = **0.9527%**

Code Compliance

Module interconnects:
Required ampacity = 8 A I_{SC} × 1.56 = 12.48 A, #10 AWG wire, type USE-2 rated 194°F/90°C has an ampacity of 40 A times the ambient temperature factor of 0.41 for its 167°F/75°C location = 16.4 A. This is acceptable because it is greater than 12.48 A.

Series string to inverter:
Required ampacity = 8 A I_{SC} × 1.56 = 12.48 A, #12 AWG wire, type XHHW-2 rated 194°F/90°C, the lugs are rated 194°F/90°C. The ampacity in the 194°F/90°C column is 30 A. The ambient design temperature of 167°F/75°C requires a temperature correction factor of 0.41 (30 × 0.41 = 12.3 A). This is not acceptable because it is less than 12.48 A. Change to a #10 wire. #10 wire ampacity is 40 A (40 A × 0.41 = 16.4 A). This is acceptable.

Inverter to service panel:
Required ampacity = inverter continuous output amps × 1.25 = 10.4 A × 1.25 = 13 A, #12 AWG wire, type THHN/THWN, rated 167°F/75°C, the lugs are rated 167°F/75°C. The ampacity in the 167°F/75°C column is 25 A. The ambient design temperature of 86°F/30°C requires a temperature correction factor of 1.0 (25 × 1.0 = 25 A). This is acceptable because it is greater than 13 A.

Corrected total voltage drop for the system is 0.00697531 (0.6975%).

Overcurrent Devices (Fuses and Breakers)

DC disconnect:

$8 \, I_{sc} \times 1.56 = 12.5 \, A = 572.88 \, V_{oc}$ corrected

Buy one 600-VDC, 15-A non-fused disconnect switch per manufacturer's recommendation.

AC disconnect:

$12 \, A_{max} \times 1.25 = 15 \, A, \, 240 \, V$

Buy one 240-VDC, 15-A AC non-fused, manual, lockable, visible contact break, disconnect switch per utility requirement.

Breaker in the service panel:

240-V inverter AC output voltage at 10.4 A

Buy one 15-A, 240-VAC breaker for service panel.

Grounding Conductors

Equipment ground wires:

Module and mount frame:
Minimum #10 (overcurrent device = 15 A) (in protected area) or #10 (same size as circuit conductor) or #6 (which does not require conduit)

Inverter: Minimum #10 (overcurrent device = 15 A) or #6

Service panel:
#12 AWG conductor from inverter. Use #12 AWG or #6.

Circuit grounds:

Inverter to service panel:
#12 AWG current-carrying conductor. Use #12 AWG in conduit.

Conduit

Series string to inverter:
2 #10 AWG XHHW-2, #10 insulated ground THHN, area = 2×0.0243 in.^{2} + 0.0211 in.^{2} = 0.0697 in.^{2}

Use ½-inch conduit.

Inverter to service panel:
2 #12 AWG THHN/THWN, #12 insulated ground THHN, area = 3×0.0133 in.^{2} = 0.0399 in.^{2}

Use ½-inch conduit.

Sample Calculation #3—On-Grid System with Batteries

Step 1: See system wiring diagram, Figure 15.1 (page 311).

Step 2: List the watts, volts, and amps of all equipment.

a. Solar modules	Watts STC	125 W	
	V_{max}	17.4 W	
	I_{max}	7.2 A	
	V_{oc}	21.7 V	
	I_{sc}	8 A	
	Nominal module volts	12 V	
	Table 15.8 temperature correction factor	1.10	
	V_{oc} corrected (see V_{oc} note)	23.87 V	

V_{oc} note: For crystalline and multi-crystalline silicon modules, multiply V_{oc} by Table15.8 correction factor (page 286). V_{oc} × correction factor × quantity modules in series = V_{oc} (corrected). Use this value for all calculations using V_{oc}. For other type modules refer to manufacturer's instructions for corrected V_{oc}.

b. Series source circuit	Quantity of modules in series	4	
	Quantity of series source circuits	3	2
	Watts STC = (module watts × quantity modules in series)	500 W	
	V_{max} = (module V_{max} × quantity modules in series)	69.6 V	
	I_{max} = (same as module I_{max})	7.2 A	
	V_{oc} (corrected) = (V_{oc} × correction factor × quantity modules in series)	95.48 V	
	I_{sc} = (same as module I_{sc})	8 A	
	Nominal system volts = (nominal module volts × quantity modules in series)	48 V	

c. Combiner box (output)	Watts STC = (series string watts × qty series source circuit strings)	1500 W	1000 W
2 combiner boxes	V_{max} = (same as series string V_{max})	69.6 V	69.6 V
	I_{max} = (series string I_{max} × quantity series strings)	21.6 A	14.4A
	V_{OC} = (same as corrected series string V_{OC})	95.48 V	95.48 A
	I_{SC} = (series string I_{SC} × quantity series strings)	24 A	16 A
	Nominal system volts	48 V	48 V
d. Charge controller (skip for batteryless systems)	Current rating	40 A	40 A
	MPPT (yes or no)	No	No
	PV V_{OC} (corrected)	95.48 V	95.48 V
2 charge controllers	Nominal battery volts	48 V	48 V
	Maximum watts = (current rating × nominal battery volts or V_{max} if MPPT)	1920 W	1920 W
e. Inverter	Continuous power rating. VA (watts)	4000 W	
	Lowest DC input volts. V	44 V	
	Output volts. VAC	120 V	
	Continuous AC current output. A	33 A	
	Maximum AC current output. A	78 A	
	Efficiency %	85%	

Do not use inverter peak efficiency. Use average inverter efficiency (usually 80 to 90%) or inverter efficiency at 75% load available on the California Energy Commission's eligible inverters list.

Step 3: List circuits. Although 4 wires run from one sub-array and 6 wires run from another sub-array, the conduits from both terminal boxes to both combiner boxes may be sized uniformly to the largest diameter.

Wire	Design Temperature	Insulation Rating	Lug Rating	Number of Wires	Conduit Yes / No
Solar module interconnects	167°F/75°C	194°F/90°C	194°F/90°C	n.a.	no
Array to terminal box	140°F/60°C	194°F/90°C	194°F/90°C	6	yes
Terminal box to combiner box	140°F/60°C	194°F/90°C	194°F/90°C	6	yes
Combiner box to batteries	86°F/30°C	167°F/75°CF	167°F/75°C	2	yes
Batteries to inverter	86°F/30°C	140°F/60°C	167°F/75°C	2	no
Inverter to service panel	86°F/30°C	140°F/60°C	167°F/75°C	2	yes

Step 4: On-Grid Systems with Batteries

	quantity	each length in feet	total length feet
a. Module interconnects	3	2	6
		length in feet	× 2 = round trip feet
a. Module interconnects		6	12
b. Series string to terminal box		10	20
c. Terminal box to combiner box		20	40
d. Combiner box to controller		—	—
e. Controller to battery bank		15	30
f. Battery bank to inverter		10	20
Total DC wire length (do not include module interconnects or battery to inverter cable)		45	90
g. AC inverter to service panel		50	100
h. Sum of DC and AC wire lengths		95	190

Calculate the proportional length of DC conductors (each conductor's round-trip length divided by the sum of the DC and AC round-trip wire lengths)		
a. Module interconnects (not included in this calculation)		
	round-trip length	÷ total length (h) = proportional length
b. Series string to T-box	20	0.10526316
c. T-box to combiner box	40	0.2105632
d. Combiner box to ~~charge controller~~	—	—
e. ~~Controller to~~ battery bank	30	0.15789474
f. Battery bank to inverter (not included in this calculation)		
g. AC inverter to service panel	100	0.52631579

Step 5: a. Calculate the voltage drop for the module interconnects.

[**1.24** ohms per 1,000 ft] ÷ 1,000] × (**12** ft$_{round\ trip}$ × **7.2** I$_{max}$)

= **0.10713600** volts lost to resistance

0.10713600 volts lost to resistance ÷ **48** system volts
= **0.00223200** voltage drop

voltage drop × 100 = **0.2232**% voltage drop

b. Calculate the voltage drop for the battery-to-inverter cable.

(**4,000** inverter rated watts ÷ **44** lowest DC input volts ÷ **0.85** inverter efficiency) × 1.25 = **133.69** amps calculated input current

The formula for voltage drop is:
[**0.0608** ohms per 1,000 ft ÷ 1,000] × (**20** ft$_{round\ trip}$ × **133.69** calculated input current) = **0.162566845** volts lost to resistance

0.162566845 volts lost to resistance ÷ **48** system volts
= **0.003386809** voltage drop

0.003386809 voltage drop × 100 = **0.33868**% voltage drop

Step 6: Specify total acceptable voltage drop: **0.03** = (**3.0**%)

Step 7: Subtract the voltage drop for module interconnects and battery
to inverter cable

0.03 voltage drop – **0.00223200** module interconnect voltage
drop – **0.003386809** battery cable voltage drop
= **0.02438119** net remaining voltage drop

Step 8: Calculate the voltage drop for each circuit.

Series string to terminal box:
0.10526316 proportion DC wire × **0.02438119** net remaining voltage drop = **0.00256644** acceptable DC voltage drop

Terminal box to combiner box:
0.21052631 proportion DC wire × **0.02438119** net remaining voltage drop = **0.00513288** acceptable DC voltage drop

Combiner box to battery bank:
0.15789474 proportion DC wire × **0.02438119** net remaining voltage
drop = **0.00384966** acceptable DC voltage drop

Inverter to service panel:
0.52631579 proportion AC wire × **0.02438119** net remaining voltage
drop = **0.01·283221** acceptable AC voltage drop

Step 9: Calculate conductor resistance.

Series string to terminal box:
(**0.00256644** DC voltage drop × **48** $V_{nominal}$ × 1,000)
÷ (**20** ft$_{round\ trip}$ × **7.2** I_{max}) = **0.855480386** ohms per 1,000 ft

Terminal box to combiner box:
(**0.00513288** DC voltage drop × **48** $V_{nominal}$ × 1,000)
÷ (**40** ft$_{round\ trip}$ × **7.2** I_{max}) = **0.855480386** ohms per 1,000 ft

Combiner box to battery bank:
(**0.00384966** DC voltage drop × **48** $V_{nominal}$ × 1,000)
÷ (**30** ft$_{round\ trip}$ × **21.6** I_{max}) = **0.285160129** ohms per 1,000 ft

Inverter to service panel:
(**0.01283221** AC voltage drop × **120** V_{out} × 1,000) ÷(**50** ft$_{round\ trip}$
× **33** $A_{AC\ output}$) = **0.466625665** ohms per 1,000 ft

Step 10: List each wire size and its actual resistance from Table 15.2 (page 281).

Series string to terminal box:
8 wire size, **0.7780** ohms per 1,000 ft

Terminal box to combiner box:
8 wire size, **0.7780** ohms per 1,000 ft

Combiner box to battery box:
3 wire size, **0.245** ohms per 1,000 ft

Inverter to service panel:
4 wire size, **0.308** ohms per 1,000 ft

Use the ohms per 1,000 feet from above and the formula from Step 5 to calculate the actual voltage drop for each circuit.

Module interconnects:
0.00223200 actual voltage drop (from Step 5)

Series string to terminal box: **0.00233400** actual voltage drop

Terminal box to combiner box: **0.00466800** actual voltage drop

Combiner box to battery bank: **0.00330750** actual voltage drop

Battery to inverter: **0.003386809** actual voltage drop (from Step 5)

Inverter to service panel: **0.00847** actual voltage drop

Total actual voltage drop:
0.2439831 (sum of conductors) = **2.439%**

Code Compliance

Module interconnects:
Required ampacity = 8 A I_{SC} × 1.56 = 12.48 A, #10 AWG wire, type USE-2 rated 194°F/90°C has an ampacity of 40 A times the ambient temperature factor of 0.41 for its 167°F/75°C location = 16.4 A. This is acceptable because it is greater than 12.48 A.

Series string to terminal box:
Required ampacity = 8 A I_{SC} × 1.56 = 12.48 A, #8 AWG wire, type THHN/THWN rated 194°F/90°C, the lugs are rated 194°F/90°C. The ampacity in the 167°F/75°C column is 50 A. The ambient design temperature of 140°F/60°C requires a tem-

perature correction factor of 0.58 (55 × 0.58 = 31.9. A). This is acceptable because it is greater than 12.48 A.

Terminal box to combiner box:
Required ampacity = 8 A I_{SC} × 1.56 = 12.5 A, #8 AWG wire, type THHN/THWN rated 194°F/90°C, the lugs are rated 194 °F/90°C. The ampacity in the 167°F/75°C column is 50 A. The ambient design temperature of 140°F/60°C requires a temperature correction factor of 0.58 (55 × 0.58 = 31.9 A). An additional correction factor of 0.80 is required for 6 current-carrying conductors in conduit. 31.9 A × 0.80 = 25.5. This is acceptable because it is greater than 12.48 A.

Combiner box to battery bank:
Required ampacity = 3 × 8 A I_{SC} × 1.56 = 37.44 A, #3 AWG wire, type THHN/THWN rated 194°F/90°C, the lugs are rated 167°F/75°C. The ampacity in the 167°F/75°C column is 100 A. The ambient design temperature of 86°F/30°C requires a temperature correction factor of 1.0. This is acceptable because 100 A is greater than 37.44 A.

Battery to inverter:
Required ampacity = [(inverter rated watts ÷ lowest DC input volts ÷ inverter efficiency) × 1.25] = (4,000 ÷ 44 ÷ 0.85) × 1.25 = 133.69 A.
   4/0 AWG wire, type THHN/THWN rated 194°F/90°C, the lugs are rated 167°F/75°C. The ampacity in the 167°F/75°C column is 230 A. The ambient design temperature of 86°F/30°C requires a temperature correction factor of 1.0. (230 A × 1.0) = 230 A. This is acceptable because it is greater than 133.69 A.

Inverter to service panel:
Required ampacity = inverter continuous output amps × 1.25 = 33 A × 1.25 = 41.25 A, #4 AWG wire, type THHN/THWN, rated 167°F/75°C, the lugs are rated 167°F/75°C. The ampacity in the 167°F/75°C column is 85 A. The ambient design temperature of 86°F/30°C requires a temperature correction factor of 1.0 (85 × 1.0 = 85 A). This is acceptable because it is greater than 41.25 A.

Overcurrent Devices (Fuses and Breakers)

Combiner box:
$8 \, I_{sc} \times 1.56 = 12.48$ A, fuse rating = 15 A = 5 each 15-A fuses

GFCI between the combiner box and the charge controller:
$3 \times 8 = 24 \, I_{sc} \times 1.56 = 37.5$, GFCI rating = 40 A
(need one for each sub-array circuit)

DC disconnect:
$3 \times 8 = 24 \, I_{sc} \times 1.56 = 37.5$ A
Buy two 125-VDC, 40-A DC circuit breakers or fused disconnect switches, one for each sub-array circuit.

AC disconnect:
33 A = inverter output current
Buy one 240-VAC, 40-A fused disconnect switch with 40-A fuse.

Fuse between the battery and the inverter:
(4,000 inverter-rated watts ÷ 44 lowest DC input volts ÷ 0.85 inverter efficiency) $\times 1.25 = 133.69$ A, fuse rating = 150 A

Breaker in the service panel:
120 V inverter AC output voltage at 33 A
Buy 120-V, 40-A breaker for service panel.

Grounding Conductors

Equipment ground wires:

Module and mount frame:
Minimum #14 (overcurrent device = 15 A) (in protected area) or #10 (same size as circuit conductor) or #6 (which does not require conduit)

GFCI box:
Minimum #10 (overcurrent device = 40 A) or same size as circuit conductor or #6

Metal battery rack:
Minimum # 6 (overcurrent device =150 A) or same size as circuit conductor

Inverter: Minimum #10 (overcurrent device = 40 A) or #6

Service panel:
#4 AWG conductor from inverter, 200-A service panel
Use #6 in conduit.

Circuit grounds:

Inverter to service panel:
#4 AWG current-carrying conductor
Use #8 or #6 AWG.

Conduit

Series string to terminal box:
2 #8 AWG XHHW-2, #10 insulated ground THHN,
area = 2×0.0437 in.^{2} + 0.0211 in.^{2} = 0.1085 in.^{2}
Use ¾-inch conduit.

Terminal box to combiner box:
6 #8 AWG THHN/THWN, #10 insulated ground THHN,
area = 6×0.0366 in.^{2} + 0.0211 in.^{2} = 0.2407 in.^{2}
Use 1-inch conduit.

Combiner box to batteries:
2 #3 AWG THHN/THWN, #10 insulated ground THHN,
area = 2×0.0973 in.^{2} + 0.0211 in.^{2} = 0.2071 in.^{2}
Use 1-inch conduit.

Batteries to inverter:
2 #4/0 AWG THHN/THWN, #8 insulated ground THHN,
area = 2×0.3237 in.^{2} + 0.0366 in.^{2} = 0.6844 in.^{2}
Use 2-inch conduit.

Inverter to service panel:
2 #4 AWG THHN/THWN, #8 insulated ground THHN,
area = 2×0.0824 in^2 + 0.0366 in^2 = 0.2014 in^2
Use 1-inch conduit.

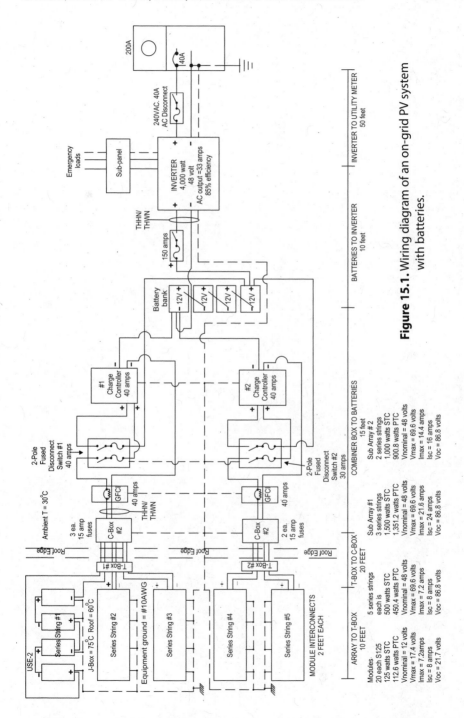

Figure 15.1. Wiring diagram of an on-grid PV system with batteries.

Chapter 16
DESIGNING YOUR PV SYSTEM: OFF-GRID

Sizing Your Off-Grid PV System

Off-grid PV systems can be either single-load or multi-load. Single-load or task-specific PV systems produce electricity on-site for a single application so sizing them is fairly straightforward. The power requirements of the load, the number of hours per day that the load needs to operate, the latitude, and sun-hours for the site are the primary factors that need to be considered. On the other hand, designing a multi-load PV system for an off-grid home demands a careful blending of science, economics, and art. Although sizing a PV system can be reduced to just the numbers, and prepackaged "PV kits" are available, the most appropriate system—the one that best meets your needs and that you will be happiest with—is the one designed specifically for your home. You must take into consideration all the electrical wants and needs that make your house your home. The different ways people use electricity, variations in climate, and the unique characteristics of available PV equipment combine to produce a wide range of system possibilities.

When designing a PV system it is important to think ahead. Anticipate your future power needs. Plan for extended periods of cloudy weather. Take the electrical consumption of the entire family into consideration. Involve everyone in the design process. We had a customer who failed to consider his teenagers' "need" to blow dry their hair every day. Consequently, his inverter was not sized to meet the family's actual load. Another customer designed and built a PV system to power a radio, telescope, and computers—his hobbies—while his wife was pumping water and washing dirty diapers by hand. Both systems had to be resized and redesigned to meet the real power demands of the entire household.

A well-designed PV system maximizes the use of equipment, performs the work you want done, wastes little, enhances the environment, is long-lived, and is within your budget.

To size your PV system, you must:

1. Determine your power requirements.
2. Match PV production in your climate to your power requirements.
3. Match battery storage for your climate and power requirements.

The steps in the PV system sizing process are interrelated and each step determines the equipment to be used. The load and your local climate conditions ultimately determine the cost of your PV system.

Determine Your Power Requirements

You have to know how much power you consume so you can know how much power your system needs to produce. The process of determining your power requirements gives you the invaluable opportunity to evaluate how you currently consume power. Are you using energy efficiently? What loads can be powered more appropriately? It is possible to maintain your standard of living and conserve energy. When you reduce power consumption, you are rewarded with a smaller PV system that costs less.

Group your electrical loads by common characteristics. List regular daily loads such as lighting, breakfast appliances (coffee maker, toaster, etc.), stereo. Make a separate list of loads used less frequently. List large power loads such as well pumps, shop tools, freezers, and other devices that have a high start-up current surge. These large loads determine the size (wattage) of your inverter. If you reduce your large loads by selecting alternative equipment, you can use a smaller inverter. A good example is using a ½-hp well pump or motor load instead of a ¾-hp pump or motor load. Or you can simply opt for a larger inverter.

One way to downsize your inverter is to stagger loads that require a lot of power. If you run the garbage disposal, the water pump, and washer while someone is using the table saw (and the refrigerator is always on), you have added at least $2,000 to the price of your inverter for the "convenience" of doing everything at once.

If you can get by with a smaller inverter, go for it. However, don't skimp on your needs and don't undersize your system. That is false economy. If you need a lot of water for your garden, overworking a small inverter and water pump to an early death is a bad bargain. Operating a marginally sized inverter at full load continuously will result in its early demise. Factor both wear-and-tear and system expansion into your design.

The Three Lists Method

An easy way to determine your power requirements is to use the "Three Lists Method." Thousands of people have used this approach, which allows you to start small, take care of essential loads, and plan for the systematic expansion of your PV system. Refer to Chapter 3, "Conservation & Energy Efficiency," for more detailed calculations.

The first list is the "Actual List," an energy audit of your present electrical power consumption. It is very important to make an actual list. Don't ignore this step, even if you plan to move from the city to the country and think you will suddenly be more energy conservative when you're off-grid. You will be the same person with the same needs and you will use similar ways to satisfy those needs wherever you live.

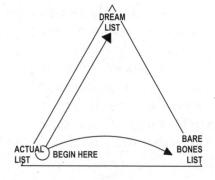

Figure 16.1. Three Lists Diagram. Make three lists of your power requirements for planning purposes. Your goal is to at least meet your Bare Bones needs as you move from your Actual (present) requirements to your Dream List. This makes upgrading your PV system orderly and helps avoid unnecessary costs.

Use the sizing worksheet on pages 337–38. Go from room to room and list every electrical device. List all the kitchen appliances. Don't forget the bathroom lights and those occasional use and seasonal appliances. List all garage and shop tools. Include well pumps and the washing machine. What about the refrigerator? Don't forget the basement—the water heater, furnace, central air conditioner, and so on.

Write down the amount of power each device consumes. You can find this information, in watts or amperes, on the identification tag or label. Also note the voltage: Some things operate at only 120 or 240 volts AC. Indicate which appliances you will use when you set up your PV system.

Write the number of minutes or hours every listed device is used each day or each week. Be accurate. You may be surprised at the numbers. Multiply the minutes or hours of use times the device's wattage (amperes × volts = watts) and put that in the watt-hours column.

Total the amounts in the watt-hours column. Multiply the daily watt-hours by 30 days and divide by 1,000 to arrive at monthly kilowatt-hours (kWh). Then compare your total to the kilowatt-hour total on your monthly electric bill. If you are not even close, check your math and check your list to see if you have missed anything, like the electric water heater or air conditioner.

When you are finished, you will have an energy audit of your electrical loads. The audit is a profile of who you are electrically. This is the first major step toward becoming your own power plant manager.

Next make your "Dream List." Now that you understand how to do an energy audit, this list should be easy, though you will need a little mental preparation. Sit in your favorite chair and close your eyes. Imagine that it is five years from now. Imagine you are in your PV dream home.

Take a mental walk through each room. List all the lights and appliances. Step outside to list power loads such as the well pump and shop tools. Be generous: Remember, this is your dream list. Be sure to include everything from your Actual List that you will still be using. If you plan to buy a new washing machine or other appliances, include them on the list, noting their actual power requirements. Your Dream List can also be used to help you shop for energy-efficient appliances. Include the number of hours you will use each device. Complete this list as you did the Actual List.

The last list is simple. It is your "Bare Bones List." Again, sit down and close your eyes. It is five years into the future, but things haven't gone so well. You are living in your PV-powered home, albeit somewhat scaled-down from your original intentions. The Bare Bones List is the least you are willing to live with. Remember that you are not on this earth to punish yourself. The Bare Bones List includes essentials and defines the lower limit of the PV system that you are going to size. Proceed as with the previous lists and total up the kilowatt-hours per day. Have every member of your household review and add to each list.

Why three lists? Few people become energy self-sufficient all at once. Change is a gradual process for most of us. With three lists, you will be able to make the transition from your present power consumption (Actual List) to future consumption and production (Dream List). The safety net of the Bare Bones List helps you set priorities: It provides a short list of what you need to do first.

With your three lists you now know your power requirements and a little more about your personal values. Are there any electrical tasks that can be done more efficiently? Perhaps some things can be done without electricity. For example, maybe it makes more sense to use a gas refrigerator or air conditioner rather than an electric appliance, especially if you plan to heat and cook with gas. View your transition to energy independence as an opportunity to investigate other renewables, such as wood heat. Learn about new energy-saving techniques and technologies. For instance, did you know that although microwave-convection ovens use more electricity when operating, their efficiency actually reduces total annual power consumption? Make an effort to learn more about energy and how to use it wisely.

Factors that Affect the Output of Your PV System

PV module output is tested at the factory under standard test conditions (STC; see page 154). The purpose of the STC rating is to provide a basis for comparison, pricing, and for the factory warranty rating. These factory test conditions do not, however, include factors such as ambient and cell temperatures and irradiance that affect the output of your modules in the real world.

Factory test temperature is 77°F/25°C. The voltage of a 36-cell crystalline solar module changes by about 0.08 volts per degree centigrade above or below 77°F/25°C. The output of amorphous solar modules does not rise and fall with temperature in the same linear manner. Still, the higher the temperature the lower the power output for all solar modules. An estimate of how your module will perform under nearly real world con-

Figure 16.2. In winter, low cell operating temperatures and sunlight reflected off snow increase array output, partly offsetting reduced output due to winter cloud cover and shorter daylight periods. (Sunnyside Solar, Inc.)

ditions is its PTC (PVUSA test conditions) test rating. PTC conditions differ slightly from STC in order to approximate "outdoor" conditions (68°F/20°C ambient temperature, and 1 meter per second wind speed at 10 meters above grade). A rule-of-thumb to estimate PTC power is to deduct 10% from the STC rating for single-crystal modules, 12% for poly- or multi-crystal modules, and 5% for amorphous modules.

Module temperature is affected by wind speed. Temperature, in turn, affects voltage. Solar modules have a greater voltage drop on windless days.

Irradiance (sunlight) also affects the output of the modules. Irradiance is affected by particulates in the air: haze, smog, clouds, and the angle of the sun on the modules. Dust and dirt on the solar array also affect the amount of solar energy that actually reaches the solar cells.

Since there are slight variations in the amperage of individual solar cells, the cells are tested, rated, and matched when they are assembled into modules. There are normal variations in the power output of individual modules. Mismatched modules have an impact on power production.

Wire losses, inverter efficiency, and battery losses (discussed in other chapters) must be taken into account when sizing your PV system.

To estimate the amount of energy that will be produced by your PV system, see the typical derating and efficiency factors in Table 16.1.

Table 16.1. Typical Derating and Efficiency Factors

Derate Factor	Derate Range	Typical Deratings
Temperature	0.95–0.80	0.90 for single-crystal Si 0.88 for polycrystalline Si 0.95 for amorphous Si
Dust and Dirt	0.98–0.90	Keep array clean for less than 5% loss
Module Mismatch	0.98–0.96	0.98 to 0.96 (2% to 4%)
DC Wire Loss	0.99–0.97	0.98 (2% or less)
Battery Conversion Efficiency	0.90–0.80	0.90 (90% coulombic efficiency)
Inverter Efficiency	0.90–0.80	0.90 for batteryless type 0.85 for battery type or manufacturer's rating at 75% load
AC Wire Loss	0.99–0.98	0.995 (1.5% or less) Keep total DC + AC wire loss below 3%

Table 16.2. PV Array Azimuth Derating Factors

		East 90.0	ESE 112.5	SE 135.0	SSE 157.5	South 180.0	SSW 202.5	SW 225.0	WSW 247.5	West 270.0
		Azimuth Degrees								
Degrees Latitude	20	0.93	0.96	0.98	0.99	1.00	0.99	0.97	0.95	0.92
	25	0.88	0.93	0.96	0.99	1.00	0.99	0.97	0.93	0.89
	30	0.85	0.91	0.96	0.99	1.00	0.98	0.97	0.89	0.83
	35	0.81	0.89	0.95	0.99	1.00	0.99	0.95	0.89	0.81
	40	0.76	0.86	0.93	0.99	1.00	0.98	0.92	0.84	0.75
	45	0.73	0.84	0.92	0.98	1.00	0.98	0.91	0.82	0.72

Table 16.3. PV Array Tilt Derating Factors

		Degrees Latitutde					
		20	25	30	35	40	45
Tilt Angle	0	0.96	0.94	0.90	0.90	0.86	0.81
	5	0.98	0.97	0.94	0.93	0.90	0.86
	10	0.99	0.98	0.96	0.96	0.93	0.89
	15	1.00	1.00	0.98	0.98	0.96	0.93
	20	1.00	1.00	0.99	1.00	0.98	0.95
	25	0.99	1.00	1.00	1.00	1.00	0.97
	30	0.98	0.99	1.00	1.00	1.00	0.99
	35	0.97	0.98	0.99	1.00	1.00	1.00
	40	0.94	0.96	0.98	0.99	1.00	1.00
	45	0.91	0.94	0.96	0.97	0.99	1.00
	50	0.88	0.90	0.95	0.95	0.98	1.00
	55	0.84	0.87	0.94	0.93	0.98	0.95
	60	0.79	0.82	0.87	0.89	0.93	0.97
	65	0.74	0.78	0.83	0.85	0.90	0.94
	70	0.69	0.73	0.78	0.80	0.86	0.91
	75	0.63	0.67	0.73	0.75	0.82	0.88
	80	0.58	0.61	0.68	0.69	0.77	0.84
	85	0.52	0.55	0.62	0.63	0.71	0.80
	90	0.47	0.50	0.56	0.57	0.65	0.75

Using these factors, a 125-watt DC STC single-crystal cell module will produce 94.38 module watts AC in a batteryless system. Thus, the DC-to-AC derating is typically 75%. In a system with batteries, the same module will be further derated to 80.2 watts AC or, typically, 65% of its STC-rated power.

Solar array orientation obviously affects the output of your PV system. The ideal array orientation in the northern hemisphere for maximum annual PV production is true south and tilted to the latitude degrees minus 5 to 15 degrees. For example, Bakersfield, California, is located at 35.4 degrees north latitude with the array tilt "sweet spot" at 20 to 30 degrees.

If your roof or solar array location is not facing true south and the pitch is not the ideal tilt, you can tilt your array to the best azimuth and angle. If you cannot ideally orient your solar array or want your array to match your roof pitch, you can factor in the loss using the charts in tables 16.2 and 16.3. A solar array that is not pointed ideally will still perform well. For example, on a house in Bakersfield that has a roof facing southwest at 225 degrees true south azimuth and tilted 20 degrees, a stand-off solar array will still produce 99% of maximum annual output. The same orientation in New York City (40.7 degrees latitude) will produce 90% of maximum output compared to the ideal south azimuth at a 25 to 35 degree tilt.

Reflected light from snow will increase array power output by approximately 10%. Lower temperature also increases module efficiency. Conversely, a solar array on a dark roof or in the middle of a black asphalt parking lot will have a lower than average output.

Match PV Production to Consumption

Now that you know your power requirements, you have to determine how much power to produce to meet your needs. This is simple math based on the output of the PV modules and the efficiency of the power conditioning and distribution equipment that you select. Here are some guidelines to help simplify your system and reduce costs.

The time of year and the time of day you use power make a difference in the size and configuration of your system. You may live in a hot climate and need cooling fans or you may have irrigation requirements.

These seasonal loads can be partly met by using a solar tracker or by manually adjusting the array. Lighting and other nighttime loads increase your dependency on batteries.

In temperate zones there is usually twice as much sun in summer as there is in winter. Tracking the winter sun in the United States only increases PV production by 10% to 15%; however, the 30% to 50% increase in production that comes from using a tracker in summer can make a difference in meeting increased summer loads and determining the size of your array.

If you live in a unique microclimate, one that differs from the general locations in this book (such as a place that gets significantly more rain than a location 50 miles away), you must allow for that difference. Coastal fog versus inland sun or increased solar radiation at higher elevations can make a significant difference.

Battery Storage

Battery storage sizing is a mathematical process, but where you locate your batteries is a major factor in determining the size of your battery bank.

If you live in a cold climate, it is worth the extra effort to put your batteries where they will stay at 70°F/21°C to reduce battery bank size. The same principle in reverse holds true in hot climates. Keep your bat-

Figure 16.3. While building your home, your temporary living quarters can be PV-powered. This small solar array provides all of the electrical needs for the trailer and the power tools.

teries cool to prolong their life. One way to do this in the desert or the tropics is to put the batteries under the house or partially buried in a well-drained battery compartment.

Your battery bank must be able to handle motor start-up current loads. The rule of thumb is to size your battery bank's ampere-hour capacity to at least four times your largest start-up current load. For example, if you have a 100-ampere start-up load, the battery bank should be at least 400 ampere-hours (20-hour rate).

Start-up current rating is usually listed on equipment. If a motor's start-up rating is not known, figure that it is at least four times the operating current.

Power Conditioning & Regulation

The type of charge controller you use depends on your array size and the special features you want. If you have DC loads, you might want a low voltage disconnect to protect your battery bank from being damaged by unintentional deep discharge.

Perhaps your PV system produces excess energy for part of the year. In this case a load diversion option may be appropriate. Most charge controller manufacturers offer meters for monitoring system voltage and charging current. Automatic starting for a generator and temperature compensation are also available. Consult with your supplier for recommendations.

DC vs. AC & Wiring

Some people prefer to use DC appliances to avoid the cost of an inverter and its associated power loss. However, DC wiring must be large enough to handle the current. Undersized wiring will rob you of power.

Current losses in the same size wire are ten times lower for 120-volt AC wiring than for 12-volt DC wiring. In all cases keep wire runs as short as possible.

If your PV array must be located far from your house, put your array, battery bank, and inverter as close together as possible and run higher voltage AC to your home.

Sizing: An Example

The power requirements used for this sizing example are similar to those of many of our clients. If the example seems either austere or overly consumptive, that's okay. We all have our own ideas of what is essential and what is excessive.

The example PV system is designed for a couple who moved from the city to the country. Their property is two miles from the nearest power lines, but this did not deter them. It is a beautiful piece of land with fine trees, an excellent house site, clean water, and the right price. The road is good and the nearest neighbors are well out of sight.

They are self-employed, both doing part-time the same work they did full-time in the city. They have also started to develop new income-earning skills. They do not need to commute. They like to garden and grow a lot of their vegetables.

In preparation for the move, over the past few years they have been replacing their appliances with sturdy, energy-efficient models. Their water well is 180 feet deep. They built a large storage building that will eventually be their barn and workshop, after their house is built. For now, they live in a trailer. They use propane for cooking, for their water heater, and for back-up heat. Wood from the property is their main heat source. A gas clothes dryer is used in winter and on rainy days. They will use a propane refrigerator for the first few years while saving for an energy-efficient electric refrigerator/freezer and additional PV modules. While they don't mind living in a cramped trailer during construction, they don't want to rough it. Rather than pack all their appliances and tools in storage they have put them to immediate use.

Let's look at their electrical power requirements. Refer to the PV System Sizing Worksheet at the end of this chapter (page 337). To calculate the daily energy consumption of each load device, use the formula:

quantity × rated watts × hours used per day
× days used per week
÷ 7 days = watt-hours per day consumption

Following the guidelines from Chapter 3, the total power consumption for the home is 3,973 watt-hours per day (or 3.973 kWh/day). This amount does not include tool use or the future electric refrigerator/freezer.

Load / Device	Number	Watts	Hours per Day Used	Days per Week Used	Daily Watt-Hours
Living Room					
Lights (reading)	2	22	2	7	88
Lights (area)	1	40	2	7	80
Television	1	100	4	7	400
Radio/stereo	1	100	2	5	142.9
DVD/VCR	1	100	2	2	57.1
Kitchen					
Lights (task)	2	40	3	7	240
Lights (area)	2	20	0.5	7	20
Microwave oven	1	1,450	0.5	3	310.7
Toaster	1	800	2 min	7	26.7
Mixer	1	150	5	5	8.9
Juicer	1	125	30 sec	7	1
Food processor	1	450	10 sec	7	1.25
Vent fan	1	88	0.5	2	12.6
Bath					
Lights	2	40	1	7	80
Bedroom					
Lights (reading)	2	20	1	3	17.1
Lights (area)	1	40	1	7	40
Appliances					
Washing machine	1	512	2	1	146.3
Dryer (motor only)	1	300	1	1	42.9
Iron	1	1000	0.5	1	71.4
Vacuum cleaner	1	600	0.25	2	42.9
Sewing machine	1	600	0.25	2	42.9
Equipment					
Water pump	1	1050	2	7	2,100
				Total	3,973
Future Needs					
Refrigerator/freezer for system planning purposes at 50% duty cycle	1	360	12	7	4,320

Tools

During construction, tools are powered by the generator providing battery charging for nighttime shop tool use. Once the house is built the number of hours of tool use is unknown.

Let's analyze power consumption by type of load. Water pumping is their largest power requirement. Using a standard ½-hp submersible pump and pressure system, this household will get 600 gallons per day. The pump draws 8.75 amperes at 120 volts or 1,050 watts. It pumps five gallons per minute into a 30-gallon pressure tank. Five gallons per minute equals 300 gallons per hour. Could they get by with less water? Possibly. Two 10-minute showers per day and a water-saving showerhead that delivers a good spray at two gallons per minute equals 40 gallons. At the washbasin they might use an additional two to five gallons. The toilet could be a water-saving unit requiring only 1.6 gallons per flush

Using two basins to wash the dishes—one for soapy water, one for rinse—consumes less water than rinsing dishes under a running facet. The garden could be gravity watered by their spring-fed pond. With proper planning and water conservation, this couple could cut their daily average water consumption—and the number of hours their pump needs to run—by half or more.

Lighting loads are the next greatest power requirement, totaling about 566 watt-hours per day. This is their average consumption because lighting use varies greatly throughout the year. In winter, more time is spent indoors; shorter days mean more artificial lighting is required in the morning and early evening. In summer, lighting needs are reduced.

When thinking of energy conservation, some people's immediate response is to get rid of kitchen appliances and gadgets. If this couple eliminated their toaster, mixer, juicer, and food processor, their daily load would be reduced by 32.55 watt-hours. Hardly worth the inconvenience since they already own and use these things. Their microwave oven could be replaced by the gas stove, although with rising propane costs and possible fuel shortages, that could mean they'll be cooking on a wood stove in the future. Since a move to the country often means a significant reduction in income, if our couple does not buy enough PV now to power the microwave, it may be several years before they will be able to afford it. They decide to keep the microwave.

Water pumping and lighting make up over two-thirds of the power requirements for this household. Conservation could reduce water consumption to the point where the combined loads are lowered to one-half the total kilowatt-hours. However, we will use their initial load calculation for this system sizing example.

The loads determine the system DC and AC voltages. Because this system's loads are all AC, the inverter DC input voltage determines the PV system primary voltage.

Inverter Sizing

A true sine wave inverter enables you to use all standard appliances as well as any computers, rechargeable tools and appliances, and stereo equipment that may not work with, or could be damaged by, a modified sine wave inverter. Selecting a 4,000-watt inverter gives our example family flexibility both in the appliances and tools they can use now, as well as in the ability to add loads to their system later.

The largest load in the example home is the water pump. The pump was selected to (1) serve their needs, (2) have readily available and serviceable parts, (3) have a low initial cost, and (4) allow for a medium-sized inverter. A 2,500-watt inverter can power a ½-hp pump motor, but a 4,000-watt inverter will handle the pump and additional loads.

The largest combined load for the example will be the water pump and washing machine. All loads will be managed to limit the largest combined or simultaneous loads from exceeding the capacity of the inverter. While it may be possible to set up a gravity water storage system instead of pumped water for the washing machine, this may not be practical or convenient. When the laundry is being done, the water pump will cycle on and off. The combined laundry and water pump starting current demands should be within the inverter's capability. Be sure that the inverter you select can handle high current loads at least four times its rating.

The inverter selected has a 48-volt DC primary input voltage. The inverter input voltage determines the DC voltage of the solar electric system. Nominal 12-volt solar modules must be used in series groups of four for 48 volts. Twelve 125-watt modules (1,500 watts) will work fine. If an inverter with a 24-volt DC input were used, the array would be wired for 24 volts.

Since a generator will be used during construction, an inverter with battery charging capability is a good choice, although a separate battery charger will also work. Be sure to check the quality and efficiency of the battery charger to optimize the generator runtime. Automatic or manual transfer between PV/inverter to generator/battery charger is required.

Solar Array Sizing

Once you know your total daily power requirements and the DC primary voltage, you can calculate the sizes of the PV array and battery bank and determine what other equipment you need.

Use the peak sun hour maps in figures 16.4 and 16.5 or the table in Appendix A to calculate PV array size. Local climate and microclimates can differ 10% or more from the charts and yearly weather may differ even more.

Joel's mountaintop property in Arkansas had three distinct microclimates. The north slope was tree covered and cold; the top of the mountain was windy with greater temperature swings; the south slope was semi-arid and hot in summer. Vegetation, soil moisture, and solar radiation differed greatly on his 32 acres. Don't let a similar situation discourage you from seeking accuracy in sizing. The validity of climatic averages has been proven over decades by solar electric homeowners.

Another source of climate data is your local PV equipment supplier. However, it is important that you do your homework and be a knowledgeable shopper. Use your supplier's information to double-check your calculations and use your numbers to double-check your supplier.

While most suppliers are ethical, there are some who will sell you what they have rather than what you need. If there is a big difference between your numbers and the supplier's, ask for an explanation. If the explanation makes sense (i.e., is based on experience and more accurate local climate data), use it. If not, buy elsewhere.

Our example's power requirements are 3,973 Wh/day. The location is 40° north, 106° west near Granby, Colorado. The peak sun-hour map indicates a yearly average of 5.5 peak sun hours per day. The winter peak sun hours are about 4.5. The 125-watt solar module used as the example in Chapter 9 has been selected because of availability, size, and price.

Sizing the system for winter with the highest consumption and the lowest insolation, the number of modules required to meet winter consumption is:

125 watts (module rating) × 4.5 (winter sun hours)
= 562.5 Wh/day per module

562.5 Wh × 0.9 (temperature derating) × 0.98 (dust and dirt derating)
× 0.98 (module mismatch derating) × 0.98 (DC wire losses)
× 0.9 (battery conversion efficiency) × 0.85 (average inverter efficiency)
× 0.99 (AC wire losses)
= 360.86 Wh/day per module actual production

3,972 Wh/day ÷ 360.86 Wh/day = 11
(which we will round up to 12 each 125-watt modules to match the
48-volt DC primary input voltage)

Figure 16.4. Peak sun hours per day.
Also see Appendix A.

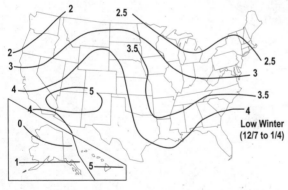

Because the homeowners use all AC appliances, inverter losses as well
as wire and battery losses have been factored in. The calculations are for
a PV array facing south with a mount that can be adjusted seasonally.
In winter the angle of the array is adjusted to degrees latitude plus 15
and in the summer it is lowered to degrees latitude minus 15. Seasonal
mount adjustments and tracking will increase annual production over
25% at this location.

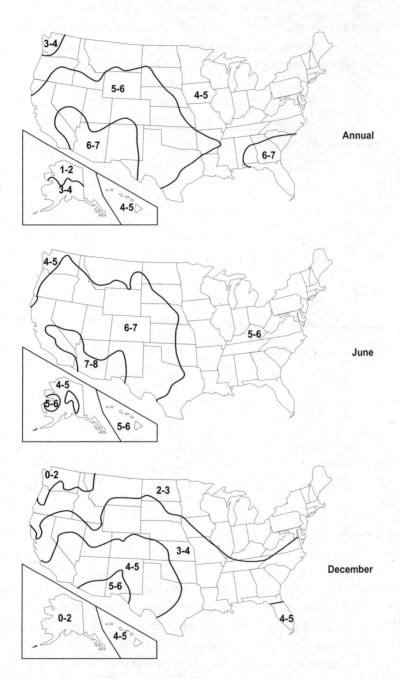

Figure 16.5. Insolation average kWh/m²/day on an array at latitude tilt. Also see Appendix A.

Sizing Your Battery Storage

Next we'll size the battery bank. There are many ways to calculate the amount of battery storage you will need. The basic rule is that you will need 10 ampere-hours of battery storage for every 1 ampere-hour of daily load requirement (a 10 to 1 ampere-hour ratio) to give you three days of battery storage.

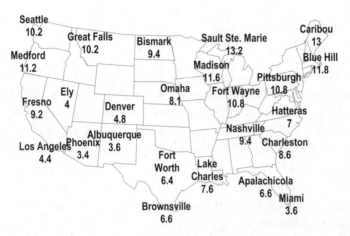

Figure 16.6. Battery storage requirements—days of autonomy.

The battery storage requirements map, Figure 16.6, shows that a system located near Denver will need 4.8 days of battery autonomy. The battery autonomy formula is:

daily watt-hours × days of autonomy
÷ 0.5 (correction for 50% depth of discharge)
× battery temperature correction factor (see Table 16.4)
= battery storage watt-hours

Use this formula for deep-cycle batteries kept at 60°F/15.6°C to 70°F/ 21.1°C. If the battery room is unheated, the battery compartment should be insulated. Normal charging produces some heat. Use a high–low thermometer (available for approximately $30) to see the temperature swings. Use the correction factor from Table 16.4 to adjust your battery bank capacity for the compartment's temperature.

Table16.4. Battery Temperature Correction Factors

Temperature °C	Correction Factor	Temperature °F
-10	1.10	+14
-15	1.55	+5
-20	2.05	-4
-25	2.75	-13
-30	3.50	-22
-35	4.25	-31

The example system, which uses 3,973 watt-hours per day, will need 38.14 kilowatt-hours of battery storage for 100% autonomy ($3,973 \times 4.8 \div 0.5 = 38,140 \div 1,000 = 38.14$). If they use golfcart batteries, like the Trojan T-105 (217 ampere-hours $\times$ 6 volts = 1,302 watt-hours), they will need 29.3 batteries. The system's primary DC voltage is 48 volts, which is provided by eight 6-volt batteries connected in series.

If there were no clouds and no seasons you would only need enough battery storage to last through the night. In the real world you need to design for more battery storage.

First, there will be cloudy days when the array will not be able to fully recharge the batteries. You must have enough capacity to last through several cloudy days in a row. Five days of autonomous storage at 50% depth of discharge is the minimum for most of the continental U.S.

Second, there are fewer hours of sunlight in winter to power your loads and keep your battery bank charged. Your battery bank should have at least as many ampere-hours of reserve capacity as it does of autonomous storage. This means you need an ampere-hour of reserve storage for each ampere-hour of autonomy capacity, which is why we recommend no more than 50% depth of discharge instead of 80%. In all cases, PV production must be sized to match consumption in order to "fill" your batteries. An oversized battery bank that cannot be fully charged on an average daily basis will become sulfated and be ruined.

Third, battery performance has its limits. Frequent deep discharging shortens battery life. Sulfation (growth of lead sulfate crystals on the battery's plates) and stress fatigue can be prevented by avoiding discharging below 50% in winter and regular equalization charging.

Your batteries will get weaker over the years, but they will last longer if protected from extremes of heat and cold. Batteries can be safely installed in a house, in an enclosed crawl space, and even in a well-drained and vented underground chamber to take advantage of the home's heat and the earth's thermal mass.

These battery bank design factors add up to a lot more than one night's storage. In sunny climates, battery capacity should be 7 to 10 times greater than the average daily load. Cloudy climates can require 20 to 30 times the daily load. If you area has a history of weeks of cloudy weather, you will need the additional battery storage.

Figure 16.7. A 4-kW PV array and battery bank provide all of the power to this custom-designed home located in eastern Massachusetts. The house also features direct solar gain, super insulation, heat mirror glazing, and active domestic water and space heating. (Photo: Solar Design Associates)

The Art of System Sizing

Designing a PV system is a process that requires matching production to consumption, equipment to task, and cost to budget. There is an art to finding the right balance. The process starts with the preliminary design. You may have to revisit and revise your original plan in order to refine your design.

The preliminary design for the example system is:

Production: 12 each 125-watt solar modules
Storage: 29.3 each golfcart batteries
Regulation: 1 each 40-ampere charge controller
Conditioning: 1 each 4,000-watt inverter

You can't have three-tenths of a battery. The system requires 48 volts; the battery selected is 6 volts. Therefore, 8 batteries in series are needed to get 48 volts. Four parallel strings of 8 each golfcart batteries in series will provide 41.66 kilowatt-hours storage. Battery manufacturers do not recommend more than four parallel strings.

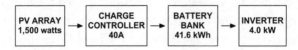

Can we increase PV production? Yes. Putting the modules on a tracker will increase winter production by 10% to 15% and summer production by as much as 55% for a yearly gain of 25%.

Sizing Your Own PV System

You can design your own PV power system based on your location, your power requirements, equipment characteristics, and your budget.

While system sizing is fresh in your mind, grab your calculator and design a PV system to suit your needs. If you are energy conservative, you will be pleased to see how few modules and batteries are needed. If you have large power requirements, do not despair. Here is some information to sleep on before you juggle the numbers again.

Budget System Sizing

Many people find that the system they "need" is beyond their budget. Rather than give up on energy independence, take another look at your budget. It should include both your initial cash outlay and what you will save over the next two, three, or more years.

Our example couple from Granby, Colorado, has budgeted $11,000 for their solar electric system. If they allow for savings gained by not using the generator as often, they can pull together $12,500.

Now look at equipment costs. Using 2006 figures, a 125-watt solar module costs approximately $500. T-105 batteries cost $100 each. A 12-module tracker costs about $1,500. The charge controller costs $150 and the inverter costs approximately $3,000. The total system will cost about $14,500 with wiring and miscellaneous hardware included.

This is over their budget. What can be done? It is false economy to downsize the inverter or charge controller or reduce the battery bank. The correct size inverter and charge controller will be needed later when they plan to expand their system. The batteries will last at least six years, probably ten. If only half the batteries are installed for the same load, they would most likely last three to five years. If it is expected that the couple's income will increase within five years, economizing now by buying fewer batteries may be okay. But if more money is not likely to be available for new batteries in five years, we recommend shopping for a better battery price. Perhaps direct purchase from a large battery distributor will save money. The 25% additional power gained by using a tracker justifies its cost.

There is a relationship between system costs and module costs. Solar modules generally account for half the cost of the major components for an off-grid PV system.

The balance of the system (all equipment except modules) costs $8,500. With $4,000 they can afford eight solar modules, which, on the tracker, will meet 80% of their winter power needs. They might consider buying an eight-module system now and adding modules later, if they are willing, initially, to use the generator a few hours a day in winter.

If our couple buys fewer modules, they might also be tempted to downsize their battery bank. Unfortunately, it's not that simple. If daily power needs are drawn from a smaller battery bank, the batteries will be cycled more often and probably more deeply. This means that the batteries will age faster and need replacement sooner. It also means greater reliance on the generator.

What is the real cost of a downsized budget system? Running the generator regularly means fuel and service expenditures (oil changes, repairs, parts), and a shorter generator life. It can also mean not having enough power if the generator fails. And fail it will. Parts may be miles or days away; repairs can be costly and time-consuming.

In review, system sizing is based on your power requirements. Climate data is used to predict average solar array performance. Battery storage is based on the number of days of autonomy required for your climate. The charge controller must be able to handle the solar array current. The inverter must be sized for the largest combined loads. The wiring must be sized to carry the current.

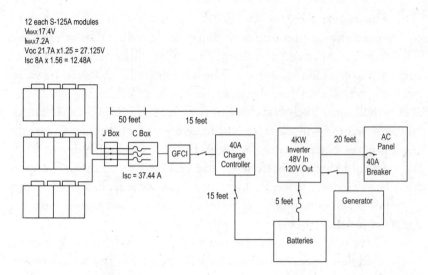

12 each S-125A modules
V$_{MAX}$ 17.4V
I$_{MAX}$ 7.2A
Voc 21.7A x1.25 = 27.125V
Isc 8A x 1.56 = 12.48A

Figure 16.8. One-line wiring diagram for the Granby, Colorado, system.

There is no simple or single answer to "What is the right design for my PV system?" Compromises may have to be made between what you want and what you can afford. It takes planning and ingenuity to get started in PV on a limited budget, but millions of people have done it. So can you.

PV SYSTEM SIZING WORKSHEET

Step One: Calculate total daily load

Device	Qty		Watts (volts × amps)		Hours per Day Used		Days per Week Used		Average Watt-Hours per Day
		×		×		×		÷7	
		×		×		×		÷7	
		×		×		×		÷7	
		×		×		×		÷7	
		×		×		×		÷7	
		×		×		×		÷7	
		×		×		×		÷7	
		×		×		×		÷7	
		×		×		×		÷7	
		×		×		×		÷7	
		×		×		×		÷7	
		×		×		×		÷7	
							Total Watt-Hours per Day		
Highest AC Surge Current									

Step Two: Adjust total watt-hours per day by derating factors (Table 16.1) and array azimuth (Table 16.2) and array tilt (Table 16.3) factors:

Example:

Watt-Hours per day ÷ 0.90 (temperature) ÷ 0.98 (dust and dirt)
÷ 0.98 (module mismatch) ÷ 0.98 (DC wire loss)
÷ 0.90 (battery conversion efficiency) ÷ 0.85(inverter efficiency)
÷ 0.99 (AC wire loss) ÷ 1.00 (array azimuth) ÷ 1.00 (array tilt)
= Adjusted Daily Watt-Hours

_____ Watt-Hours per day
÷ 0.90 ÷ 0.98 ÷ 0.98 ÷ 0.98 ÷ 0.90 ÷ 0.85 ÷ 0.99 ÷ 1.00 ÷ 1.00
= _____ Adjusted Daily Watt-Hours

Step Three: Calculate the required array peak watts using yearly average or worst month peak sun hours. See peak sun hour maps (Figure 16.4) or insolation chart (Appendix A).

Adjusted Daily Watt-Hours ÷ Peak Sun Hours = Array Peak Watts

_____ ÷ _____ = _____ Array Peak Watts

Step Four: Calculate the number of solar modules required. For 24-volt DC systems using 12-volt modules round up to the nearest number divisible by two. For 48-volt DC systems using 12-volt modules round up to the nearest number divisible by four.

Array Peak Watts ÷ Module Wattage Rating = Number of Modules

_____ ÷ _____ = _____ Modules

Step Five: Calculate the battery bank size. Use the Days Autonomy multiplier for battery storage requirements from the map in Figure 16.6.

Daily Watt-Hours x Autonomy Multiplier
÷ 0.5 (correction for 50% depth of discharge)
× Battery Temperature Correction Factor (Table16.4)
= Watt-Hours of Storage Needed

_____ × _____ ÷ 0.5 × _____ = _____Watt-Hours

The battery's rated storage capacity (in ampere-hours) times the battery voltage equals the battery watt-hours.

Battery Ampere-Hours × Battery Voltage = Battery Watt-Hours

_____ × _____ = _____Battery Watt-Hours

Watt-Hours of Storage Needed ÷ Battery Watt-Hours
= Number of Batteries Needed

_____ ÷ _____ = _____Batteries

(Round up to the nearest even number of batteries in a series string to equal the system voltage.)

Chapter 17
DESIGNING YOUR PV SYSTEM: ON-GRID

Sizing Your On-Grid PV System

There are two types of on-grid PV systems: systems without battery storage (batteryless) and systems with batteries. Batteryless PV systems connect in parallel through your utility service panel to the electrical grid. The PV electricity they produce displaces the utility electricity you consume. On-grid PV systems with batteries are connected to the utility grid in the same manner. They also displace electrical consumption and the batteries function as an "uninterruptible power supply" (UPS) to provide electricity when grid power is interrupted. Both systems automatically disconnect from the electrical grid when a grid power failure occurs and both can be net metered.

Here are eight questions to answer before you start to design your on-grid PV system.

1. What do I want my PV system to do for me?
2. What are my daily energy requirements (kWh/day)?
3. Where will I install my PV system?
4. How much room do I have for the solar array?

5. What is my budget?
6. Given my budget and available space, what will this PV system do for me?
7. Do I need an uninterruptible power system (UPS)?
8. Will my PV system be net metered?

Defining why you want a PV system and what you expect it to do are the first steps in sizing your on-grid PV system. It is very tempting to say "I want to eliminate my entire utility electric bill." This may not be a realistic goal when you consider your available space, budget, and energy requirements. Once you decide the type and size of system you want (based on your space, budget, and energy requirements) you will be able to determine its impact on your electric bill.

Knowing your power requirements is as important for an on-grid PV system as it is for an off-grid system. Determining your daily energy consumption is explained in chapters 3 and 15. Even though you may not be relying on your PV system for 100% of your energy, you will be choosing the percentage of your grid consumption you want to offset with PV.

You probably won't want to produce more electricity than you consume. Each state handles excess generation differently. In some, excess generation is purchased by the utility, but at the wholesale rate. In other states excess energy is forfeited to the utility either on a monthly or an annual basis. A few utility companies pay a premium for PV power.

Look at your property and select a location to install your solar array and balance of system (BOS) equipment. The same guidelines for site selection outlined in Chapter 10 apply. Ideally, the array location should face true south, be unshaded and without obstructions, and allow for an optimum array tilt. You need about 110 square feet for each kW of crystalline silicon PV.

PV modules come in fixed sizes. If you plan to put the PV array on your roof, look at what is already there. Vents, vent stacks, skylights, and other rooftop fixtures may have to be moved. Roofs with multiple planes and pitches might have only a small area suitable for PV even if the house is very large.

If the roof of your house is unsuitable, is there space for a south-facing pergola, awning, or carport? What about the garage roof? Could you install your PV array on a ground or pole mount?

Even if you do not have a suitable south-facing site at the optimum tilt you can still install a PV array. Chapter 16 explains how to calculate the energy production for a less than ideal location.

Once you decide where to install your PV array and how many PV modules will fit in the space, you can estimate the system's output. The calculation is similar to the system sizing calculation in the previous chapter.

Module wattage rating (from manufacturer)
× number of modules
× peak sun hours (from Figure 16.4 or the data in Appendix A)
= watt-hours per day

Watt-hours per day
× 0.90 (single crystal cell temperature derating)
× 0.98 (dust and dirt derating)
× 0.98 (module mismatch derating)
× 0.98 (DC wire loss)
× 0.9 (battery conversion efficiency, use only for systems with batteries)
× 0.85 (inverter efficiency for systems with batteries or
0.90 for systems without batteries,
or use manufacturer's rated efficiency at 75% load)
× 0.99 (AC wire loss)
= Adjusted daily watt-hours

Adjusted daily watt-hours
× correction factor for azimuth (from Table 16.2)
× correction for sun angle (from Table 16.3)
= watt-hours per day estimated energy production

Example 1

A homeowner decides to install a 2.5-kW PV system that has twenty 125-watt solar modules (described in Chapter 9). The home is located in Los Angeles. The array will be installed on a carport and face true south at 20° tilt. It will be a net-metered, batteryless system. Yearly average insolation at latitude -15° is 5.5 sun hours (from Appendix A).

125 W × 20 modules × 5.5 sun hours = 13,750 watt-hours per day

13,750 watt-hours per day × 0.90 (temperature derating)
× 0.98 (dust and dirt derating)
× 0.98 (module mismatch derating)
× 0.98 (DC wire losses)
× 0.90 (inverter efficiency for batteryless type)
× 0.99 (AC wire loss)
= 10,377 watt-hours per day

The system has no batteries and is a good tilt angle and ideal azimuth orientation, so these factors do not have to be considered. The average estimated output of the system will be 10,377 watt-hours per day (10.37 kWh/day).

Example 2

Example 1's next-door neighbor decides to install the exact same batteryless system. But this system will go on the roof of the house, which faces southwest at a 15° pitch. Use tables 16.2 and 16.3 for array azimuth and tilt factors.

10,377 watt-hours per day
× 0:95 (correction for southwest orientation)
× 0.98 (correction for tilt angle)
= 9,660 watt-hours per day (9.66 kWh/day)

While mechanically identical to the system in Example 1, this system will produce about 9.3% less electricity annually.

Example 3

Another neighbor installs the exact same array (their roof also faces southwest at a 15° tilt), but unlike the other two, they include batteries for a UPS.

125 W × 20 modules × 5.5 sun hours = 13,750 watt-hours per day

13,750 watt-hours per day × 0.9 (temperature derating)
× 0.98 (dust and dirt derating)

$\times$ 0.98 (module mismatch derating)
$\times$ 0.98 (DC wire losses)
$\times$ 0.90 (battery conversion loss)
$\times$ 0.85 (inverter efficiency for battery type)
$\times$ 0.99 (AC wire loss)
$\times$ 0.95 (orientation correction)
$\times$ 0.98 (tilt angle correction)
= 8.21 kWh/day

Even though this system is identical to the systems in examples 1 and 2, its annual production will be nearly 21% less electricity than system 1 and 15% less than system 2.

If the array were mounted vertically on the southwest face of the home, this system would produce 4.67 kWh per day.

8.21 kWh/day $\times$ 0.57 (orientation correction) = 4.67 kWh /day

Mounting an array vertically is not unheard of. A building in Santa Monica, California, that has won architectural awards for outstanding ecological design has panels mounted vertically on a west wall that is partially shaded by palm trees. The Condé Nast building in New York City also has vertically mounted solar arrays (see photo page 388).

System design compromises are often unavoidable, but they can be quantified. If the electrical rate for these three homes is 15¢ per kWh, in a 30-day billing cycle, the home in Example 1 produces (10.37 kWh $\times$ 15¢ $\times$ 30 days) = $46.66 worth of electricity. Example 2 produces $43.47 worth of electricity, and Example 3 produces $36.94 worth of electricity. The design concessions made on the second home (not factoring in inflation or interest) only cost about $38 per year in lost production of electricity.

Do I Really Need an Uninterruptible Power Supply (UPS)?

Batteries are necessary for an off-grid PV home for two reasons. The most obvious is that batteries will provide electricity on cloudy days and at night. The second reason is that inverters designed for off-grid use require batteries to operate. Some off-grid inverters can also be used for on-grid systems. The wiring diagram in Figure 17.4 shows the design for a residential, on-grid, net-metered system with battery backup in Los Angeles. This is our PV-powered home.

All grid-connected PV systems must automatically disconnect from the grid during utility power outages. This anti-islanding safety feature prevents your power from feeding into utility lines that utility personnel may be working on. All UL-listed inverters designed for utility interconnection meet the anti-islanding requirement. If you have a batteryless inverter your PV system will not operate during power outages. A grid-connected PV system with batteries will isolate the array, inverter, and batteries from the power grid while household circuits routed through a subpanel will continue to receive power.

We've noticed that the first thing people tend to do when their electricity goes off is to look out the window to see if the neighbors are also in the dark. We confess to smiling during power outages while our neighbors flip circuit breakers on and off and look up their electrician's telephone number until we remind them that we have a PV power system with battery backup. Our small battery bank is enough to power four circuits overnight if we limit our electrical consumption. In the morning, when the sun rises, our PV system powers our home and recharges the batteries.

There is a downside to using a grid-connected PV system with an inverter and batteries. You can lose as much as 10% of your PV production due to battery conversion efficiency. If you feel that utility power outages are frequent and inconvenient enough to require a UPS, or if you want to protect sensitive equipment from power outages and poor quality utility electricity, the added cost of a battery system and lower PV production is worth it. There is, however, an alternative.

Your UPS can be freestanding and independent of the PV system. A separate inverter and set of batteries can be sized to meet the needs of selected loads. The batteries can be trickle-charged by the electrical grid. The battery bank sizing worksheet at the end of Chapter 16 (start at Step 5 on page 338) will help you size your UPS.

One reason our PV system has batteries is that it was one of the first systems installed in California when the first rebates went into effect. Affordable batteryless inverters for grid-tied systems were not available at the time. If our PV system were designed today it would have a batteryless inverter and a separate UPS. When we expanded our PV system in 2007, we used a batteryless inverter (see pages 84 to 90).

Will My System Be Net Metered?

Whether your system is net metered or not will depend on the laws in your state. Net metering makes a difference in system design. If your system is not net metered, all of your PV power will feed into a subpanel that is completely separate from your grid-connected metered panel. The load circuits powered by a PV system that is not net metered must be isolated from the wiring in the rest of your house. The isolated electrical loads will not be powered by the utility, only by the PV system and batteries.

Putting Your PV System Together

Designing a code-compliant PV system is fairly straightforward if you follow the flow of electricity from the photovoltaic array to its point of use.

Example 1: A Batteryless PV System

This system is located in Los Angeles, California, at 34° north latitude. The house is 2,000 square feet with a 1,000-square foot south-facing roof at a 20° pitch. The homeowners were using 20 kWh/day, but through conservation they reduced their consumption to 16 kWh/day. The system is net metered and has no batteries. They would like to produce 100% of their electrical consumption.

The inverter selected requires a DC input of 250 to 600 volts; the maximum AC output is 2,500 watts. The PV module is the 125-watt module used in the examples throughout this book (see pages 159–60). The maximum input power for this inverter is 3,000 watts DC. This works out to 24 modules (3,000 watts ÷ 125 watts). The 24 modules cost more than the homeowners had budgeted, so they decide to try to use 22 modules, which still meets the inverter DC voltage input requirement. The V_{max} voltage for a 22-module series string is 376.2 volts (22 modules × 17.1 volts each).

There are 5.5 peak sun-hours per day (yearly average) at this location. The 22 modules will produce:

22 modules × 125 watts × 5.5 (peak sun hours)
= 15,125 watt-hours per day (15.13 kWh)

15.13 kWh × 0.90 (temperature derating)
× 0.98 (dust and dirt derating)
× 0.98 (module mismatch derating) × 0.98 (DC wire losses)
× 0.90 (inverter efficiency for batteryless type)
× 0.99 (AC wire losses)
= 11.42 kWh per day

The 11.42 kWh per day falls short of their 16 kWh per day goal, so the homeowners decide to install 24 modules (12.45 kWh/day) as determined by the original calculations. Though still short of their goal to be 100% solar-powered, this does encourage them to consider buying more energy-efficient appliances to replace their old ones.

The next step is to draw a block diagram of the system (see Figure 17.1). This will help you visualize what equipment and materials are needed and to calculate wire and fuse sizes.

The homeowners are planning to install the array on the roof, which has space for two rows of twelve modules. The inverter will be mounted on an outside wall of the house, next to the utility service panel.

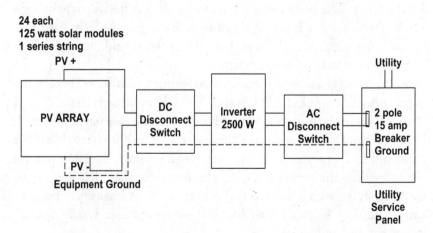

Figure 17.1. Block diagram of an on-grid batteryless PV system.

Note that there are both DC and AC disconnect switches, as required by the *National Electrical Code* and their local utility, as well as a 15-amp 250-VAC breaker in their service panel. There was not enough space or capacity for the dedicated PV system breaker in their old 100-amp service panel so they hired an electrician to replace it with a 200-amp service panel.

There is no requirement for lightning protection in this area. The wiring diagram (Figure 17.2) includes all the equipment needed for their system. Next they have to determine the wire and conduit sizes required to meet the *National Electrical Code* and include them on the wiring diagram in their building permit package. Refer to Chapter 15 for this information.

Example 2. A Grid-Connected System with Batteries

This system has a few more parts. Although net metered, it has battery backup for emergency power. For the sake of simplicity and comparison, we will use the same house, but a slightly different approach to calculating the system design. The homeowners want to produce 16 kWh per day on an annualized basis.

16 kWh per day ÷ 0.90 (temperature derating)
÷ 0.98 (dust and dirt derating)
÷ 0.98 (module mismatch derating)
÷ 0.98 (DC wire losses)
÷ 0.90 (battery conversion loss)
÷ 0.85 (average inverter efficiency for battery type)
÷ 0.99 (AC wire losses)
= 24.94 kWh/day

24.94 kWh/day ÷ 5.5 peak sun hours = 4.53 kW STC

4.53 kW STC ÷ 0.125 kilowatts per 125-watt module
= 36.24 modules (rounded down to 36 modules
to match the 48-volt inverter input of nine parallel strings of
four 12-volt modules in series)

An alternative method for an approximate calculation is to divide the amount of electricity they want to produce by the overall efficiency of a

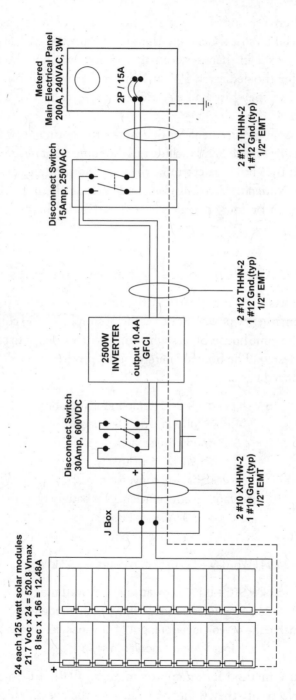

Figure 17.2. Wiring diagram of an on-grid batteryless PV system.

PV system (65% for a system with batteries or 75% for a system without batteries) and then divide by the number of sun hours at their location to arrive at a system size that meets their requirements.

$$16 \text{ kWh/day} \div 0.65 = 24.6 \text{ kWh/day}$$

$$24.6 \text{ kWh/day} \div 5.5 \text{ peak sun hours} = 4.47 \text{ kW DC STC}$$

$$4.47 \text{ kW DC STC} \div 0.125 \text{ STC kilowatts per 125 watt module} = 36 \text{ modules}$$

The owners decide to use 32 modules, operating at 48 volts DC to match inverter input voltage and power rating as well as their budget. The modules will be connected in eight parallel strings of four modules in series.

$$32 \text{ modules} \times 125 \text{ watts} = 4{,}000 \text{ watts STC}$$

The total watts (STC) is 4,000. This will require a 4,000-watt inverter that can also handle up to 4 kW of emergency loads when utility power fails and their system switches to off-grid mode. (Had they decided to use 36 modules they would have needed a larger inverter.) Each series string will be 500 watts STC (440 watts PTC) with a $V_{nominal}$ of 48 volts, V_{max} of 69.6 volts, and I_{max} of 7.2 amps.

The eight combined parallel series strings have an I_{max} of 57.6 amps. The charge controller is rated at 60 amps. This includes the NEC 125% safety factor and the 125% for conditions when irradiance is greater than 1,000 W/m².

The homeowners need to decide how much back-up battery storage they want. They decide to move two circuits to a new subpanel and make a list of the devices on these circuits. The loads total almost 5 kWh per day. They decide that one day of back-up battery storage for power outages is adequate in their area.

Using Step Five from the PV System Sizing Worksheet in Chapter 16 (page 338), they calculate:

$$5 \text{ kWh daily watt-hours} \times 1 \text{ day}$$
$$\div 0.98 \text{ (DC wire loss)}$$
$$\div 0.90 \text{ (battery efficiency)}$$
$$\div 0.85 \text{ (inverter efficiency)}$$
$$\div 0.99 \text{ (AC wire loss)}$$
$$= 6.73 \text{ kWh}$$

6.73 kWWh ÷ 0.5 (for 50% depth of discharge)
= 13.46 kWh desired storage

They select sealed gel cell, 12-volt, 86-ampere-hour batteries.

86 amp hours × 12 volts
= 1,032 watt-hours storage capacity per battery
= 1.032 kWh

13.46 kWh needed storage ÷ 1.032 kWh per battery = 13.04 batteries

This is a 48-volt DC system so the 12-volt batteries must be in series strings of four. They buy 16 batteries, which also gives them additional autonomy (16 × 1.032 = 16.5 kWh).

The next step is to draw the block diagram and the one-line diagram. The homeowners are now ready to apply for their building permit, contact their utility for an interconnect agreement, and order their equipment. Every residential PV system design follows these same basic steps.

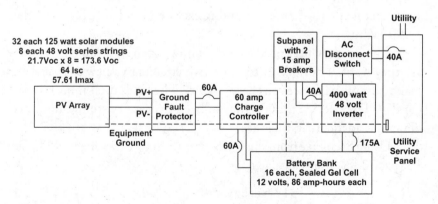

Figure 17.3. Block diagram of an on-grid PV system with battery backup.

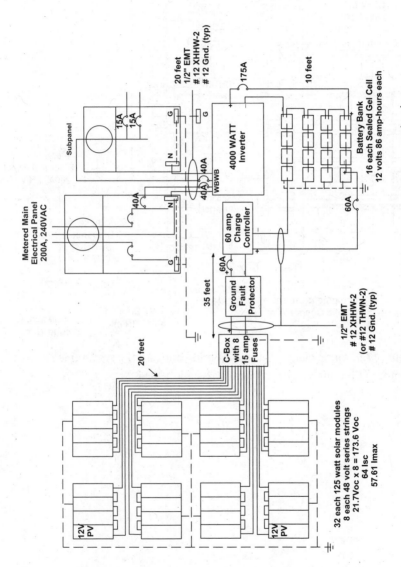

Figure 17.4. Wiring diagram of an on-grid PV system with battery backup.

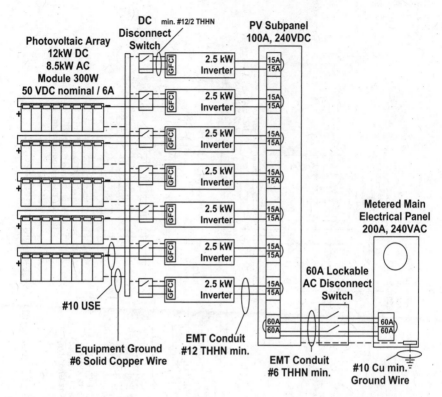

Figure 17.5. Wiring diagram of a 12-kW grid-connected PV system. (See system photos in Chapter 6, figures 6.1 and 6.2, page 108.)

Chapter 18
COMBINING PV & OTHER ELECTRIC GENERATORS

Many people who now use solar electricity previously used other electric power sources. When first moving to a new location with no utility power, some homeowners used generators or even automobile or truck batteries to provide power for appliances, tools, lighting, and music.

Several years ago, there was a growing cottage industry that supplied dual battery kits for remote cabins. Daily driving charged both the vehicle starter battery and a separate deep-cycle battery by splitting the output of the vehicle alternator with a battery isolator. The kit also included a junction box and plug so that the home loads could be plugged into the second battery. Do-it-yourselfers devised similar dual-battery systems (see Figure 18.1).

Eventually solar modules were added to the vehicle charging system so that the time between trips to charge the battery could be extended. Then the batteries were moved from the vehicle to a vented battery box in the home as the addition of more PV modules eliminated the need for the dual battery system.

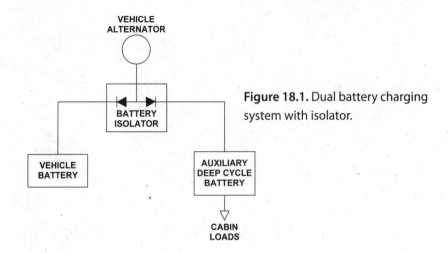

Figure 18.1. Dual battery charging system with isolator.

Some people who move away from the grid use a gasoline, diesel, or pro-pane generator (genset). The generator powers tools during construction as well as large appliances and water pumps. Generator capacity can be utilized more efficiently by adding a battery charger, battery bank, and inverter. Now the generator can be used to charge batteries while power-ing other things.

Charging batteries while using a lightly loaded generator increases fuel efficiency. At night or when loads are light, the generator can be turned off for some quiet time because the battery bank and inverter can power the loads. The result is less generator running time and less frequent generator servicing and repairs. Solar modules can be added as the budget permits to further reduce generator runtime.

Eventually, solar modules replace the generator altogether. If your timing is good, the generator will last as long as it is needed. However, many people use a generator until it becomes a financial burden or they get frustrated with the noise, exhaust fumes, and repairs.

A few people have dabbled with thermoelectric generators. These solid-state devices, a semiconductor thermocouple, have been used by homeowners with high heating requirements. Some thermoelectric units operate on propane and can be modified to run on biogas. Some low watt-age units attach to the firebox of a wood, coal, oil, or gas heater.

In water-abundant mountain locations, hydropower can be used in conjunction with photovoltaics. The hydroelectric power plant can be

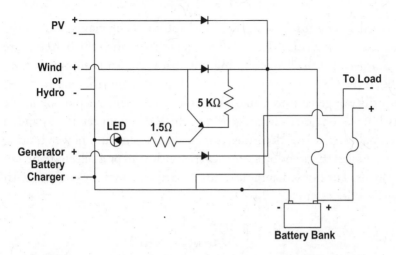

Figure 18.2. This circuit diagram shows three 12-volt inputs feeding into a single battery bank. The blocking diodes isolate each power source to prevent backfeeding. The LED comes on when power is being delivered.

fairly large with the familiar dam and spillover, or it can be a small genera-tor attached to a pipe from a spring or even a kitchen faucet. For those who have gravity-fed water and modest power requirements, a faucet-attached generator makes sense. Imagine turning on the tap to produce electricity. Of course, this system will work only when the home is located near a controlled flow of water.

Small human-powered generators resembling exercise bicycles have been linked to a PV-charged battery bank. You can build a pedal-powered generator from spare bicycle parts and a small generator. But be forewarned: If you use a car alternator or generator, you've got a steep hill to pedal. A half-hour on an electric exercise cycle may be all you can do in a day. The average human can generate up to one-third horsepower for a relatively short period of time. At 100% efficiency, this means you could pedal over 200 watts into your battery bank.

The Standby PV System

An autonomous PV system, also called a standby PV system or unin-terruptible power supply (UPS), can be used in conjunction with either utility power or a generator. PV charges batteries that are also charged by the utility or a fossil-fuel generator. Automatic or manual switching can be used to prioritize the PV. This means that when the batteries are fully charged and the sun is shining, any loads will get their power from the sun. If the loads are greater than the PV production, they are switched to the utility or generator. Thus, PV can be used to offset a portion of the utility bill. An obvious advantage is that the PV system can be expanded to eventually replace the dependency on utility or generator power. During power outages, the standby system carries either all or a selected portion of the loads. The block diagram, Figure 18.3, shows a standby system with a transfer safety switch.

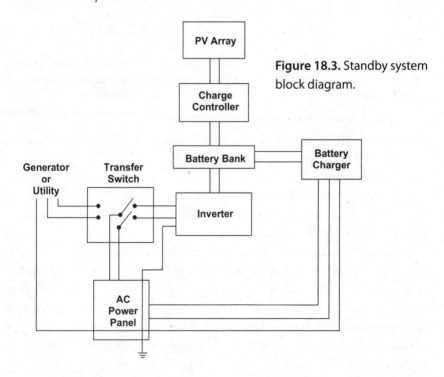

Figure 18.3. Standby system block diagram.

Operation of a standby PV system is straightforward. Power comes from two sources—utility (or genset) and PV. When the storage batteries have been fully charged by PV and the other source—utility (or genset), power will come from PV if the sun is present. The system acts like a UPS in this mode by "simulating" a utility (or genset) outage, and PV takes over. In the event of a real utility (or genset) power failure, the system automatically switches over to PV for the duration of the outage or the capacity of the battery storage. A properly designed standby PV system should have adequate battery storage. If the battery bank is too large to be fully charged by the PV system, the batteries will sulfate and can be ruined. If the battery bank is too small, the standby system will not optimize generator runtime or provide the required stand-alone autonomy.

Installation of the standby PV system is straightforward as well. Figures 18.6 and 18.7 show the equipment room and block diagram for a standby system that provides power for a home in California. PV is set up as usual. Either manual or automatic switching can be used to tie the PV system to the generator. A battery charger, either built into the inverter or separate, completes the necessary equipment.

Some inverters and power panels can take power inputs from multiple sources. They can also automatically start a fossil-fuel generator if the battery voltage falls below a preset point. Preassembled or partially assembled power centers simplify the installation of PV systems tied to a utility or a generator making it possible for a handy person with some electrical skills to install a PV and utility or generator PV system.

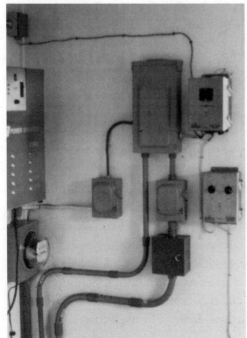

Figure 18.4. Wiring and controls for this system provide safety cut-outs, charge control, and monitoring. (Talmage Engineering, Kennebunkport, ME)

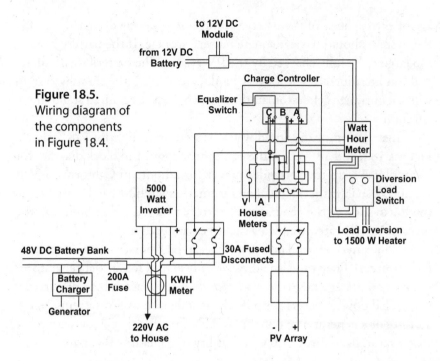

Figure 18.5.
Wiring diagram of
the components
in Figure 18.4.

We have coached thousands of do-it-yourselfers by e-mail, mail, and telephone. If you buy quality equipment from a knowledgeable supplier and are willing to take the time to do a neat, safe, and thorough job, you can install a preassembled power center or individual components. Doing it yourself will give you a better understanding of the equipment and its capabilities.

It doesn't matter if your set-up is called a standby PV system, photo/genset, UPS, or genset plus; they are all essentially the same multi-source power supply.

Some remote power system designers justify using a fossil-fuel generator with PV by stating cost-effectiveness and greater generator efficiency, but this practice is only a temporary expediency. The justification does not stem from the efficiency or cost-effectiveness of the genset, but from the peripheral equipment (battery charger, battery bank, and inverter) used with the generator to reduce its running time. Genset noise, service, repairs, fossil fuel consumption, and the resulting costs in dollars, resource depletion, and pollution must be considered. The most cost-effective and efficient approach is to totally eliminate fossil-fuel generators.

It should be clear that we are against the use of fossil fuel genera-tors. However, we have included this section because we do not live in a perfect world. We all must go through a transition period on our way to becoming 100% solar. We hope you will use this information not to justify the use of a generator, but to help in your transition to clean, renewable power.

Genset Costs

Initially, a generator can be less expensive than a solar electric system. A 4.5-kilowatt generator costs less than $3,000. A small 2.5-kilowatt generator costs less than $800 plus a few hundred dollars for a fuel stor-age tank and wiring.

How much does it cost to own and operate a generator? Depending on the circumstances, output from a 2.5-kilowatt generator—if operated efficiently at or near peak loading—will cost $1.50 to over $2.00 per kilowatt-hour based on life-cycle cost of fuel, maintenance, repairs, and replacement over a 20-year period. In comparison, an off-grid PV system costs between 30¢ and $1.50 per kilowatt-hour.

Occasionally, a PV system designer is asked to design a PV and generator power system when the power requirements are not known. This is often unavoidable in new house construction. That was the case with a home located in a utility-serviced suburban community. The home-owners wanted to produce their own power (Figure 18.6). An 800-watt tracking solar array was mounted on the roof. Storage consists of an 800-ampere-hour industrial-grade battery bank. A 2.5-kW inverter with load demand provides household power. For back-up power, a 4-kW propane generator and battery charger were also installed.

As it turned out, the homeowners' power needs fall almost within the system's production. From March to September, solar power pro-vides for all their electrical needs. In December the generator operates two hours per week to keep the batteries topped off. During January, February, October, and November, only one hour per week of generator runtime is needed.

The homeowners could have eliminated the generator and battery charger altogether if they had analyzed their power requirements. That money could have been used to expand the solar array and battery bank.

In locations where winter cloud cover is extreme, a small back-up generator may be needed.

Consider your present and future power requirements carefully. You may find that a fossil-fuel generator is unnecessary if you add a few more solar modules and pay attention to your power consumption. You can go from an energy user to an energy producer and skip the fossil-fuel generator step entirely.

Figure 18.6. The equipment room of a Fresno, California, home. From left to right: generator-powered battery charger, 2.5-kW inverter, charge controller, DC panel, AC panel, and array disconnect switch. Twelve 2-volt forklift battery cells are located in the battery box below the AC breaker panel. (Don Loweburg, Offline Independent Energy Systems)

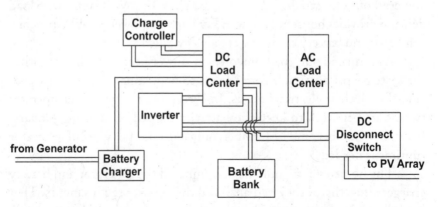

Figure 18.7. Block diagram of equipment shown in Figure 18.6.

Wind & Solar

When Joel first started using homemade electricity, small homestead-sized windchargers, similar to windchargers used in the U.S. before rural electrification, were starting to reappear. Although controls and construction materials had been improved, these windchargers were basically the same as those made in the 1930s.

For a short time, Joel tried to use wind power, but he didn't like climbing the tower to service the moving parts. Good windchargers are dependable, but cheap chargers are pure aggravation. Owning a windcharger always means more work than PV. Wind generators need regular servicing and repair, the frequency correlates to the weather and the quality of the wind generator.

If you are considering wind power and your budget is limited, you can buy a new or used unit for under $2,000 or, if you are good with your hands, you could build your own. The windcharger at Anderson's General Store, Guemes Island, Washington (see figures 18.8, 18.9, and 18.10), was built by Solar Energy International students.

Finding a good site for a windcharger can be difficult. Most people who think that they have "lots of wind" do not. Buy or lease wind monitoring equipment to evaluate your site. Windcharger dealers often provide monitoring service for a nominal fee. Some state energy offices or colleges will assist you as well.

Table 18.1. Wind Speed Indicator

Wind Speed (mph)	Wind Effect
0–1	Smoke rises vertically.
2–3	Direction of wind shown by smoke drift but not by wind vanes.
4–7	Wind felt on face, leaves rustle. Ordinary wind vane moves.
8–12	Leaves and twigs in constant motion. Wind extends a light flag.
13–18	Raises dust and loose paper. Small branches are moved.
19–24	Small trees in leaf begin to sway. Crested wavelets form on inland waters.
25–31	Large branches in motion, whistling heard in power lines. Umbrella use is difficult.

Figure 18.8. Anderson's General Store, Guemes Island, Washington, gets its power from PV, wind, batteries, and the utility grid. PV production from eight ground-mounted 140-watt modules averages 90 DC kWh/month. Wind power from the home-built 500-watt generator on an 80-foot tower adds another 59 DC kWh/month. (Anderson's General Store, Solar Energy International, Sagrillo Power & Light, and *Home Power*. Photo: Jonathan Prescott)

Figure 18.9.
A 3.6-kW inverter provides net-metered PV and wind power to Anderson's General Store. In addition to PV and windpower charge controllers, above the DC and AC power panels is an electric heating element connected through a load controller to use excess power during windy periods (Anderson's General Store, OutBack Power Systems, and Solar Energy International. Photo: Ian Woofenden)

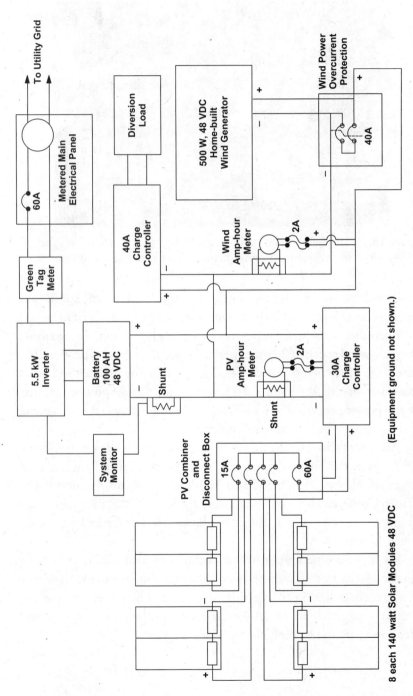

Figure 18.10. Wiring diagram of Anderson's General Store PV and wind power systems (figures 18.8 and 18.9).

Figure 18.11. The Talmage home in Maine has a hybrid system using wind and PV for all of its electrical needs. (Talmage Engineering, Kennebunkport, ME)

Wind speed at 30 feet and higher is often 1½ times the speed measured at ground level. At 120 feet, wind speed can be twice that measured on the ground. Most wind generators are designed to deliver maximum power at a wind speed of 30 miles per hour and the power is proportional to the cube of the wind speed. This means that the power production curve is very steep. At 15 miles per hour, the amount of power produced is ⅛ the power produced at 30 miles per hour. The lowest starting wind speed for most windchargers is 7 miles per hour.

Another factor in selecting a windcharger site is the proximity and height of surrounding buildings and trees. To avoid turbulence, a windcharger should be mounted 30 feet higher than any obstruction within 300 feet. Commercially available towers are 31 to 90 feet in height. Some building codes require that the tower be situated twice its height from any property lines.

PV has become a mainstream technology and wind power is big business, but using renewable power sources is about individuals seeking alternatives to pollution and resource depletion. There are many paths to becoming energy self-reliant. A combination of solar power, wind power, hydropower, biogas or biofuels can work for you.

Chapter 19
INSTALLATION, TESTING & MAINTENANCE

Installation Guidelines

Every PV installation is different, but some basic guidelines will help you think through your installation and avoid problems. As in all building projects, it is important to list each step before starting work. Keep an ongoing list of ideas and suggestions. Finalize your materials list so that when you start you have all the hardware, fasteners, wires, and parts that you need at hand. This is particularly important if stores are a long drive away. An unfinished installation is an invitation to trouble. You can be sure that rain or snow will catch you at the most inopportune time.

The key rule is keep it simple and uncluttered. Your system will change with your needs, and a straightforward, accessible installation will make future modifications easier. As always, safety first. Make electrical wiring and connections in accordance with codes and proper practices. Make sure connections are mechanically tight and that you use the right fasteners. Soldering requires special skill, but don't be afraid to learn. Wire nuts and terminals will also work if used properly. Visit an electrical supply house and see their selection of wire connectors, stand-offs, and terminal blocks. Become familiar with what is available in your area. Ask questions and tell them about your PV system. Most people will be curious and they will want to help.

Equipment Location

The first step in the installation procedure is selecting the location for your equipment (review the site selection section in Chapter 10). Most manufacturers provide free installation guidelines and instructions. Read their recommendations carefully. It is in everyone's interest that the equipment you buy be used to its best advantage. Most companies offer assistance by telephone or email. Distributors and dealers will also help. It is wise to buy your equipment from someone who actually uses the hardware that they sell.

Site constraints require additional consideration. If the electricity from the PV array is low-voltage DC, you must take into account the resistance in long wire runs from the array to your home. In general, setting up a pole or ground mount a distance from your house and running bigger cable is a better choice than cutting down a shade tree. However, there are practical limits to array location and point of use versus cable size and cost. Check the wire size chart (Chapter 15) and local wire prices to determine the most practical distance between your array and your house. Consider underground cable instead of unsightly and potentially unsafe overhead wire runs.

You will have to work on your installed equipment periodically, so keep easy access in mind when siting the array, battery bank, and other equipment. A short wire run to a dark corner of a crowded closet is false economy. Accident, injury, and just plain aggravation will make you wish you had spent the extra money for longer and bigger wire to reach a place where you can move freely around your battery bank and power center. Allow at least three feet of open workspace in front of equipment. See Chapter 12 for a complete discussion of battery location.

Again, safety first! Be sure your array and batteries are disconnected when you work on your system.

The location of the charge controller, inverter, and controls is important. Controllers and inverters require proper ventilation to keep the internal parts cool. Dirt and dust on controls and contacts can be a problem. Your charge controller and inverter must "breathe." Install them in a dust-free area. Vented and rainproof electrical enclosures are available in all sizes for equipment mounting and protection. Don't put your controls in or near a closet where someone might use them for a clothes hanger. Never locate your controls and regulator directly above a heat source or near flammable materials.

Don't forget to protect your batteries, open wires, and fasteners from falling objects. Children and the curious can be all hands. Keep things out of reach, safe, and secure.

A note on mounting: Simple temporary ground mounts—legs attached to the array frame—may be needed for testing purposes. It is extremely risky to use a temporary mount for too long. We have used fixed and adjustable ground and roof mounts as temporary testing mounts. The fixed roof mounts were attached to wood stand-offs of 2"×4" pressure-treated lumber blocks or to channel steel or aluminum angle bolted to the roof rafters. Then the array was fastened with bolts to the stand-offs. The adjustable array was mounted in the same way except that metal legs at the sides of the array were added to make tilt adjustments. Look at a few billboards and signs to see what kinds of metal frames, angle iron, tubing, and hardware are used to withstand wind and snow loads in your area.

This is obvious, but of utmost importance for mounts: Be sure to fasten the array securely to the roof, pole, or ground (Chapter 10). The best way to check the array mount is to grab it and pull with all your weight. Try hanging on it to test the strength. A periodic inspection of the fastenings is essential.

Off-Grid PV Installation: An Example

The example PV system in Chapter 16 consists of:

> 12 each 125-watt solar modules
> 32 each 6-volt batteries
> 40-ampere charge controller with digital meter
> 4,000-watt inverter

Frame, Modules & Mounts

The solar array will be mounted on the roof of the house. The primary voltage to power the inverter is 48 volts DC. The batteries and other equipment are in a shed on the north side of the house. There are no local restrictions or zoning regulations regarding height, size, or appearance of arrays. This house is in a snowy region, with a roof pitch of 55° angled to the south with all-day access to the sun.

The dimensions of the modules are 24.5"×56.3". The modules have weatherproof interconnect boxes on the back, which make wiring easy. Since the modules have aluminum frames, stainless-steel hardware and aluminum angle are selected for the array frame to prevent electrolysis corrosion caused by dissimilar metals. Given the choice of wood or metal array framing, always choose metal. Wood mounts do not provide an earth ground or lightning protection path and are subject to weathering, which weakens them. The array may not be seasonally adjusted, but the mounting hardware is made to be adjustable because the cost difference is small and the increased energy production may be wanted later.

Here a stand-off mount is used, the simplest and least costly configuration for a roof of the proper tilt. To allow for a summer tilt of 25° (40° latitude -15°), the top of the array is hinged with the adjustable lower legs bolted and stowed under the array frame. The array is situated on

Figure 19.1. Solar array assembly of prewired modules.

the roof so that two people on separate ladders can reach it safely from the eaves to raise the array to 25°. In doing so, the adjustable lower legs drop into position to be fastened to the angle aluminum stand-offs.

All of the stand-offs are securely fastened to the roof framing members at the array corners and at no more than 4-foot centers, using ⅜-inch stainless-steel lag bolts. Silicone caulk is used to make the roof penetrations watertight. The stand-offs are 2 inches high, enough to provide underside module cooling in warm weather without creating a large snow dam in winter.

The mounting rack can be made in a shop and all hardware (hinges, stand-offs, and fasteners) assembled on the ground to ensure fit and proper alignment. While on the ground, check the modules for proper fit. Redrilling module mounting holes when you are on the roof can be dangerous. Test the modules electrically on a sunny day while they are still on the ground. Write down each module's V_{OC} and I_{SC} and the test date, along with other pertinent information, such as weather conditions.

With all the hardware and tools on hand and at least two helpers, mount the array on the roof. It is a good idea to have a few extra people nearby when the array is being installed, but wait until the roof and

electrical work is finished before having a party.

The modules can be wired after they are on the roof because the upper edge of this frame is hinged, providing access to the underside of the modules. If your array mount is not hinged, the modules already fastened to the mount have to be prewired on the ground. A preassembled array can be very heavy and difficult to handle on ladders and on the roof.

Figure 19.2. A preassembled solar array being moved into place. (Sunnyside Solar, Inc.)

Equipment Room

With the frame and modules safely stowed, awaiting the day they will be mounted on the roof, it is time to work in the equipment room. First, make a drawing and lay out the location of the equipment. The inside wall adjoining the house is selected for the sturdy battery rack made of angle steel. Wood racks can be used, but they must be strong enough to support the weight of the batteries. The first shelf of batteries is a few inches off the ground to make cleaning underneath easy. Each shelf is high enough to allow a hydrometer to be used to check the batteries. Always make it easy to service your equipment.

Batteries are placed carefully on the shelves with proper polarity so that wire runs are as simple and as short as possible. Batteries are heavy. These thirty-two 6-volt batteries weigh over 4,000 pounds and are configured in 4 parallel strings of 8 in series. To leave room for servicing, the homeowner put 16 batteries on the two lower shelves and 16 batteries on the top shelf. This means that each shelf must be strong enough to hold a ton.

Another option for stowing batteries is to build an enclosure like the battery box detailed in Chapter 12 and described on page 222.

The battery bank can now be wired. Take extra precautions when working with batteries. Keep tools clear of the terminals and wires. The batteries are first wired in series groups using #4/0 AWG USE cable with the proper crimped and soldered connectors. This means each string of eight batteries will have seven 1-foot long battery interconnects connecting positive to negative. Next, the series strings are wired in parallel. Make short parallel connections so that all four strings are paralleled.

These four sets of eight 6-volt batteries connected in parallel make one large 48-volt battery. This battery bank has enough power to run an all-electric home for one or two days. It is also enough power to weld thick steel plate and cause severe burns. Use insulated tools or wrap wrenches with insulating electrical tape. A smoke detector in the equipment room is a good investment. (See Chapter 12 for information on venting and other precautions.)

Next, the combiner box, GFCI, charge controller, and disconnect switches are mounted on the wall between the battery bank and where the wires from the solar array enter the room. The inverter is mounted on the wall or secured to a shelf as close to the batteries as possible. If AC power is needed to finish the installation, connect the wire from the battery bank to the inverter to get power from the battery bank. (Be sure to recharge the batteries daily.) The inverter must be turned off before you make these connections. Follow inverter installation instructions to the letter and test the unit before you put it to use. A safety fuse or breaker between the inverter positive output and the battery bank is required.

Installing the Array

With the battery bank, charge controller, switches, and fuses installed, you are ready to work on the roof array and put the system into operation. Prior to installation, check the roof framing to make sure it is adequately braced. Although this array weighs less than 400 pounds, additional rafter braces were nailed in the attic to support the weight of the workers.

On a nice windless day, when the roof is dry, assemble the mounting hardware, frame, modules, and necessary tools. Use wooden or plastic ladders, scaffolding, and extreme caution if other live wires are near the work area. Take special care when working on the roof. Use safety ropes. For the groundworkers' sake, don't toss things from the roof. Avoid damaging the roofing material.

Locate the roof framing members and mark the placement for the stand-offs. If you are not sure how to do this, hire someone skilled in roofing, carpentry, and wiring. When the stand-offs are secured, make sure all roof penetrations are sealed and waterproofed.

PV modules produce power as soon as the sun strikes them. Securely cover the modules with blankets or heavy paper and keep them covered until all connections have been made. Carefully pass the modules one by one up to the roof and bolt each one into place, noting layout and polarity. If the array is preassembled on the ground, use ropes and plenty of help to lift the heavy array onto the roof and hold it in place while fastening it down. Do not twist or bend modules or panels when handling them.

After the #10 AWG module interconnect wires are connected, the next step is to hook up the array wire. As the module junction box terminals cannot accommodate larger than #10 AWG wire, a separate junction box is located on the roof directly above or to the side of the array, fastened to the roof and sealed to prevent leaks. A fused combiner box, required by code, can also serve as the separate junction point. The modules wired in series strings are connected to the combiner box where the series strings are paralleled. In the combiner box, put a neat one-inch wire loop in each string wire positive lead to allow DC current readings to be made with an amp clamp meter. Use wire and conduit clamps to securely fasten the interconnect wires and conduit running to the combiner box to the roof or the array mount. Most inspectors require conduit to be fastened on blocks.

The array mount is earth-grounded with #6 AWG wire and the ground wire is run into the combiner box, which should have three separate array terminal blocks for array positive (+), array negative (-), and ground. Wire or conduit can be run either through the inside or outside of the house to the equipment room in accordance with the National Electrical Code and local codes.

In the equipment room, connect the array to the wall-mounted fused array disconnect switch. A #6 AWG ground wire is run, securely fastened along its length, to the ground rod located at the house's electrical service panel. Chassis ground wires from the controller, the inverter, and other equipment are also run to the ground rod.

A second 40-ampere fused disconnect switch from the charge controller to the battery bank is also wall-mounted. Cables from the battery bank are run to the inverter. With the array switch off, you are ready to wire the array through the controller and the disconnect switches to the battery bank. Even though the array is covered, you should double-check voltage and polarity at each step. The voltage at the array disconnect switch should be the array's open-circuit voltage of 84 volts nominal (4 5 approximately 21 volts). The battery's nominal voltage is 48 volts.

Follow the charge controller instructions. The connection sequence is typically:

1. Connect battery negative to controller
2. Connect battery positive to controller
3. Connect array negative to controller
4. Connect array positive to controller

Now you can remove the coverings from the array. You have installed a safe, well-grounded solar electric system, and the batteries are now being charged. The example system's monitoring meter will read between 17 and 21 amperes on a sunny day.

With the inverter off, the next step is to connect the house to the solar electric system. AC wiring from the inverter to the home distribution circuit breaker box should be in conduit and wired in accordance to standard house wiring codes and practices. With the house service panel breaker off, turn on the inverter according to the manufacturer's instructions. Then turn on the house panel breaker and test the house circuits for proper AC voltage and polarity before plugging in appliances. The installation is now complete.

Tracker & Generator Installation

The homeowners could decide to install the solar modules on a tracker or add a generator. At the selected tracker site site, dig a hole, according to the manufacturer's recommendations, at least three feet deep. Use a galvanized-steel pipe for the pole. It is a good idea to insert a metal cross-brace through the pole below grade to keep the pole from turning in high winds. Place a flat rock or concrete pad footing at the bottom of the hole. Position the pole in the hole and brace it to keep it straight and plumb. Then set in the concrete.

Fasten conduit to the side of the pole and down into the ground for array positive and negative wires and a ground wire. A weatherproof junction box on the pole simplifies fastening module interconnects to the array wire. Use USE-2 wire between the j-box and tracker-mounted modules with enough slack to allow for tracker movement. Bury the conduit to the equipment room to the locally required depth. When the pole mount cement has hardened, mount the tracker on the pole. Cover and mount the modules and fasten the interconnects. Connect the array to the junction box the same way the roof-mounted array was wired.

In the equipment room connections must be made from the standby generator located in a shed safely away from the house. Generator location, housing, and fuel storage must be in compliance with safety and fire prevention regulations. In the absence of local regulations, use good judgment and check with your generator supplier, fire department, or forest service. Be extremely careful with generator fuel and lubricants. Do not use the generator shed for storage.

Run conduit from the generator shed to the equipment room. A transfer switch is required to select either generator or solar. The simplest transfer switch is a manual 3-pole safety switch rated for the generator and inverter. AC hot, neutral, and ground wires are brought into the box. A generator transfer switch could also be used in grid-connect systems, if allowed by local building codes.

To make the manual transfer, bring the generator up to speed for the proper voltage. Then switch from Inverter-to-Distribution panel across Center-Off to Generator-to-Distribution panel. Motor loads and sensitive equipment should be turned off before making the transfer. Manual transfer switches are available for under $500. There are more expensive transfer switches that are adjustable and have monitoring and control circuitry. Some transfer switches and generators with remote starters allow switching to occur at the house. Automatic transfer switches cost ten times more than manual switches. Some automatic transfer switches have time delays to allow for genset startup.

Some inverters with a built-in battery charger have an automatic transfer option. By wiring the inverter/battery charger input to the generator through a time delay, you can start the generator while the inverter automatically switches to battery charger mode. There are cautions to observe with this set-up. First, when running, the power from the generator will be routed through the inverter, so be sure that loads through the inverter are within its capacity and protected with a dedicated circuit breaker. Second, be sure you have installed a time delay on the battery charger to prevent the standby transfer relay from chattering and burning out. Some homeowners dedicate large electrical loads to the generator only.

For systems with a separate generator- or grid-powered battery charger, the charger is hardwired into the house circuits just like a major appliance. In all cases with inverters, generators, transfer switches, and battery chargers, the manufacturer's instructions must be followed.

Hiring a PV Installer

While installing a simple PV system may be within the capabilities of an experienced handy person, installing a PV system with transfer switches and a generator and a utility connection is beyond the scope of most do-it-yourselfers. When you hire someone to install all or part of your PV system, expect professional performance, prompt attention to your needs, timely service, and reasonable fees. In many locations an electrician's license is required to perform these tasks. If a permit is needed for the work, the installer is responsible for obtaining it and meeting all inspection requirements.

Check with your state's contractor licensing board to ensure that the installer's licenses, insurance, and bonds are up-to-date. Ask your electrician for references. Talk to other customers and look at their installations. If you like what you see and hear, you will probably be happy with the contractor's work.

An experienced electrician should be able to understand and install your PV system. However, working with new technology and working with DC will slow down anyone unfamiliar with PV systems.

If you want to assist in the installation, be sure the installer understands what you want to do. Some technicians will not work with the homeowner because of the hindrance factor or insurance liability; others won't because their lack of skill will become obvious.

The quality of work in the installations illustrated in this book ranges from very good to excellent. These systems were selected because they are typical and show the thought and attention that the designers and installers put into their work.

The knowledgeable do-it-yourselfer knows that well-planned, organized work done safely will result in a quality installation. Wise do-it-yourselfers know their own limitations. When in doubt, call in a professional.

Monitoring & Testing

In the normal course of events you will hardly notice your PV power system. If a problem does occur, it won't be a surprise because you have been keeping a relaxed eye on things.

There are basic tests that should be performed when your system is first installed, whenever it is modified, and at regular intervals throughout its life. Some owners test their systems twice a year when they make seasonal array adjustments. Others test their systems once a year. In all cases, wiring, components, and array fasteners should be inspected at least once a year for wear and tear. Performing these tests is the responsibility of the owner.

Figure 19.3. Testing the voltage and current of an array on a tracking mount. (Photo: Bob McGill)

Test your modules when you first receive them. Prior to testing, always inspect all equipment for manufacturing defects and shipping damage. File all papers, receipts, and warranties in a safe place.

Solar array performance is tested by putting a DC ammeter amp-clamp around the positive leads from your solar array to read the solar array output under load. In full sun at noon standard time or 1 p.m. daylight savings time, the current reading should equal or be greater than 70% of the array's maximum current rating.

If you measure less than 70% of a series string's maximum current, disconnect the series string's positive and negative leads. Connect an ammeter in series to the leads. Be careful. DC current will spark and burn on contact. Read the short-circuit amperage for no less than 70% of the rated output. Then with your voltmeter across the array leads, take an open-circuit voltage reading. This reading should be at least 80% of the array open-circuit voltage.

If the readings are less than specified, isolate the problem. An open circuit or short circuit in the wiring or a bad fuse will be easy to isolate as the readings will differ dramatically from the expected. If readings are low, isolate the problem string and then the problem module, modules, wiring, or fuse. A nicked wire, intermittent short, or open wire can be very difficult to isolate. If the modules are faulty and under warranty, contact your supplier and ask for instructions about which tests to perform before returning them.

If your array and controller are performing properly and your power production matches your consumption, your battery bank specific gravity should remain at or near a full state of charge. If not, it may be necessary to test your battery bank's self-discharge rate. This requires taking the batteries out of service.

Monitoring equipment is essential to assess the status of your PV system. You need to know your system voltage, current, and the state of charge of your battery bank. These monitoring devices will give you the whole picture:

1. array ammeter (charging current)
2. DC voltmeter (DC system voltage)
3. production and consumption ampere-hour, watt-hour, or kilowatt-hour meters
4. AC voltmeter (AC system voltage)
5. hydrometer (to measure battery state of charge)
6. multitester (analog or digital) to test your system
7. load resistor (to test module output) (optional)

A good digital multitester, available for about $100, is essential to test resistance, voltage, and current. You can also use it to check home lighting circuits and wires for shorts or opens, and even perform electrical tests on your vehicle. The multitester you select should have several ranges so you can check up to 240 AC volts, system DC volts, at least 20 amperes DC current, resistance, and continuity. In addition to a multimeter, clamp-on ammeters or amp-clamps that read 50 amps DC, AC current, and voltage are available for under $75. An amp-clamp will give you a quick indication of how your system is performing. The array output ammeter on some charge controllers shows array charging current and also lets you know when the controller is on, off, or in trickle-charge mode.

The results of DC voltmeter and hydrometer testing will indicate the condition of your batteries. After you have tested your battery bank a few times and you understand the charge/discharge characteristics of your system, an occasional glance at your voltmeter will give you a good indication of the average specific gravity of the battery bank. This, in turn, lets you know your battery's depth of discharge.

Figure 19.4. Testing wire and insulation with a high voltage "megger." Insulated gloves and only using one hand while testing prevents heart-stopping currents from flowing across the chest. (Solar Electrical Systems)

You can calculate the average specific gravity of the cells in your battery bank by dividing your battery voltage by the number of series cells in the battery bank string minus 0.84. For example, a 48-volt (nominal) battery bank is 49.5 volts. There are twenty-four 2-volt cells used to make up the 48-volt bank: $49.5 \div 24 = 2.0625 - 0.84 = 1.2225$. Referring to depth of discharge vs. specific gravity table in Chapter 12 (page 228), the battery bank is at 70% capacity (discharged 30%).

To see how much energy you consume, install a kilowatt-hour meter in the AC circuit between the inverter and the AC distribution breaker box. This is the familiar glass-encased counter that utility companies use. Kilowatt-hour meters sell for under $100. Digital meters are also available but cost more.

A DC ampere-hour meter allows you to monitor and match production and consumption of a DC-only system. Some charge controllers have built in amp-hour meters.

It is possible to operate a PV system without metering and monitoring equipment hardwired directly into the circuit if you occasionally take measurements with a good multitester. If you decide to install a number of meters on your system, be sure they can be switched off. These parasitic loads consume precious PV power.

Although not essential, a resistor dummy load can be used to test each PV module for output under simulated in-service conditions. A wire-wound variable resistor capable of handling the modules' watts is used. Refer to the Ohm's Law chart in Chapter 8 (page 140) to determine the resistor size. Put your load resistor across each separate PV module's positive and negative leads or terminals to measure load voltage (V_{max}). To measure load current (I_{max}), attach one side of the load resistor to each module's positive lead or terminal. Attach your ammeter to the other side of the resistor and then to the module's negative lead or terminal so that the resistor and ammeter are in series.

It should take about an hour to test your installed PV system the first time, depending on the size of your PV array. Regular yearly testing of your system will take even less time because you do not have to test and record each individual module's output as you did the first time. Measure the voltage and current at noon on a sunny day with no haze. Record your readings for each module. Readings on a sunny day should be at least 70% of the manufacturer's rating for current and 80% for voltage. Keep test result data in your equipment log book.

Test your batteries with a hydrometer following manufacturer's instructions. (Be careful not to drip any battery acid on yourself.) Measure the specific gravity of each cell every two months during the first year. Your batteries should be fully charged initially.

Next, test the entire PV system. Measure and record the current and voltage output. Test the entire system four to six times during the first year; after that testing once or twice a year is sufficient.

You may want to perform a self-discharge test on your batteries. This test takes a week, but gives a good indication of the remaining useful life of older batteries. Disconnect the batteries and charge them with an industrial charger at a current rate that does not exceed the battery's capacity in ampere-hours divided by 20 hours (C/20). For example, a 200-ampere-hour battery should be charged at 10 amps or less. When all cell voltages at rest are 2.3 volts or the manufacturer's rating, the battery is fully charged. Record the specific gravity for each cell. Keep the batteries out of use for a week at 60 to 70°F/15.5 to 21.1°C and test for specific gravity again. Good cells will have a self-discharge rate less than specific gravity 0.015. It is advisable to perform a self-discharge test before buying used batteries.

Maintenance

Maintaining your PV system is very easy. Check your battery electrolyte level and replenish with distilled water as necessary. Don't overfill. Clean and tighten battery connectors. Keep your batteries clean since dust and grime can conduct current between the positive and negative terminals and the case. To retard corrosion, use the treated felt rings or spray-on protectants sold in auto parts stores on the posts and connectors. Once a year visually inspect all equipment, fasteners, wires, cords, connectors, plugs, and outlets. Repair or replace as necessary.

Check your PV array for dirt build-up. Arrays with 15° or more tilt usually stay clean with rain. If you live in a city, in a pollution zone, or on a dusty road, you may have to wash your array a few times a year with a mild soap or plain water and a soft cloth. Do not use solvents or strong detergents. Most PV systems only need to be rinsed off with a garden hose. Do not wash your array in the middle of the day as cold water can cause the hot glass to crack.

How does dirt on solar arrays affect electrical production? The New Mexico Solar Energy Institute at Las Cruces reported a 6% increase in performance after detergent washing a portion of a 17-kW array near El Paso, Texas. A few days later a gentle rain washed the rest of the array. The difference in output between the two was undetectable. The conclusion was that detergent washing is a waste of time and money, but the dust, dirt, and pollutants at other locations may require detergent.

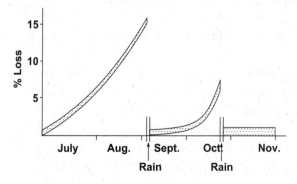

Figure 19.5. Power loss from dirt on a Southern California array increased in the springtime. A summer rain rinsed the array. Solar arrays are self-cleaning at locations with regular rainfall.

Think Safety

Familiarize yourself with procedures for putting out an electrical fire. Make sure your fire extinguisher is in good working order and that it has a multi-class rating that includes "Class C," which is suitable for use on electrical fires. If your have a fire, immediately turn off the electricity. Grid-connected homes using high DC voltage are required to have disconnect switches from the array to the controller and to the utility intertie inverter. Access to wires and quick disconnect switches or plugs is essential.

Dangerous currents are present on both the DC and AC sides of the PV system. Always assume that a wire is hot unless you have personally disconnected and tested it. Cut-out switches, circuit breakers, and safety switches should be plainly marked and tagged.

Your PV system is an electric power plant and you are the power plant safety engineer. Make your job easier with proper planning and preparation. Think safety.

PV Installation Checklist

This checklist is adapted from the California Energy Commission's *Guide to Photovoltaic (PV) System Design and Installation*, provided by Endecon Engineering.

A hard hat, insulated gloves, and eye protection are recommended for all testing.

Before starting any PV system testing:
- Check that non-current-carrying metal parts are grounded properly and that array frames, metal boxes, etc. are connected to the grounding system.
- Ensure that all labels and safety signs are in place.
- Verify that all disconnect switches (from the main AC disconnect all the way through to the combiner fuse switches) are in the open position. Tag each box with a warning sign to show that work on the system is in progress.

PV Array

- Verify that all combiner fuses are removed and that no voltage is present at the output of the combiner box.
- Visually inspect any plug and receptacle connectors between the modules and panels to ensure they are fully engaged.
- Check that strain reliefs/cable clamps are properly installed on all cables and cords by pulling on cables to verify.
- Check to make sure all panels are attached properly to their mounting brackets and that nothing is misaligned.
- Visually inspect the array for cracked modules.
- Check to see that all wiring is neat and well supported.

PV Array Circuit Wiring

- Check homerun wires (from PV modules to combiner box) at DC string combiner box to ensure there is no voltage.
- Connect the homerun wires to the DC string combiner box terminals in the proper order and make sure labeling is clearly visible.

Source Circuit String Wiring

The next three steps must be followed for each source circuit string using a systematic approach—east to west or north to south. Ideal testing conditions are midday on cloudless days March through October.

- Check open-circuit voltage of each module in the string to verify that it provides the manufacturer's specified voltage in full sun. Modules under the same sunlight conditions should have similar voltages. Beware of a two volt or more difference under the same sunlight conditions.
- Verify that both the positive and negative string connectors are identified properly with permanent wire marking.
- Repeat this sequence for all source circuit strings.

Continuation of PV Array Circuit Wiring

- Verify the polarity of each source circuit string in the DC string combiner box. Place the common test lead on the negative grounding block and the positive test lead on each string connection—make sure there is never a negative measurement. Verify that open-circuit voltage is within proper range according to module specifications. Number each string and note string position on as-built drawing.

A fire can start in the fuse block if the polarity of one source circuit string is reversed. This can result in the destruction of the combiner box and adjacent equipment. Reverse polarity on an inverter can cause damage that is not covered under the equipment warranty.

- Retighten all terminals in the DC string combiner box.

Wiring Tests

- Verify that the only place where the AC neutral is grounded is at the main service panel.
- Check that the AC line voltage at main AC disconnect is within proper limits (115–125 volts AC for 120 volts and 230–250 for 240 volts). If installation contains additional AC disconnect switches, repeat this step. Perform a voltage check on each switch working from the main service entrance to the inverter AC disconnect switch. Close each switch after the test is made, except for the final switch before the inverter.

Inverter Start-up Tests

Be sure the inverter is off before proceeding with this section.

- Test the continuity of all DC fuses to be installed in the DC string combiner box, install all string fuses, and close fused switches in the combiner box.
- Check open-circuit voltage at DC disconnect switch to ensure it is within proper limits. If the installation contains additional DC disconnect switches, repeat this step for each switch, working from the PV array to the inverter DC disconnect switch. Close each switch after the test is made, except for the final switch before the inverter.
- Consult the inverter manual for proper start-up procedure.
- Confirm that the inverter is operating and record the DC operating voltage in the equipment log book.
- Confirm that the operating voltage is within proper limits according to the manufacturer's installation manual.
- After recording the operating voltage at the inverter, close any open boxes related to the inverter system.
- Confirm that the inverter is producing the expected power.

System Acceptance Test

Ideal testing conditions are midday on cloudless days March through October. However, this test procedure accounts for less than ideal conditions and can be conducted on a sunny day.

- Check to make sure that the PV array is in full sun with no shading whatsoever. If it is impossible to find a time during the day when the whole array is in full sun, only that portion that is in full sun will be able to be accepted.
- If the system is not operating, turn it on and allow it to run for 15 minutes before taking any performance measurements.
- Obtain the solar irradiance measurement by one of two methods below and record the irradiance (in W/m^2) in the log book. To obtain percentage of peak sun, divide irradiance by 1,000 W/m^2 and record the value.
  Example: 692 W/m^2 ÷ 1,000 W/m^2 = 0.692 or 69.2%
  Use the decimal percent value to calculate watts $AC_{corrected}$.

Method 1:
Take measurement from calibrated solar meter or pyranometer.

Method 2:
Place a properly operating PV module of the same model found in the array in full sun in the exact same orientation as the array being tested. After 15 minutes of full exposure, test the short-circuit current with a digital multimeter.

- Divide this number into the short-circuit current (I_{SC}) value of the module. Multiply this by 1,000 W/m². Record this value.

Example: $I_{SC\text{-measured}}$ = 3.6 amps; I_{SC} printed on module = 5.2 amps
Irradiance = 3.6 amps ÷ 5.2 amps × 1,000 W/m² = 692 W/m²

- Sum the total of the module watts STC. Multiply this by 0.7 to obtain expected peak AC output. Record as watts$_{AC\text{-estimated}}$.
- Record AC watts output from the inverter or system meter and record as watts$_{AC\text{-measured}}$.
- Divide watts$_{AC\text{-measured}}$ by the decimal percent peak irradiance. Record as watts$_{AC\text{-corrected}}$. This "AC-corrected" value is the rated output of the the PV system. This number must be within 90% or higher of watts$_{AC\text{-estimated}}$. If it is less than 90%, the PV system is either shaded, dirty, miswired, the fuses are blown, or the modules or inverter are not operating properly.

Sample calculation: A PV system is composed of 20 PV modules rated 100 watts STC operating at an estimated irradiance of 692 W/m². The measured power output is 1,000 watts$_{AC\text{-measured}}$ at the time of the test.

Sum of module ratings = 100 watts$_{STC}$ per module
× 20 modules = 2,000 watts$_{STC}$

Estimated AC power output = 2,000 watts$_{STC}$ × 0.7
= 1,400 watts$_{AC\text{-estimated}}$

Measured AC output = 1,000 watts$_{AC\text{-measured}}$

Corrected AC output = 1,000 watts$_{AC\text{-measured}}$ ÷ 0.692
= 1,445 watts$_{AC\text{-corrected}}$

Comparison of corrected and estimated outputs:

1,445 watts$_{AC\text{-corrected}}$ ÷ 1,400 watts$_{AC\text{-estimated}}$
= 1.03 ≥ 0.9 (acceptable performance)

Chapter 20
THE FUTURE OF PV

PV has evolved from a scientific curiosity to a reliable power source for orbiting satellites and space stations to cost-effective power for remote homes and villages and to renewable energy distributed generation in cities and towns around the world. PV use will continue to rise rapidly as the need to replace fossil fuel and nuclear power plants becomes increasingly critical. The change to PV will accelerate as new materials and production methods come on-line. As more people understand how easy, practical, and rewarding it is to use PV, electric utilities will find ways to integrate privately owned PV into the grid power mix. PV is already becoming an important addition to utility companies' portfolios.

Building-integrated PV (BIPV) will soon be commonplace. Megawatts of PV roofs, curtain walls, and windows are being installed every day. PV shade structures and carports are turning parking lots into solar power plants. Tomorrow, BIPV and PV shade structures, which are simply PV arrays with long legs, will be built on polluted urban wasteland as part of the "brownfields" clean-up process.

Figure 20.1. The Condé Nast building in Manhattan heralds a future when all buildings may get their power from the sun. Vertical PV arrays on the south and east facades from the 37th to the 43rd floors offset part of the building's electric load. (Kiss + Cathcart, Architects)

In the not-too-distant future—less than 10 years if we focus our resources on peace and plenty instead of war and waste—buildings will be rated as either net energy consumers or net energy producers. Older, high density, multi-story buildings like skyscrapers and inner-city hospitals with limited roof space or inadequate solar access will remain net energy consumers, while new buildings will be designed to produce all the energy that they consume. Low energy use buildings like warehouses will have PV roofs that produce excess electricity for credit or to sell to their neighbors.

Homes of the future will be super energy efficient and have PV roofs that provide all the home's electricity plus power for the homeowner's electric or hydrogen fuel cell automobile. PV modules will be available in many shapes, sizes, and colors to fit any building application. Mounting hardware is already more user-friendly. Building-integrated mounting and wiring will be inconspicuous. Multi-purpose inverters that operate with or without batteries will allow you to choose battery backup now, later, or not at all and back-up power will be stored as hydrogen and utilized through fuel cells.

Electric service panels will be made solar-ready with built-in connections and circuit breakers for on-site power production. Electricians and tradespeople are learning how to install PV, just as they learned to install heating and air conditioning units a few decades ago. Someday all electricians will be as familiar with PV as they are with lighting equipment.

PV will change the way people think about energy production. Most of us are already concerned about continuously increasing consumption, power plant pollution, diminishing energy resources, and buying oil and gas from nations that harbor terrorists. On-site PV equipment costs are now included in homeowner's mortgage payments in lieu of paying a separate monthly electric bill. Mortgage-financed PV is already cheaper in many places than utility power. In Japan, Germany, and the United States every year hundreds of thousands of people buy PV for their homes for economic and environmental reasons. This trend is expanding throughout Europe and to other countries.

In the future, politicians will be required to provide energy solutions that reduce the threat of war, improve the environment, and enhance individual freedom. Individuals and corporations will be held responsible for the energy resources they consume and the health and environmental impacts of their actions. PV will help everyone to meet these goals.

Figure 20.2. In the future, more people will work from their homes. This large house and home-based business has a 12-kW solar array that produces over 50 kWh/day. (Solar Electrical Systems)

Figure 20.3. This comfortable mountain cabin has a 3.4-kW building-integrated solar array. Snow quickly slides off the BIPV array. (Stephen Heckeroth)

Figure 20.4. Filtered light shines through the clear encapsulant of twenty-four 300-watt solar modules that feed into three 2.5 kW net-metered inverters. The trellis framing is recycled steel pipe from an oil refinery. (groSolar)

On-site PV use will continue to grow by thousands of megawatts every year. Large-scale centralized PV power plants will still be required to power urban areas that house over 50% of the world's population, but plans are on the drawing board for a global network of PV arrays that will feed into a global utility network. The Genesis Project is one such plan. Genesis, which stands for Global Energy Network Equipped with Solar cells and International Superconductor grids, is the brainchild of Dr. Yukinori Kuwano, one PV's most innovative scientists and the founder of Sanyo's PV division. His 200-year plan is to make PV the planet's prime energy source.

A look at Earth from space shows that rainy and cloudy regions cover less than 30% of the total land mass and that it is always daylight on one side of the globe. A global PV power grid connected with super-conducting cables would enable areas in daylight to provide clean solar energy around-the-clock.

When Kuwano first proposed the Genesis Project, global energy demands were equivalent to 14 billion barrels of crude oil per year. About 4% of the world's desert areas (800 square kilometers) covered with 10% efficient solar cells could meet this requirement.

Figure 20.5. PV systems like this—twenty-one 185-watt PV modules feeding into a 3.5-kW inverter—are becoming commonplace in the U.S., Japan, and Germany. (Gaiam Real Goods)

Is a global PV grid possible? The individuals who install PV on their grid-connected homes and businesses are building the global PV network right now. As this trends grows, national networks can then be connected by superconductor cables now in development.

The United States and Canada are already interconnected, as are the electric power grids of some European countries. The "Silk Road Genesis," proposed by the Tokyu Construction Company, calls for the construction of PV power plants along that ancient trade route. The first stage of the project, which has the support of Sanyo and other companies, involves building over one hundred 100-megawatt PV plants along the Silk Road by 2030.

Even if a global electric grid does not materialize, hundreds of megawatts of PV are already connected to local and regional grids while hundreds of thousands of households in developing countries are getting their first electricity from PV each year.

While you were reading this short chapter, more than 1,000 watts of PV were installed somewhere. In 2007, over 60,000 watts of PV were installed every 10 minutes. We hope you will join the PV Revolution and go solar now.

Appendix A
SOLAR RADIATION DATA
& PEAK SUN HOURS

Solar radiation received on a surface is insolation (not to be confused with insulation). The units for irradiance are watt-hours per square meter (Wh/m²). The units for insolation are irradiance per unit of time, such as watt-hours per square meter per day (Wh/m²/day) or kilowatt-hours per square meter per day (kWh/m²/day). Peak Sun Hours is another term for kWh/m²/day.

The following information is from the U.S. Department of Energy database. The same data is used by PVWATTS, an online calculator used to estimate PV performance. See http://rredc.nrel.gov/solar/calculators/PVWATTS/.

The data in average kWh/m²/day for each location is given for four solar array tilt angles:
1. horizontal (flat)
2. the location's latitude degrees
3. the latitude plus 15° (optimal for winter PV production when the sun is low to the horizon)
4. the latitude minus 15° (optimal for summer PV production when the sun is higher in the sky)

To estimate a solar electric system's average annual performance, use the following formula:

solar module peak watts DC STC

× quantity of modules in the array

× Peak Sun Hours

= kWh/day DC × derating factors (see Table 16.1, page 319)

= kWh/day AC

For example, a batteryless net-metered PV system in Lexington, Kentucky, has a solar array with twenty 200-watt solar modules at latitude tilt and the Peak Sun Hours are 4.5:

200 watts DC STC module

× 20 modules

× 4.5 Peak Sun Hours

= 18 kWh/day DC × 0.9 × 0.98 × 0.98 × 0.98 × 0.9 × 0.99
(typical derating factors for a batteryless net-metered system)

= 13.58 kWh/day AC

Use the yearly Peak Sun Hours for quick calculations, but note that winter Sun Hours at many locations may be less than half as much as summer. Yearly Peak Sun Hours are useful if you plan a grid-connected, net-metered PV system with annual utility company billing. If you plan an off-grid PV system, use December or January Peak Sun Hours to ensure that you have enough energy during cloudy periods and longer nights that require more artificial lighting. Use the two-month average to calculate daily PV production for a day at the beginning or end of a month. For example, use average June and July Peak Sun Hours to estimate PV production for July 4. Also see the Sun Hours Maps on pages 329 and 330.

	Jan	Feb	Mar	Apr	May	Jun	Jul	Aug	Sep	Oct	Nov	Dec	Year
BIRMINGHAM, AL—Latitude 33.57													
0	2.5	3.3	4.4	5.5	6.0	6.2	5.9	5.6	4.8	4.0	2.8	2.3	4.4
Lat - 15	3.3	4.0	4.9	5.7	6.0	6.1	5.8	5.8	5.2	4.8	3.6	3.1	4.9
Lat	3.7	4.4	5.1	5.6	5.6	5.6	5.4	5.6	5.3	5.2	4.1	3.5	4.9
Lat + 15	3.9	4.5	5.0	5.2	5.0	4.9	4.8	5.1	5.1	5.2	4.3	3.8	4.7
HUNTSVILLE, AL—Latitude 34.65													
0	2.4	3.1	4.1	5.3	5.9	6.3	6.1	5.7	4.7	3.9	2.7	2.1	4.4
Lat - 15	3.1	3.8	4.7	5.6	5.9	6.2	6.0	5.9	5.2	4.7	3.5	2.9	4.8
Lat	3.5	4.2	4.8	5.5	5.6	5.7	5.6	5.7	5.3	5.0	3.9	3.3	4.8
Lat + 15	3.7	4.3	4.7	5.1	5.0	4.9	4.9	5.2	5.0	5.1	4.1	3.5	4.6
MOBILE, AL—Latitude 30.68													
0	2.7	3.5	4.4	5.4	5.9	5.9	5.6	5.2	4.7	4.2	3.1	2.5	4.4
Lat - 15	3.3	4.1	4.9	5.6	5.8	5.8	5.5	5.3	5.1	4.9	3.8	3.2	4.8
Lat	3.7	4.5	5.0	5.6	5.5	5.4	5.1	5.2	5.1	5.2	4.3	3.6	4.9
Lat + 15	4.0	4.6	4.9	5.2	4.9	4.7	4.5	4.7	4.9	5.3	4.5	3.9	4.7
MONTGOMERY, AL—Latitude 32.30													
0	2.7	3.5	4.5	5.7	6.2	6.4	6.1	5.7	4.9	4.2	3.0	2.5	4.6
Lat - 15	3.4	4.2	5.0	5.9	6.2	6.3	6.0	5.9	5.3	4.9	3.8	3.3	5.0
Lat	3.8	4.6	5.2	5.8	5.8	5.8	5.6	5.7	5.4	5.3	4.3	3.7	5.1
Lat + 15	4.0	4.7	5.1	5.4	5.2	5.1	4.9	5.2	5.2	5.4	4.5	4.0	4.9
ANCHORAGE, AK—Latitude 61.17													
0	0.3	1.0	2.3	3.6	4.6	4.9	4.6	3.5	2.2	1.1	0.4	0.2	2.4
Lat - 15	0.9	2.1	3.8	4.7	4.9	5.0	4.8	4.1	3.1	2.0	1.1	0.5	3.1
Lat	1.0	2.2	3.9	4.6	4.6	4.5	4.4	3.8	3.1	2.1	1.2	0.6	3.0
Lat + 15	1.0	2.3	3.9	4.3	4.0	3.9	3.8	3.4	2.9	2.0	1.3	0.6	2.8
FAIRBANKS, AK—Latitude 64.82													
0	0.1	0.8	2.3	4.0	5.1	5.6	5.1	3.7	2.3	1.0	0.3	0.0	2.5
Lat - 15	0.7	2.2	4.5	5.6	5.7	5.7	5.4	4.5	3.4	1.9	1.0	0.2	3.4
Lat	0.7	2.4	4.7	5.6	5.3	5.2	4.9	4.2	3.4	2.0	1.1	0.3	3.3
Lat + 15	0.8	2.5	4.7	5.3	4.6	4.5	4.3	3.8	3.2	2.0	1.1	0.3	3.1
NOME, AK—Latitude 64.50													
0	0.2	0.8	2.3	4.3	5.3	5.5	4.6	3.3	2.1	1.0	0.3	0.1	2.5
Lat - 15	0.8	2.5	4.5	5.9	5.9	5.5	4.7	3.7	3.2	2.2	1.0	0.4	3.4
Lat	0.9	2.8	4.7	5.9	5.6	5.0	4.3	3.5	3.2	2.4	1.1	0.5	3.3
Lat + 15	0.9	2.9	4.7	5.6	5.0	4.3	3.7	3.1	3.0	2.4	1.2	0.5	3.1

	Jan	Feb	Mar	Apr	May	Jun	Jul	Aug	Sep	Oct	Nov	Dec	Year
FLAGSTAFF, AZ—Latitude 35.13													
0	3.1	4.0	5.1	6.3	7.2	7.7	6.4	5.9	5.4	4.4	3.3	2.8	5.1
Lat - 15	4.4	5.2	5.9	6.8	7.2	7.4	6.2	6.1	6.1	5.6	4.7	4.2	5.8
Lat	5.2	5.8	6.2	6.7	6.7	6.7	5.8	5.9	6.3	6.1	5.4	4.9	6.0
Lat + 15	5.6	6.1	6.2	6.2	5.9	5.7	5.0	5.4	6.0	6.3	5.8	5.4	5.8
PHOENIX, AZ—Latitude 33.43													
0	3.2	4.3	5.5	7.1	8.0	8.4	7.6	7.1	6.1	4.9	3.6	3.0	5.7
Lat - 15	4.4	5.4	6.4	7.5	8.0	8.1	7.5	7.3	6.8	6.0	4.9	4.2	6.4
Lat	5.1	6.0	6.7	7.4	7.5	7.3	6.9	7.1	7.0	6.5	5.6	4.9	6.5
Lat + 15	5.5	6.2	6.6	6.9	6.6	6.3	6.0	6.4	6.7	6.7	5.9	5.3	6.3
PRESCOTT, AZ—Latitude 34.65													
0	3.1	3.9	5.1	6.6	7.5	8.0	6.9	6.3	5.7	4.6	3.4	2.8	5.3
Lat - 15	4.4	5.1	5.9	7.0	7.5	7.7	6.7	6.5	6.5	5.8	4.8	4.1	6.0
Lat	5.1	5.7	6.2	6.9	7.0	7.0	6.2	6.3	6.6	6.4	5.5	4.9	6.1
Lat + 15	5.5	5.9	6.1	6.4	6.1	6.0	5.4	5.7	6.4	6.5	5.9	5.4	5.9
TUCSON, AZ—Latitude 32.12													
0	3.4	4.4	5.6	7.1	7.9	8.1	7.1	6.7	6.0	5.0	3.8	3.2	5.7
Lat - 15	4.6	5.5	6.4	7.5	7.8	7.8	6.9	6.9	6.6	6.1	5.0	4.3	6.3
Lat	5.4	6.2	6.7	7.3	7.3	7.1	6.4	6.6	6.8	6.6	5.8	5.1	6.5
Lat + 15	5.9	6.4	6.6	6.8	6.4	6.1	5.6	6.0	6.6	6.8	6.2	5.6	6.3
FORT SMITH, AR—Latitude 35.33													
0	2.6	3.4	4.4	5.4	6.0	6.5	6.6	6.0	4.8	3.9	2.8	2.3	4.6
Lat - 15	3.6	4.2	5.0	5.7	6.0	6.3	6.5	6.2	5.3	4.8	3.6	3.2	5.1
Lat	4.1	4.6	5.2	5.6	5.7	5.8	6.0	6.0	5.4	5.1	4.1	3.7	5.1
Lat + 15	4.3	4.8	5.1	5.2	5.0	5.0	5.3	5.5	5.2	5.2	4.3	4.0	4.9
LITTLE ROCK, AR—Latitude 34.73													
0	2.5	3.3	4.3	5.3	6.1	6.5	6.4	5.9	4.8	3.9	2.7	2.2	4.5
Lat - 15	3.4	4.1	4.9	5.6	6.1	6.4	6.3	6.1	5.3	4.8	3.5	3.0	5.0
Lat	3.8	4.5	5.1	5.5	5.7	5.9	5.9	5.9	5.4	5.1	4.0	3.5	5.0
Lat + 15	4.1	4.6	5.0	5.1	5.1	5.1	5.2	5.4	5.2	5.2	4.2	3.7	4.8
ARCATA, CA—Latitude 40.98													
0	1.8	2.5	3.6	5.0	5.8	6.0	5.9	5.0	4.4	3.1	2.0	1.6	3.9
Lat - 15	2.7	3.3	4.3	5.4	5.9	5.9	5.8	5.3	5.1	3.9	2.9	2.5	4.4
Lat	3.0	3.5	4.4	5.3	5.5	5.4	5.4	5.0	5.1	4.1	3.2	2.8	4.4
Lat + 15	3.2	3.6	4.3	4.9	4.9	4.7	4.7	4.6	4.9	4.1	3.3	3.0	4.2

BAKERSFIELD, CA—Latitude 35.42													
0	2.3	3.3	4.7	6.2	7.4	8.1	8.0	7.2	5.9	4.4	2.9	2.1	5.2
Lat - 15	3.0	4.2	5.4	6.6	7.4	7.8	7.8	7.5	6.8	5.5	3.8	2.8	5.7
Lat	3.3	4.5	5.6	6.5	6.9	7.1	7.2	7.3	6.9	6.0	4.3	3.2	5.7
Lat + 15	3.5	4.7	5.5	6.0	6.1	6.1	6.2	6.6	6.7	6.1	4.5	3.4	5.4
DAGGETT, CA—Latitude 34.87													
0	3.2	4.2	5.5	7.0	7.9	8.4	8.0	7.3	6.3	4.9	3.6	2.9	5.8
Lat - 15	4.6	5.4	6.5	7.5	7.9	8.1	7.8	7.6	7.1	6.2	5.0	4.4	6.5
Lat	5.3	6.0	6.8	7.4	7.4	7.4	7.2	7.3	7.3	6.8	5.8	5.2	6.6
Lat + 15	5.7	6.2	6.7	6.8	6.5	6.3	6.2	6.6	7.0	6.9	6.2	5.6	6.4
FRESNO, CA—Latitude 36.77													
0	2.1	3.2	4.7	6.3	7.5	8.1	8.0	7.2	5.9	4.3	2.7	1.9	5.2
Lat - 15	2.8	4.1	5.5	6.8	7.6	7.8	7.9	7.5	6.8	5.5	3.6	2.5	5.7
Lat	3.1	4.4	5.7	6.7	7.1	7.2	7.3	7.3	6.9	6.0	4.1	2.8	5.7
Lat + 15	3.2	4.5	5.6	6.2	6.3	6.1	6.3	6.6	6.7	6.1	4.2	3.0	5.4
LONG BEACH, CA—Latitude 33.82													
0	2.8	3.6	4.7	6.0	6.4	6.7	7.3	6.7	5.4	4.2	3.1	2.6	5.0
Lat - 15	3.8	4.5	5.4	6.4	6.4	6.5	7.2	6.9	6.0	5.0	4.1	3.6	5.5
Lat	4.3	4.9	5.6	6.3	6.1	6.0	6.7	6.7	6.1	5.4	4.7	4.2	5.6
Lat + 15	4.6	5.1	5.5	5.8	5.4	5.2	5.8	6.1	5.8	5.5	5.0	4.5	5.4
LOS ANGELES, CA—Latitude 33.93													
0	2.8	3.6	4.8	6.1	6.4	6.6	7.1	6.5	5.3	4.2	3.2	2.6	4.9
Lat - 15	3.8	4.5	5.5	6.4	6.4	6.4	7.1	6.8	5.9	5.0	4.2	3.6	5.5
Lat	4.4	5.0	5.7	6.3	6.1	6.0	6.6	6.6	6.0	5.4	4.7	4.2	5.6
Lat + 15	4.7	5.1	5.6	5.9	5.4	5.2	5.8	6.0	5.7	5.5	5.0	4.5	5.4
SACRAMENTO, CA—Latitude 38.52													
0	1.9	3.0	4.3	5.9	7.2	7.9	7.9	7.0	5.7	4.0	2.4	1.7	4.9
Lat - 15	2.6	3.9	5.2	6.5	7.3	7.6	7.8	7.5	6.7	5.3	3.3	2.4	5.5
Lat	2.9	4.2	5.4	6.3	6.8	7.0	7.2	7.2	6.9	5.7	3.7	2.7	5.5
Lat + 15	3.1	4.3	5.2	5.9	6.0	6.0	6.3	6.5	6.6	5.8	3.9	2.9	5.2
SAN DIEGO, CA—Latitude 32.73													
0	3.1	3.9	4.9	6.1	6.3	6.5	6.9	6.5	5.4	4.4	3.4	2.9	5.0
Lat - 15	4.1	4.8	5.6	6.4	6.3	6.3	6.8	6.7	6.0	5.3	4.5	3.9	5.6
Lat	4.7	5.3	5.8	6.3	5.9	5.8	6.4	6.5	6.1	5.7	5.1	4.6	5.7
Lat + 15	5.1	5.5	5.7	5.9	5.2	5.1	5.6	5.9	5.8	5.8	5.4	5.0	5.5

	Jan	Feb	Mar	Apr	May	Jun	Jul	Aug	Sep	Oct	Nov	Dec	Year
SAN FRANCISCO, CA—Latitude 37.90													
0	2.2	3.0	4.2	5.7	6.7	7.2	7.3	6.5	5.4	3.9	2.5	2.0	4.7
Lat - 15	3.1	3.9	5.0	6.2	6.8	7.0	7.3	6.9	6.2	5.0	3.5	2.9	5.3
Lat	3.5	4.2	5.2	6.1	6.4	6.5	6.8	6.7	6.4	5.4	3.9	3.4	5.4
Lat + 15	3.7	4.4	5.1	5.6	5.7	5.6	5.9	6.1	6.1	5.5	4.1	3.6	5.1
SANTA MARIA, CA—Latitude 34.90													
0	2.8	3.7	4.9	6.2	7.0	7.4	7.5	6.8	5.6	4.4	3.2	2.7	5.2
Lat - 15	4.0	4.7	5.7	6.6	7.0	7.2	7.4	7.1	6.3	5.4	4.4	3.9	5.8
Lat	4.6	5.2	5.9	6.5	6.6	6.6	6.9	6.9	6.4	5.8	5.0	4.5	5.9
Lat + 15	4.9	5.4	5.8	6.0	5.8	5.7	6.0	6.2	6.2	6.0	5.3	4.9	5.7
ALAMOSA, CO—Latitude 37.45													
0	3.0	4.0	5.2	6.4	7.1	7.7	7.2	6.5	5.6	4.5	3.3	2.7	5.3
Lat - 15	4.7	5.5	6.2	6.9	7.1	7.4	7.0	6.8	6.5	5.9	4.9	4.4	6.1
Lat	5.5	6.2	6.5	6.8	6.6	6.8	6.5	6.5	6.7	6.4	5.6	5.2	6.3
Lat + 15	6.0	6.5	6.4	6.3	5.8	5.7	5.6	5.9	6.4	6.5	6.0	5.7	6.1
COLORADO SPRINGS, CO—Latitude 38.82													
0	2.5	3.4	4.5	5.7	6.2	6.9	6.7	6.0	5.1	4.0	2.8	2.3	4.7
Lat - 15	4.0	4.7	5.5	6.2	6.2	6.7	6.6	6.4	6.0	5.4	4.2	3.7	5.5
Lat	4.6	5.2	5.7	6.1	5.9	6.2	6.1	6.1	6.1	5.8	4.8	4.4	5.6
Lat + 15	5.0	5.4	5.6	5.6	5.2	5.3	5.3	5.6	5.9	5.9	5.1	4.8	5.4
BOULDER, CO—Latitude 40.02													
0	2.4	3.3	4.4	5.6	6.2	6.9	6.7	6.0	5.0	3.8	2.6	2.1	4.6
Lat - 15	3.8	4.6	5.4	6.1	6.2	6.6	6.6	6.3	5.9	5.1	4.0	3.5	5.4
Lat	4.4	5.1	5.6	6.0	5.9	6.1	6.1	6.1	6.0	5.6	4.6	4.2	5.5
Lat + 15	4.8	5.3	5.6	5.6	5.2	5.2	5.3	5.5	5.8	5.7	4.8	4.5	5.3
EAGLE, CO—Latitude 39.65													
0	2.4	3.3	4.4	5.6	6.4	7.2	6.9	6.1	5.1	3.9	2.5	2.1	4.7
Lat - 15	3.7	4.6	5.3	6.1	6.4	7.0	6.9	6.5	6.1	5.2	3.8	3.4	5.4
Lat	4.3	5.2	5.6	6.0	6.0	6.4	6.3	6.3	6.2	5.6	4.3	3.9	5.5
Lat + 15	4.6	5.4	5.5	5.6	5.3	5.5	5.5	5.7	5.9	5.7	4.5	4.3	5.3
GRAND JUNCTION, CO—Latitude 39.12													
0	2.5	3.5	4.6	6.0	7.0	7.7	7.4	6.6	5.5	4.1	2.7	2.2	5.0
Lat - 15	3.8	4.7	5.5	6.5	7.0	7.5	7.3	7.0	6.5	5.4	4.1	3.6	5.7
Lat	4.4	5.2	5.7	6.4	6.6	6.8	6.7	6.7	6.6	5.9	4.6	4.1	5.8
Lat + 15	4.7	5.4	5.6	6.0	5.8	5.8	5.8	6.1	6.4	6.0	4.9	4.5	5.6

PUEBLO, CO—Latitude 38.28													
0	2.7	3.6	4.7	6.0	6.7	7.4	7.2	6.5	5.4	4.2	2.9	2.4	5.0
Lat-15	4.1	4.9	5.7	6.5	6.7	7.2	7.1	6.9	6.3	5.6	4.4	3.9	5.8
Lat	4.8	5.4	6.0	6.4	6.3	6.6	6.6	6.6	6.4	6.0	5.0	4.6	5.9
Lat+15	5.2	5.6	5.9	6.0	5.6	5.6	5.7	6.0	6.2	6.2	5.3	5.0	5.7
BRIDGEPORT, CT—Latitude 41.17													
0	1.9	2.7	3.7	4.7	5.4	5.9	5.8	5.4	4.2	3.1	2.0	1.6	3.8
Lat-15	2.9	3.7	4.4	5.1	5.5	5.8	5.8	5.5	4.9	4.0	2.8	2.4	4.4
Lat	3.3	4.0	4.6	5.0	5.2	5.3	5.4	5.3	4.9	4.3	3.1	2.8	4.4
Lat+15	3.5	4.1	4.5	4.6	4.6	4.6	4.7	4.8	4.7	4.3	3.3	2.9	4.2
HARTFORD, CT—Latitude 41.93													
0	1.9	2.7	3.7	4.6	5.4	5.9	5.9	5.1	4.1	3.0	1.9	1.5	3.8
Lat-15	2.9	3.7	4.4	5.1	5.5	5.8	5.8	5.4	4.8	3.9	2.7	2.3	4.4
Lat	3.3	4.1	4.6	4.9	5.2	5.3	5.4	5.2	4.8	4.1	2.9	2.7	4.4
Lat+15	3.5	4.2	4.5	4.6	4.6	4.6	4.8	4.7	4.6	4.2	3.0	2.8	4.2
WILMINGTON, DE—Latitude 39.67													
0	2.0	2.9	3.9	4.9	5.6	6.2	6.1	5.4	4.4	3.3	2.2	1.7	4.1
Lat-15	3.0	3.8	4.6	5.3	5.7	6.1	6.0	5.8	5.0	4.3	3.1	2.6	4.6
Lat	3.4	4.2	4.8	5.2	5.4	5.6	5.6	5.5	5.1	4.5	3.5	3.0	4.6
Lat+15	3.6	4.3	4.7	4.8	4.7	4.9	4.9	5.0	4.9	4.6	3.6	3.2	4.4
DAYTONA BEACH, FL—Latitude 29.18													
0	3.1	3.9	5.0	6.2	6.4	6.1	6.0	5.7	4.9	4.2	3.4	2.9	4.8
Lat-15	3.8	4.5	5.5	6.4	6.4	6.0	5.9	5.8	5.2	4.7	4.1	3.6	5.2
Lat	4.3	4.9	5.7	6.3	6.0	5.5	5.5	5.6	5.3	5.0	4.6	4.1	5.2
Lat+15	4.6	5.1	5.6	5.9	5.4	4.8	4.9	5.1	5.1	5.1	4.8	4.4	5.1
JACKSONVILLE, FL—Latitude 30.50													
0	2.9	3.7	4.7	5.9	6.1	6.0	5.8	5.4	4.6	4.0	3.2	2.7	4.6
Lat-15	3.6	4.3	5.2	6.1	6.1	5.8	5.7	5.5	5.0	4.5	3.9	3.4	4.9
Lat	4.2	4.7	5.5	6.0	5.7	5.4	5.4	5.3	5.0	4.9	4.4	3.9	5.0
Lat+15	4.4	4.9	5.4	5.6	5.1	4.7	4.7	4.9	4.8	4.9	4.7	4.2	4.9
KEY WEST, FL—Latitude 24.55													
0	3.7	4.4	5.5	6.3	6.3	6.1	6.1	5.8	5.2	4.6	3.8	3.4	5.1
Lat-15	4.2	4.9	5.8	6.5	6.3	6.0	6.0	5.9	5.4	5.0	4.4	4.0	5.4
Lat	4.9	5.5	6.1	6.4	6.0	5.5	5.6	5.7	5.5	5.4	5.0	4.7	5.5
Lat+15	5.3	5.7	6.0	6.0	5.3	4.8	5.0	5.2	5.3	5.5	5.3	5.1	5.4

	Jan	Feb	Mar	Apr	May	Jun	Jul	Aug	Sep	Oct	Nov	Dec	Year
MIAMI, FL—Latitude 25.80													
0	3.5	4.2	5.2	6.0	6.0	5.6	5.8	5.6	4.9	4.4	3.7	3.3	4.8
Lat - 15	4.1	4.7	5.5	6.2	5.9	5.5	5.7	5.6	5.1	4.7	4.2	3.9	5.1
Lat	4.7	5.2	5.7	6.1	5.6	5.1	5.4	5.5	5.1	5.1	4.7	4.5	5.2
Lat + 15	5.0	5.4	5.6	5.7	5.0	4.5	4.8	5.0	4.9	5.1	4.9	4.9	5.1
TALLAHASSEE, FL—Latitude 30.38													
0	2.9	3.7	4.7	5.9	6.3	6.1	5.8	5.5	4.9	4.3	3.3	2.7	4.7
Lat - 15	3.6	4.3	5.2	6.1	6.2	6.0	5.7	5.6	5.3	5.0	4.1	3.4	5.0
Lat	4.0	4.7	5.4	6.0	5.9	5.6	5.4	5.4	5.3	5.4	4.6	4.0	5.1
Lat + 15	4.3	4.9	5.3	5.6	5.2	4.9	4.7	4.9	5.1	5.5	4.8	4.2	5.0
TAMPA, FL—Latitude 27.97													
0	3.2	4.0	5.1	6.2	6.4	6.1	5.8	5.5	4.9	4.4	3.6	3.1	4.9
Lat - 15	3.9	4.6	5.5	6.4	6.4	5.9	5.7	5.5	5.2	5.0	4.2	3.8	5.2
Lat	4.5	5.1	5.8	6.3	6.0	5.5	5.3	5.4	5.2	5.4	4.8	4.4	5.3
Lat + 15	4.8	5.3	5.7	5.9	5.3	4.8	4.7	4.9	5.0	5.5	5.1	4.7	5.1
WEST PALM BEACH, FL—Latitude 26.68													
0	3.3	4.0	5.0	5.9	6.0	5.7	5.9	5.6	4.8	4.2	3.4	3.1	4.7
Lat - 15	3.8	4.5	5.3	6.1	5.9	5.6	5.8	5.6	5.1	4.6	4.0	3.7	5.0
Lat	4.4	5.0	5.6	6.0	5.6	5.2	5.4	5.4	5.1	4.9	4.5	4.3	5.1
Lat + 15	4.7	5.1	5.5	5.6	5.0	4.5	4.8	5.0	4.9	5.0	4.7	4.7	5.0
ATLANTA, GA—Latitude 33.65													
0	2.6	3.4	4.5	5.7	6.2	6.4	6.2	5.7	4.8	4.1	2.9	2.4	4.6
Lat - 15	3.4	4.2	5.1	6.0	6.2	6.3	6.1	5.9	5.3	4.9	3.8	3.2	5.0
Lat	3.8	4.6	5.3	5.8	5.8	5.8	5.7	5.7	5.4	5.2	4.2	3.7	5.1
Lat + 15	4.1	4.7	5.1	5.4	5.2	5.1	5.0	5.2	5.1	5.3	4.5	3.9	4.9
AUGUSTA, GA—Latitude 33.37													
0	2.6	3.5	4.5	5.7	6.1	6.3	6.1	5.5	4.8	4.1	3.0	2.4	4.6
Lat - 15	3.4	4.3	5.1	6.0	6.1	6.2	6.0	5.7	5.2	4.9	3.8	3.3	5.0
Lat	3.9	4.7	5.3	5.9	5.8	5.7	5.6	5.5	5.3	5.3	4.3	3.8	5.1
Lat + 15	4.1	4.8	5.2	5.5	5.1	5.0	4.9	5.0	5.1	5.3	4.6	4.1	4.9
MACON, GA—Latitude 32.70													
0	2.7	3.5	4.6	5.7	6.2	6.3	6.0	5.6	4.8	4.1	3.0	2.5	4.6
Lat - 15	3.4	4.3	5.1	6.0	6.2	6.2	5.9	5.8	5.2	4.9	3.9	3.3	5.0
Lat	3.9	4.7	5.3	5.9	5.8	5.7	5.5	5.6	5.3	5.3	4.4	3.8	5.1
Lat + 15	4.1	4.8	5.2	5.5	5.2	5.0	4.9	5.1	5.1	5.4	4.6	4.0	4.9

SAVANNAH, GA—Latitude 32.13													
0	2.8	3.5	4.7	5.8	6.2	6.3	6.1	5.5	4.7	4.1	3.1	2.6	4.6
Lat - 15	3.5	4.3	5.2	6.1	6.2	6.1	6.0	5.6	5.1	4.8	3.9	3.4	5.0
Lat	4.0	4.7	5.4	6.0	5.8	5.7	5.6	5.4	5.1	5.1	4.4	3.9	5.1
Lat + 15	4.3	4.8	5.3	5.6	5.2	4.9	4.9	5.0	4.9	5.2	4.6	4.2	4.9
HILO, HI—Latitude 19.72													
0	3.8	4.3	4.6	4.8	5.2	5.4	5.2	5.3	5.0	4.3	3.7	3.5	4.6
Lat - 15	4.0	4.4	4.7	4.8	5.1	5.3	5.1	5.3	5.1	4.5	3.9	3.7	4.7
Lat	4.5	4.9	4.8	4.7	4.8	4.9	4.8	5.1	5.2	4.8	4.3	4.3	4.8
Lat + 15	4.9	5.0	4.7	4.4	4.3	4.3	4.3	4.7	5.0	4.8	4.5	4.6	4.6
HONOLULU, HI—Latitude 21.33													
0	3.9	4.7	5.4	5.9	6.4	6.5	6.6	6.5	5.9	5.0	4.1	3.7	5.4
Lat - 15	4.3	5.0	5.6	5.9	6.3	6.4	6.5	6.5	6.1	5.3	4.5	4.1	5.5
Lat	4.9	5.5	5.8	5.9	5.9	5.9	6.0	6.2	6.2	5.7	5.1	4.8	5.7
Lat + 15	5.3	5.8	5.8	5.5	5.3	5.1	5.3	5.7	6.0	5.8	5.4	5.2	5.5
BOISE, ID—Latitude 43.92													
0	1.6	2.5	3.8	5.3	6.5	7.2	7.6	6.6	5.9	3.4	1.9	1.4	4.4
Lat - 15	2.5	3.5	4.7	5.9	6.7	7.1	7.6	7.1	6.3	4.9	2.9	2.3	5.1
Lat	2.8	3.8	4.9	5.8	6.2	6.5	7.0	6.8	6.5	5.2	3.2	2.6	5.1
Lat + 15	2.9	3.9	4.8	5.4	5.5	5.5	6.0	6.2	6.0	5.3	3.3	2.8	4.8
POCATELLO, ID—Latitude 42.92													
0	1.7	2.6	3.8	5.1	6.2	7.0	7.3	6.3	5.0	3.5	2.0	1.5	4.3
Lat - 15	2.6	3.6	4.7	5.6	6.3	6.8	7.3	6.8	6.1	4.9	2.9	2.3	5.0
Lat	2.9	3.9	4.9	5.5	5.9	6.2	6.7	6.6	6.2	5.3	3.2	2.6	5.0
Lat + 15	3.0	4.0	4.8	5.1	5.2	5.4	5.8	6.0	6.0	5.3	3.4	2.8	4.7
CHICAGO, IL—Latitude 41.78													
0	1.8	2.6	3.5	4.6	5.7	6.3	6.1	5.4	4.2	3.0	1.8	1.5	3.9
Lat - 15	2.7	3.5	4.1	5.0	5.8	6.1	6.1	5.7	4.9	3.9	2.5	2.2	4.4
Lat	3.1	3.8	4.2	4.9	5.4	5.7	5.6	5.5	4.9	4.2	2.8	2.4	4.4
Lat + 15	3.3	3.9	4.1	4.5	4.8	4.9	4.9	5.0	4.7	4.2	2.9	2.6	4.1
MOLINE, IL—Latitude 41.45													
0	1.9	2.7	3.6	4.7	5.7	6.4	6.3	5.5	4.3	3.2	2.0	1.6	4.0
Lat - 15	2.9	3.8	4.3	5.1	5.8	6.2	6.2	5.9	5.1	4.2	2.8	2.4	4.6
Lat	3.3	4.1	4.4	5.0	5.4	5.7	5.8	5.6	5.1	4.5	3.1	2.7	4.6
Lat + 15	3.5	4.2	4.3	4.6	4.8	4.9	5.1	5.1	4.9	4.5	3.3	2.9	4.4

	Jan	Feb	Mar	Apr	May	Jun	Jul	Aug	Sep	Oct	Nov	Dec	Year
PEORIA, IL—Latitude 40.67													
0	2.0	2.8	3.6	4.8	5.8	6.4	6.3	5.5	4.4	3.2	2.0	1.6	4.0
Lat - 15	2.9	3.7	4.2	5.2	5.8	6.3	6.2	5.9	5.2	4.3	2.9	2.4	4.6
Lat	3.3	4.1	4.4	5.1	5.5	5.8	5.8	5.7	5.2	4.6	3.2	2.7	4.6
Lat + 15	3.5	4.2	4.2	4.7	4.8	5.0	5.1	5.1	5.0	4.6	3.3	2.9	4.4
ROCKFORD, IL—Latitude 42.20													
0	1.9	2.7	3.5	4.6	5.7	6.3	6.1	5.4	4.2	3.0	1.8	1.5	3.9
Lat - 15	2.9	3.8	4.3	5.0	5.7	6.1	6.1	5.7	5.0	4.0	2.6	2.3	4.5
Lat	3.3	4.1	4.4	4.9	5.4	5.6	5.7	5.5	5.0	4.2	2.9	2.6	4.5
Lat + 15	3.6	4.2	4.3	4.5	4.8	4.9	4.9	5.0	4.8	4.3	3.0	2.8	4.3
SPRINGFIELD, IL—Latitude 39.83													
0	2.1	2.9	3.7	5.0	6.0	6.5	6.4	5.7	4.6	3.4	2.2	1.7	4.2
Lat - 15	3.1	3.9	4.4	5.4	6.0	6.4	6.4	6.0	5.4	4.5	3.1	2.6	4.8
Lat	3.5	4.2	4.5	5.3	5.6	5.9	5.9	5.8	5.4	4.8	3.4	2.9	4.8
Lat + 15	3.8	4.4	4.4	4.9	5.0	5.1	5.1	5.3	5.2	4.8	3.5	3.1	4.5
FORT WAYNE, IN—Latitude 41.00													
0	1.8	2.6	3.5	4.6	5.6	6.2	6.1	5.3	4.3	3.0	1.8	1.4	3.9
Lat - 15	2.5	3.4	4.1	5.0	5.7	6.1	6.0	5.7	5.0	3.9	2.5	2.0	4.3
Lat	2.8	3.7	4.2	4.9	5.4	5.6	5.6	5.4	5.0	4.1	2.7	2.2	4.3
Lat + 15	3.0	3.8	4.1	4.5	4.8	4.9	4.9	4.9	4.8	4.1	2.8	2.3	4.1
INDIANAPOLIS, IN—Latitude 39.73													
0	2.0	2.8	3.7	4.9	5.9	6.5	6.3	5.6	4.6	3.3	2.1	1.6	4.1
Lat - 15	2.8	3.6	4.3	5.2	5.9	6.3	6.2	5.9	5.2	4.2	2.8	2.3	4.6
Lat	3.1	3.9	4.4	5.1	5.6	5.8	5.8	5.7	5.3	4.5	3.1	2.6	4.6
Lat + 15	3.3	4.0	4.3	4.7	4.9	5.0	5.1	5.2	5.1	4.5	3.2	2.7	4.3
SOUTH BEND, IN—Latitude 41.70													
0	1.7	2.5	3.4	4.6	5.6	6.2	6.0	5.3	4.1	2.9	1.7	1.4	3.8
Lat - 15	2.4	3.3	4.0	5.0	5.7	6.1	6.0	5.6	4.8	3.7	2.3	1.9	4.2
Lat	2.7	3.5	4.1	4.8	5.4	5.6	5.6	5.4	4.8	3.9	2.5	2.0	4.2
Lat + 15	2.8	3.6	4.0	4.5	4.7	4.8	4.9	4.9	4.6	3.9	2.6	2.1	4.0
DES MOINES, IA—Latitude 41.53													
0	2.0	2.8	3.8	4.9	5.8	6.5	6.5	5.7	4.4	3.2	2.1	1.7	4.1
Lat - 15	3.2	3.9	4.6	5.3	5.9	6.4	6.5	6.0	5.2	4.4	3.1	2.6	4.8
Lat	3.6	4.3	4.7	5.2	5.5	5.8	6.0	5.8	5.3	4.7	3.4	3.0	4.8
Lat + 15	3.9	4.4	4.7	4.8	4.9	5.1	5.2	5.3	5.1	4.7	3.6	3.2	4.6

MASON CITY, IA—Latitude 43.15													
0	1.9	2.7	3.7	4.7	5.8	6.3	6.3	5.5	4.3	3.0	1.8	1.5	4.0
Lat - 15	3.1	3.9	4.6	5.1	5.9	6.2	6.3	5.9	5.1	4.1	2.8	2.5	4.6
Lat	3.5	4.3	4.7	5.0	5.5	5.7	5.8	5.6	5.1	4.4	3.1	2.8	4.6
Lat + 15	3.8	4.4	4.7	4.6	4.9	4.9	5.1	5.1	4.9	4.4	3.2	3.0	4.4
SIOUX CITY, IA—Latitude 42.40													
0	1.9	2.8	3.8	4.9	5.8	6.6	6.5	5.7	4.4	3.2	2.0	1.6	4.1
Lat - 15	3.1	3.9	4.7	5.4	5.9	6.4	6.5	6.1	5.2	4.4	3.0	2.6	4.8
Lat	3.6	4.3	4.9	5.3	5.5	5.9	6.0	5.8	5.3	4.7	3.4	3.0	4.8
Lat + 15	3.9	4.4	4.8	4.9	4.9	5.1	5.3	5.3	5.1	4.7	3.5	3.2	4.6
DODGE CITY, KS—Latitude 37.77													
0	2.7	3.6	4.7	5.9	6.5	7.2	7.2	6.3	5.1	4.0	2.8	2.4	4.9
Lat - 15	4.0	4.8	5.5	6.3	6.5	7.0	7.1	6.6	5.9	5.2	4.0	3.6	5.5
Lat	4.6	5.3	5.8	6.2	6.1	6.4	6.5	6.4	6.0	5.6	4.6	4.2	5.6
Lat + 15	5.0	5.5	5.7	5.8	5.4	5.5	5.7	5.8	5.7	5.7	4.8	4.6	5.4
TOPEKA, KS—Latitude 39.07													
0	2.3	3.0	4.0	5.1	5.9	6.5	6.6	5.8	4.6	3.5	2.4	1.9	4.3
Lat - 15	3.4	4.0	4.7	5.5	5.9	6.3	6.5	6.1	5.3	4.6	3.4	2.9	4.9
Lat	3.9	4.4	4.9	5.4	5.5	5.8	6.0	5.9	5.4	4.9	3.8	3.4	4.9
Lat + 15	4.2	4.5	4.7	5.0	4.9	5.0	5.3	5.3	5.2	4.9	4.0	3.6	4.7
WICHITA, KS—Latitude 37.65													
0	2.5	3.3	4.3	5.4	6.1	6.7	6.8	6.1	4.9	3.8	2.6	2.2	4.6
Lat - 15	3.6	4.3	5.0	5.8	6.1	6.5	6.8	6.4	5.5	4.8	3.7	3.3	5.2
Lat	4.2	4.7	5.2	5.7	5.7	6.0	6.3	6.1	5.6	5.2	4.2	3.8	5.2
Lat + 15	4.5	4.9	5.1	5.3	5.1	5.2	5.5	5.6	5.4	5.3	4.4	4.1	5.0
LEXINGTON, KY—Latitude 38.03													
0	2.0	2.8	3.7	4.9	5.7	6.2	6.0	5.5	4.4	3.4	2.2	1.7	4.1
Lat - 15	2.8	3.5	4.3	5.2	5.7	6.0	5.9	5.7	5.0	4.3	2.9	2.4	4.5
Lat	3.1	3.8	4.4	5.1	5.4	5.6	5.5	5.5	5.1	4.6	3.2	2.7	4.5
Lat + 15	3.3	3.8	4.3	4.7	4.8	4.8	4.8	5.0	4.8	4.6	3.4	2.9	4.3
LOUISVILLE, KY—Latitude 38.18													
0	2.0	2.8	3.8	5.0	5.8	6.3	6.1	5.6	4.5	3.5	2.2	1.7	4.1
Lat - 15	2.8	3.6	4.4	5.3	5.8	6.2	6.0	5.9	5.1	4.4	3.0	2.4	4.6
Lat	3.1	3.9	4.5	5.2	5.5	5.7	5.6	5.7	5.2	4.7	3.3	2.7	4.6
Lat + 15	3.3	4.0	4.4	4.8	4.9	4.9	4.9	5.1	5.0	4.7	3.4	2.9	4.4

	Jan	Feb	Mar	Apr	May	Jun	Jul	Aug	Sep	Oct	Nov	Dec	Year
BATON ROUGE, LA—Latitude 30.53													
0	2.6	3.5	4.4	5.4	5.9	6.0	5.7	5.4	4.8	4.3	3.0	2.5	4.5
Lat - 15	3.2	4.0	4.8	5.6	5.9	5.9	5.6	5.6	5.2	5.0	3.7	3.1	4.8
Lat	3.6	4.4	5.0	5.5	5.5	5.5	5.3	5.4	5.2	5.3	4.1	3.6	4.9
Lat + 15	3.8	4.5	4.9	5.1	4.9	4.7	4.6	4.9	5.0	5.4	4.4	3.8	4.7
LAKE CHARLES, LA—Latitude 30.12													
0	2.7	3.6	4.5	5.4	6.0	6.3	6.0	5.6	5.0	4.3	3.2	2.6	4.6
Lat - 15	3.3	4.1	4.9	5.5	6.0	6.2	5.9	5.7	5.4	5.0	3.9	3.2	4.9
Lat	3.7	4.5	5.1	5.4	5.6	5.7	5.5	5.6	5.4	5.4	4.3	3.7	5.0
Lat + 15	3.9	4.6	4.9	5.1	5.0	5.0	4.9	5.1	5.2	5.4	4.6	3.9	4.8
NEW ORLEANS, LA—Latitude 29.98													
0	2.7	3.6	4.5	5.5	6.1	6.1	5.7	5.5	4.9	4.3	3.1	2.6	4.6
Lat - 15	3.3	4.2	4.9	5.7	6.0	6.0	5.7	5.6	5.3	5.0	3.8	3.2	4.9
Lat	3.7	4.5	5.0	5.6	5.7	5.5	5.3	5.4	5.3	5.3	4.3	3.7	5.0
Lat + 15	3.9	4.6	4.9	5.3	5.1	4.8	4.7	4.9	5.1	5.4	4.5	3.9	4.8
SHREVEPORT, LA—Latitude 32.47													
0	2.6	3.4	4.4	5.4	6.0	6.4	6.4	6.0	5.0	4.1	3.0	2.5	4.6
Lat - 15	3.4	4.1	4.9	5.6	6.0	6.3	6.3	6.2	5.4	4.9	3.7	3.2	5.0
Lat	3.8	4.5	5.1	5.5	5.7	5.8	5.9	6.0	5.5	5.2	4.2	3.7	5.1
Lat + 15	4.0	4.6	5.0	5.1	5.0	5.0	5.2	5.5	5.3	5.3	4.4	3.9	4.9
CARIBOU, ME—Latitude 46.87													
0	1.6	2.6	3.8	4.6	5.2	5.7	5.6	4.8	3.6	2.3	1.4	1.2	3.6
Lat - 15	2.9	3.9	5.0	5.1	5.3	5.6	5.6	5.2	4.3	3.1	2.2	2.2	4.2
Lat	3.3	4.3	5.2	5.0	4.9	5.1	5.2	4.9	4.3	3.3	2.4	2.5	4.2
Lat + 15	3.5	4.5	5.2	4.7	4.4	4.5	4.6	4.5	4.1	3.3	2.5	2.7	4.0
PORTLAND, ME—Latitude 43.65													
0	1.9	2.8	3.8	4.7	5.6	6.1	6.0	5.4	4.2	2.9	1.8	1.5	3.9
Lat - 15	3.1	4.1	4.8	5.2	5.7	6.0	6.0	5.8	5.1	4.0	2.8	2.6	4.6
Lat	3.6	4.5	5.0	5.1	5.3	5.5	5.6	5.5	5.1	4.3	3.1	3.0	4.6
Lat + 15	3.9	4.7	5.0	4.7	4.7	4.7	4.9	5.0	4.9	4.3	3.2	3.2	4.4
BALTIMORE, MD—Latitude 39.18													
0	2.1	2.9	3.9	4.9	5.6	6.2	6.0	5.3	4.4	3.3	2.2	1.8	4.0
Lat - 15	3.1	3.8	4.6	5.3	5.7	6.0	6.0	5.6	5.0	4.3	3.2	2.7	4.6
Lat	3.5	4.2	4.8	5.2	5.3	5.6	5.5	5.4	5.1	4.6	3.6	3.1	4.6
Lat + 15	3.7	4.3	4.7	4.8	4.7	4.8	4.9	4.9	4.8	4.6	3.7	3.3	4.4

BOSTON, MA—Latitude 42.37													
0	3.9	1.5	1.9	3.0	4.3	5.4	6.1	6.1	5.6	4.7	3.7	2.7	1.9
Lat - 15	4.5	2.5	2.8	4.1	5.0	5.7	6.0	6.0	5.7	5.2	4.6	3.8	3.0
Lat	4.6	2.9	3.1	4.3	5.1	5.5	5.6	5.5	5.3	5.0	4.7	4.2	3.4
Lat + 15	4.4	3.1	3.3	4.4	4.9	5.0	4.9	4.8	4.7	4.7	4.6	4.3	3.6

WORCHESTER, MA—Latitude 42.37													
0	3.9	1.5	1.9	3.0	4.2	5.2	6.0	6.0	5.5	4.7	3.8	2.8	1.9
Lat - 15	4.5	2.4	2.8	4.0	4.9	5.6	5.9	5.8	5.5	5.1	4.6	3.8	3.0
Lat	4.5	2.8	3.0	4.3	5.0	5.3	5.5	5.4	5.2	5.0	4.8	4.2	3.4
Lat + 15	4.3	3.0	3.2	4.3	4.7	4.8	4.9	4.6	4.6	4.6	4.7	4.4	3.6

ALPENA, MI—Latitude 45.07													
0	3.7	1.2	1.5	2.5	3.8	5.1	6.1	6.2	5.7	4.7	3.7	2.5	1.6
Lat - 15	4.3	1.8	2.1	3.3	4.5	5.5	6.1	6.1	5.8	5.2	4.7	3.6	2.5
Lat	4.2	2.0	2.3	3.5	4.5	5.3	5.7	5.5	5.4	5.1	4.9	3.9	2.8
Lat + 15	4.0	2.1	2.3	3.5	4.3	4.8	4.9	4.8	4.8	4.7	4.9	4.0	2.9

DETROIT, MI—Latitude 42.42													
0	3.8	1.3	1.7	2.8	4.1	5.3	6.1	6.2	5.6	4.6	3.4	2.5	1.6
Lat - 15	4.3	1.9	2.4	3.7	4.8	5.6	6.1	6.1	5.7	5.0	4.1	3.3	2.4
Lat	4.2	2.1	2.6	3.9	4.8	5.4	5.6	5.6	5.4	4.9	4.2	3.6	2.7
Lat + 15	4.0	2.2	2.6	3.9	4.6	4.9	4.9	4.9	4.8	4.5	4.1	3.7	2.8

FLINT, MI—Latitude 42.97													
0	3.7	1.3	1.6	2.7	4.0	5.2	6.0	6.1	5.6	4.6	3.4	2.5	1.6
Lat - 15	4.2	1.8	2.2	3.5	4.7	5.6	6.0	6.0	5.7	5.0	4.1	3.3	2.3
Lat	4.1	2.0	2.4	3.7	4.7	5.3	5.5	5.5	5.3	4.8	4.2	3.6	2.6
Lat + 15	3.9	2.1	2.5	3.7	4.5	4.8	4.8	4.7	4.7	4.5	4.1	3.7	2.7

GRAND RAPIDS, MI—Latitude 42.88													
0	3.8	1.3	1.6	2.7	4.1	5.3	6.2	6.3	5.7	4.7	3.5	2.5	1.6
Lat - 15	4.3	1.8	2.2	3.6	4.8	5.7	6.1	6.2	5.8	5.1	4.2	3.3	2.3
Lat	4.2	2.0	2.4	3.8	4.8	5.4	5.7	5.7	5.4	5.0	4.3	3.5	2.5
Lat + 15	4.0	2.0	2.5	3.8	4.6	4.9	5.0	4.9	4.8	4.6	4.2	3.6	2.6

HOUGHTON, MI—Latitude 47.17													
0	3.6	1.1	1.3	2.3	3.6	5.0	6.0	6.0	5.5	4.6	3.5	2.2	1.3
Lat - 15	4.1	1.6	1.9	3.2	4.4	5.5	6.0	5.9	5.6	5.2	4.5	3.2	2.1
Lat	4.1	1.8	2.1	3.4	4.4	5.2	5.6	5.4	5.3	5.2	4.7	3.5	2.3
Lat + 15	3.8	1.9	2.1	3.4	4.2	4.7	4.8	4.7	4.7	4.8	4.7	3.6	2.5

	Jan	Feb	Mar	Apr	May	Jun	Jul	Aug	Sep	Oct	Nov	Dec	Year
LANSING, MI—Latitude 42.78													
0	1.6	2.5	3.5	4.6	5.6	6.2	6.1	5.2	4.0	2.7	1.7	1.3	3.8
Lat - 15	2.4	3.3	4.2	5.0	5.7	6.0	6.1	5.6	4.7	3.6	2.3	1.8	4.2
Lat	2.6	3.6	4.3	4.9	5.3	5.6	5.6	5.4	4.8	3.8	2.4	2.0	4.2
Lat + 15	2.8	3.7	4.2	4.5	4.7	4.8	4.9	4.9	4.5	3.7	2.5	2.1	4.0
SAULT STE. MARIE, MI—Latitude 46.73													
0	1.6	2.6	3.9	4.8	5.7	6.1	6.0	5.0	3.5	2.2	1.4	1.2	3.7
Lat - 15	2.5	3.8	5.1	5.4	5.8	5.9	6.0	5.4	4.2	3.0	1.9	1.9	4.3
Lat	2.8	4.2	5.3	5.3	5.4	5.4	5.6	5.2	4.2	3.1	2.1	2.2	4.2
Lat + 15	2.9	4.3	5.3	5.0	4.8	4.7	4.9	4.7	4.0	3.1	2.1	2.3	4.0
TRAVERSE CITY, MI—Latitude 44.73													
0	1.5	2.4	3.5	4.6	5.6	6.2	6.1	5.1	3.7	2.4	1.4	1.2	3.6
Lat - 15	2.1	3.3	4.4	5.0	5.8	6.1	6.1	5.4	4.4	3.2	2.0	1.6	4.1
Lat	2.3	3.5	4.6	4.9	5.4	5.6	5.6	5.2	4.4	3.4	2.1	1.8	4.1
Lat + 15	2.4	3.6	4.5	4.6	4.8	4.8	4.9	4.7	4.2	3.3	2.1	1.8	3.8
DULUTH, MN—Latitude 46.83													
0	1.6	2.6	3.8	4.8	5.6	6.0	6.1	5.1	3.7	2.5	1.5	1.2	3.7
Lat - 15	2.8	4.0	5.0	5.5	5.7	5.9	6.1	5.5	4.5	3.4	2.4	2.2	4.4
Lat	3.2	4.4	5.2	5.4	5.4	5.4	5.6	5.3	4.5	3.6	2.6	2.5	4.4
Lat + 15	3.4	4.5	5.2	5.1	4.7	4.7	4.9	4.8	4.3	3.6	2.7	2.7	4.2
INTERNATIONAL FALLS, MN—Latitude 48.57													
0	1.4	2.4	3.7	4.8	5.5	5.8	5.8	4.9	3.5	2.2	1.4	1.1	3.6
Lat - 15	2.6	3.9	4.9	5.5	5.6	5.7	5.9	5.4	4.3	3.1	2.1	2.1	4.3
Lat	3.0	4.3	5.1	5.4	5.2	5.2	5.4	5.1	4.3	3.2	2.3	2.4	4.3
Lat + 15	3.2	4.5	5.1	5.1	4.6	4.5	4.7	4.6	4.1	3.2	2.4	2.6	4.0
MINNEAPOLIS, MN—Latitude 44.88													
0	1.8	2.7	3.8	4.7	5.7	6.3	6.3	5.4	4.1	2.8	1.7	1.4	3.9
Lat - 15	3.1	4.1	4.8	5.3	5.8	6.1	6.4	5.8	4.9	3.9	2.6	2.3	4.6
Lat	3.5	4.5	5.0	5.1	5.5	5.6	5.9	5.6	5.0	4.1	2.9	2.7	4.6
Lat + 15	3.8	4.7	4.9	4.8	4.8	4.9	5.1	5.1	4.7	4.2	3.0	2.9	4.4
ROCHESTER, MN—Latitude 43.92													
0	1.8	2.7	3.7	4.6	5.6	6.2	6.2	5.3	4.0	2.8	1.7	1.4	3.8
Lat - 15	2.9	3.9	4.6	5.1	5.7	6.0	6.2	5.7	4.8	3.8	2.6	2.3	4.5
Lat	3.3	4.2	4.7	4.9	5.3	5.5	5.7	5.5	4.8	4.1	2.8	2.6	4.5
Lat + 15	3.6	4.4	4.7	4.6	4.7	4.8	5.0	4.9	4.6	4.1	2.9	2.8	4.3

SAINT CLOUD, MN—Latitude 45.55

0	1.7	3.8	4.7	5.6	6.2	6.3	5.4	4.0	2.7	1.7	1.3	3.8
Lat - 15	3.0	4.8	5.3	5.8	6.1	6.3	5.8	4.8	3.8	2.6	2.3	4.6
Lat	3.4	5.0	5.1	5.4	5.6	5.8	5.6	4.9	4.1	2.9	2.6	4.6
Lat + 15	3.7	5.0	4.7	4.8	4.8	5.1	5.0	4.6	4.1	3.0	2.8	4.4

JACKSON, MS—Latitude 32.32

0	2.6	4.5	5.5	6.1	6.4	6.2	5.8	4.9	4.2	3.0	2.4	4.6
Lat - 15	3.5	5.0	5.8	6.1	6.3	6.1	6.0	5.4	5.0	3.7	3.2	5.0
Lat	4.1	5.2	5.7	5.8	5.8	5.7	5.8	5.5	5.3	4.2	3.6	5.1
Lat + 15	4.5	5.1	5.3	5.1	5.0	5.0	5.3	5.2	5.4	4.4	3.9	4.9

MERIDIAN, MS—Latitude 32.33

0	2.6	4.4	5.4	5.9	6.2	5.9	5.6	4.8	4.1	2.9	2.4	4.5
Lat - 15	3.2	4.8	5.7	5.9	6.0	5.8	5.7	5.2	4.8	3.7	3.1	4.8
Lat	3.6	5.0	5.6	5.6	5.6	5.4	5.5	5.2	5.2	4.1	3.5	4.9
Lat + 15	3.8	4.9	5.2	5.0	4.9	4.8	5.0	5.0	5.2	4.3	3.8	4.7

COLUMBIA, MO—Latitude 38.82

0	2.2	4.0	5.2	6.0	6.6	6.6	6.6	5.9	4.6	3.5	2.3	4.3
Lat - 15	3.3	4.7	5.6	6.0	6.4	6.6	6.5	6.2	5.3	4.5	3.2	4.9
Lat	3.8	4.9	5.5	5.6	5.9	6.1	6.0	5.9	5.4	4.9	3.6	4.9
Lat + 15	4.0	4.8	5.1	5.0	5.1	5.3	5.3	5.4	5.2	4.9	3.8	4.7

KANSAS CITY, MO—Latitude 39.30

0	2.2	3.9	5.1	5.9	6.5	6.6	6.6	5.8	4.6	3.6	2.3	4.3
Lat - 15	3.4	4.7	5.5	5.9	6.3	6.5	6.5	6.1	5.3	4.6	3.4	4.9
Lat	3.8	4.8	5.4	5.6	5.8	6.0	6.0	5.9	5.4	5.0	3.8	4.9
Lat + 15	4.1	4.7	5.0	4.9	5.0	5.3	5.3	5.2	5.2	5.0	4.0	4.7

SPRINGFIELD, MO—Latitude 37.23

0	2.4	4.1	5.2	5.9	6.4	6.6	6.6	5.9	4.7	3.7	2.5	4.4
Lat - 15	3.4	4.7	5.6	5.9	6.2	6.5	6.5	6.2	5.3	4.6	3.4	4.9
Lat	3.8	4.9	5.4	5.6	5.7	6.0	6.0	5.9	5.4	4.9	3.8	4.9
Lat + 15	4.1	4.8	5.0	4.9	5.0	5.2	5.3	5.4	5.1	5.0	3.9	4.7

ST. LOUIS, MO—Latitude 38.75

0	2.2	3.9	5.0	5.9	6.4	6.4	6.6	5.7	4.6	3.5	2.3	4.2
Lat - 15	3.2	4.6	5.4	5.9	6.3	6.3	6.5	6.0	5.3	4.5	3.2	4.8
Lat	3.6	4.2	5.3	5.6	5.8	5.9	6.0	5.7	5.3	4.8	3.5	4.8
Lat + 15	3.8	4.3	4.9	4.9	5.0	5.1	5.2	5.2	5.1	4.8	3.7	4.6

	Jan	Feb	Mar	Apr	May	Jun	Jul	Aug	Sep	Oct	Nov	Dec	Year
BILLINGS, MT—Latitude 45.80													
0	1.7	2.6	3.8	5.0	5.9	6.7	7.0	6.1	4.5	3.1	1.9	1.4	4.1
Lat - 15	2.9	3.9	4.9	5.6	6.0	6.6	7.1	6.7	5.6	4.5	3.2	2.7	5.0
Lat	3.3	4.2	5.1	5.5	5.6	6.0	6.5	6.5	5.7	4.8	3.5	3.1	5.0
Lat + 15	3.5	4.4	5.1	5.1	5.0	5.2	5.7	5.9	5.5	4.9	3.7	3.3	4.8
CUT BANK, MT—Latitude 48.60													
0	1.4	2.2	3.5	4.9	5.9	6.6	6.9	5.8	4.2	2.8	1.6	1.1	3.9
Lat - 15	2.6	3.7	4.8	5.6	6.1	6.5	7.0	6.5	5.4	4.3	2.9	2.4	4.8
Lat	3.0	4.0	4.9	5.5	5.7	5.9	6.5	6.2	5.4	4.6	3.3	2.7	4.8
Lat + 15	3.2	4.2	4.9	5.1	5.1	5.1	5.6	5.6	5.2	4.6	3.5	2.9	4.6
GLASGOW, MT—Latitude 48.22													
0	1.5	2.3	3.6	4.7	5.7	6.5	6.7	5.7	4.1	2.7	1.6	1.2	3.9
Lat - 15	2.7	3.7	4.8	5.4	5.8	6.4	6.8	6.4	5.2	4.1	2.8	2.3	4.7
Lat	3.1	4.1	5.0	5.3	5.5	5.9	6.3	6.1	5.3	4.3	3.1	2.6	4.7
Lat + 15	3.3	4.2	4.9	4.9	4.8	5.0	5.5	5.5	5.0	4.4	3.3	2.8	4.5
GREAT FALLS, MT—Latitude 47.48													
0	1.4	2.4	3.7	4.9	5.8	6.7	7.1	5.9	4.3	2.8	1.7	1.2	4.0
Lat - 15	2.6	3.7	4.8	5.5	6.0	6.6	7.2	6.5	5.4	4.3	2.9	2.3	4.8
Lat	3.0	4.0	5.0	5.4	5.6	6.0	6.6	6.3	5.5	4.6	3.3	2.7	4.8
Lat + 15	3.2	4.2	5.0	5.0	4.9	5.2	5.7	5.7	5.3	4.6	3.4	2.9	4.6
HELENA, MT—Latitude 46.60													
0	1.5	2.3	3.5	4.8	5.8	6.5	7.0	5.9	4.4	2.9	1.7	1.2	4.0
Lat - 15	2.5	3.6	4.6	5.4	5.9	6.4	7.1	6.5	5.5	4.3	2.8	2.2	4.7
Lat	2.8	3.9	4.8	5.3	5.5	5.8	6.5	6.2	5.6	4.6	3.1	2.6	4.7
Lat + 15	3.0	4.0	4.7	4.9	4.9	5.0	5.7	5.7	5.4	4.7	3.3	2.8	4.5
KALISPELL, MT—Latitude 48.30													
0	1.2	2.0	3.1	4.3	5.4	6.1	6.7	5.6	4.0	2.5	1.3	1.0	3.6
Lat - 15	1.9	2.9	4.0	4.9	5.5	6.0	6.7	6.2	5.1	3.6	1.9	1.5	4.2
Lat	2.1	3.1	4.1	4.8	5.1	5.5	6.2	5.9	5.2	3.8	2.0	1.7	4.1
Lat + 15	2.2	3.2	4.0	4.4	4.6	4.7	5.4	5.4	5.0	3.8	2.1	1.7	3.9
LEWISTOWN, MT—Latitude 47.05													
0	1.5	2.3	3.6	4.8	5.7	6.4	6.8	5.8	4.2	2.8	1.7	1.2	3.9
Lat - 15	2.6	3.6	4.7	5.4	5.8	6.3	6.9	6.4	5.3	4.2	2.9	2.3	4.7
Lat	3.0	3.9	4.9	5.3	5.4	5.8	6.4	6.1	5.4	4.5	3.2	2.7	4.7
Lat + 15	3.2	4.1	4.8	4.9	4.8	5.0	5.5	5.5	5.1	4.6	3.4	2.8	4.5

MILES CITY, MT—Latitude 46.43													
0	1.7	2.6	3.8	4.9	5.9	6.8	7.0	6.0	4.4	3.0	1.8	1.4	4.1
Lat - 15	3.0	3.9	5.0	5.6	6.0	6.7	7.1	6.7	5.6	4.4	3.0	2.6	5.0
Lat	3.4	4.3	5.2	5.5	5.7	6.1	6.5	6.5	5.7	4.7	3.4	3.0	5.0
Lat + 15	3.6	4.5	5.1	5.0	5.0	5.3	5.6	5.8	5.4	4.8	3.6	3.2	4.8
MISSOULA, MT—Latitude 46.92													
0	1.3	2.1	3.2	4.5	5.5	6.3	6.9	5.8	4.2	2.7	1.4	1.1	3.8
Lat - 15	2.0	3.0	4.0	5.0	5.6	6.1	7.0	6.4	5.3	3.9	2.1	1.7	4.4
Lat	2.2	3.2	4.1	4.9	5.2	5.6	6.4	6.1	5.4	4.1	2.3	1.9	4.3
Lat + 15	2.3	3.2	4.0	4.5	4.6	4.8	5.6	5.5	5.1	4.1	2.4	2.0	4.0
GRAND ISLAND, NE—Latitude 40.97													
0	2.2	3.0	4.1	5.3	6.1	6.9	6.8	6.0	4.7	3.5	2.3	1.9	4.4
Lat - 15	3.6	4.2	5.0	5.8	6.1	6.7	6.8	6.4	5.5	4.7	3.5	3.1	5.1
Lat	4.1	4.6	5.2	5.7	5.7	6.2	6.3	6.1	5.6	5.1	4.0	3.6	5.2
Lat + 15	4.4	4.8	5.2	5.3	5.1	5.3	5.5	5.6	5.4	5.2	4.2	3.9	5.0
OMAHA, NE—Latitude 41.37													
0	2.1	2.9	3.9	5.0	5.9	6.7	6.6	5.7	4.5	3.3	2.1	1.7	4.2
Lat - 15	3.3	4.0	4.7	5.5	6.0	6.5	6.5	6.1	5.2	4.4	3.2	2.7	4.9
Lat	3.8	4.4	4.9	5.3	5.6	6.0	6.0	5.8	5.3	4.7	3.5	3.2	4.9
Lat + 15	4.1	4.6	4.8	5.0	5.0	5.2	5.3	5.3	5.1	4.7	3.7	3.4	4.7
SCOTTSBLUFF, NE—Latitude 41.87													
0	2.1	3.0	4.1	5.3	6.0	6.9	7.0	6.2	4.9	3.5	2.3	1.9	4.4
Lat - 15	3.5	4.3	5.1	5.8	6.0	6.7	7.0	6.7	5.9	4.9	3.6	3.2	5.2
Lat	4.0	4.8	5.3	5.7	5.7	6.2	6.5	6.4	6.0	5.3	4.1	3.8	5.3
Lat + 15	4.3	4.9	5.2	5.3	5.0	5.3	5.6	5.8	5.7	5.3	4.3	4.1	5.1
ELKO, NV—Latitude 40.83													
0	2.1	2.9	4.0	5.3	6.3	7.1	7.4	6.6	5.4	3.8	2.3	1.9	4.6
Lat - 15	3.3	4.1	4.9	5.7	6.4	6.9	7.3	7.0	6.5	5.2	3.5	3.1	5.3
Lat	3.8	4.5	5.1	5.6	6.0	6.3	6.8	6.8	6.6	5.7	4.0	3.6	5.4
Lat + 15	4.1	4.7	5.0	5.2	5.3	5.4	5.9	6.1	6.4	5.8	4.2	3.9	5.2
ELY, NV—Latitude 39.28													
0	2.6	3.4	4.5	5.8	6.6	7.5	7.3	6.5	5.6	4.1	2.8	2.2	4.9
Lat - 15	4.0	4.7	5.5	6.3	6.6	7.2	7.2	6.9	6.6	5.5	4.1	3.6	5.7
Lat	4.6	5.2	5.7	6.2	6.2	6.6	6.6	6.6	6.7	6.0	4.7	4.3	5.8
Lat + 15	5.0	5.5	5.6	5.7	5.5	5.6	5.7	6.0	6.5	6.1	5.0	4.7	5.6

	Jan	Feb	Mar	Apr	May	Jun	Jul	Aug	Sep	Oct	Nov	Dec	Year
LAS VEGAS, NV—Latitude 36.08													
0	3.0	4.0	5.4	6.9	7.8	8.4	7.9	7.2	6.2	4.7	3.4	2.8	5.7
Lat - 15	4.4	5.3	6.4	7.5	7.8	8.1	7.7	7.5	7.1	6.1	4.8	4.2	6.4
Lat	5.1	5.9	6.7	7.4	7.3	7.4	7.1	7.2	7.2	6.6	5.5	4.9	6.5
Lat + 15	5.6	6.1	6.6	6.8	6.5	6.3	6.2	6.5	7.0	6.8	5.9	5.4	6.3
RENO, NV—Latitude 39.50													
0	2.3	3.2	4.5	5.9	7.0	7.6	7.8	6.9	5.7	4.1	2.6	2.1	5.0
Lat - 15	3.6	4.4	5.5	6.5	7.1	7.4	7.7	7.4	6.8	5.6	3.9	3.3	5.8
Lat	4.1	4.9	5.7	6.4	6.6	6.8	7.1	7.1	6.9	6.1	4.4	3.9	5.8
Lat + 15	4.4	5.1	5.6	5.9	5.8	5.8	6.1	6.4	6.7	6.2	4.6	4.2	5.6
TONOPAH, NV—Latitude 38.07													
0	2.7	3.6	4.8	6.2	7.1	7.9	7.8	7.0	5.9	4.4	3.0	2.4	5.2
Lat - 15	4.1	4.9	5.8	6.7	7.1	7.6	7.7	7.3	6.9	5.9	4.4	3.8	6.0
Lat	4.8	5.4	6.1	6.6	6.7	6.9	7.1	7.1	7.1	6.4	5.0	4.5	6.1
Lat + 15	5.1	5.6	6.0	6.1	5.9	5.9	6.1	6.4	6.8	6.5	5.3	4.9	5.9
WINNEMUCCA, NV—Latitude 40.90													
0	2.1	2.9	4.1	5.5	6.6	7.4	7.7	6.7	5.5	3.8	2.3	1.9	4.7
Lat - 15	3.3	4.1	5.0	6.0	6.7	7.1	7.6	7.2	6.6	5.3	3.5	3.0	5.5
Lat	3.7	4.5	5.2	5.9	6.2	6.5	7.0	6.9	6.7	5.7	3.9	3.5	5.5
Lat + 15	4.0	4.6	5.1	5.5	5.5	5.6	6.0	6.3	6.5	5.8	4.1	3.8	5.2
CONCORD, NH—Latitude 43.20													
0	1.9	2.8	3.9	4.7	5.6	6.1	6.1	5.3	4.2	2.9	1.8	1.5	3.9
Lat - 15	3.1	4.1	4.8	5.2	5.7	5.9	6.0	5.7	5.0	3.9	2.7	2.5	4.6
Lat	3.5	4.5	5.0	5.1	5.3	5.4	5.6	5.5	5.0	4.2	3.0	2.8	4.6
Lat + 15	3.8	4.7	5.0	4.7	4.7	4.7	4.9	4.9	4.8	4.2	3.1	3.1	4.4
ATLANTIC CITY, NJ—Latitude 39.45													
0	2.0	2.8	3.9	4.9	5.6	6.1	5.9	5.3	4.4	3.3	2.2	1.8	4.0
Lat - 15	3.0	3.8	4.6	5.3	5.7	6.0	5.9	5.6	5.0	4.3	3.2	2.7	4.6
Lat	3.5	4.1	4.8	5.2	5.3	5.5	5.5	5.4	5.1	4.6	3.6	3.1	4.6
Lat + 15	3.7	4.3	4.7	4.8	4.7	4.8	4.8	4.9	4.9	4.6	3.8	3.3	4.4
NEWARK, NJ—Latitude 40.70													
0	1.9	2.7	3.8	4.8	5.5	6.0	5.9	5.2	4.3	3.2	2.0	1.6	3.9
Lat - 15	2.9	3.7	4.5	5.2	5.5	5.8	5.8	5.5	4.9	4.1	2.9	2.4	4.4
Lat	3.3	4.0	4.6	5.1	5.2	5.4	5.4	5.3	5.0	4.4	3.2	2.8	4.5
Lat + 15	3.5	4.1	4.5	4.7	4.6	4.7	4.7	4.8	4.7	4.4	3.3	3.0	4.3

ALBUQUERQUE, NM—Latitude 35.05

0	3.2	4.2	5.4	6.8	7.7	8.1	7.5	6.9	5.9	4.7	3.5	2.9	5.6
Lat - 15	4.6	5.4	6.3	7.3	7.7	7.8	7.4	7.2	6.6	5.9	4.8	4.3	6.3
Lat	5.3	6.0	6.5	7.2	7.2	7.1	6.9	6.9	6.8	6.5	5.5	5.0	6.4
Lat + 15	5.8	6.2	6.5	6.6	6.3	6.1	6.0	6.3	6.5	6.6	5.9	5.5	6.2

TUCUMCARI, NM—Latitude 35.18

0	3.0	3.9	5.1	6.4	7.0	7.5	7.2	6.5	5.5	4.5	3.3	2.7	5.2
Lat - 15	4.3	5.1	5.9	6.8	7.0	7.2	7.1	6.8	6.2	5.6	4.5	4.0	5.9
Lat	5.0	5.6	6.2	6.7	6.6	6.6	6.6	6.5	6.3	6.1	5.2	4.8	6.0
Lat + 15	5.4	5.9	6.1	6.2	5.8	5.7	5.7	5.9	6.1	6.2	5.5	5.2	5.8

ALBANY, NY—Latitude 42.75

0	1.8	2.6	3.6	4.7	5.5	6.0	6.1	5.2	4.1	2.8	1.7	1.4	3.8
Lat - 15	2.7	3.6	4.4	5.0	5.5	5.8	6.0	5.5	4.8	3.7	2.4	2.1	4.3
Lat	3.0	3.9	4.5	4.9	5.1	5.4	5.5	5.2	4.8	3.9	2.6	2.4	4.3
Lat + 15	3.2	4.1	4.4	4.5	4.6	4.6	4.8	4.8.	4.6	3.9	2.7	2.5	4.1

BINGHAMTON, NY—Latitude 42.22

0	1.7	2.5	3.5	4.5	5.3	5.8	5.8	5.0	3.9	2.7	1.7	1.4	3.7
Lat - 15	2.5	3.3	4.2	4.8	5.3	5.6	5.7	5.3	4.5	3.5	2.3	1.9	4.1
Lat	2.8	3.5	4.3	4.7	5.0	5.2	5.3	5.1	4.5	3.7	2.4	2.1	4.1
Lat + 15	2.9	3.6	4.2	4.4	4.4	4.5	4.6	4.6	4.3	3.7	2.5	2.2	3.8

BUFFALO, NY—Latitude 42.93

0	1.6	2.4	3.4	4.5	5.5	6.1	6.0	5.2	3.9	2.6	1.6	1.3	3.7
Lat - 15	2.2	3.1	4.1	5.0	5.6	6.0	6.0	5.5	4.6	3.4	2.1	1.8	4.1
Lat	2.4	3.3	4.2	4.8	5.2	5.5	5.5	5.3	4.6	3.6	2.3	1.9	4.1
Lat + 15	2.5	3.4	4.1	4.5	4.6	4.8	4.8	4.8	4.3	3.6	2.3	2.0	3.8

NEW YORK CITY, NY—Latitude 40.78

0	1.9	2.7	3.8	4.9	5.7	6.1	6.0	5.4	4.3	3.2	2.0	1.6	4.0
Lat - 15	2.9	3.7	4.6	5.3	5.8	6.0	6.0	5.7	5.0	4.1	2.9	2.4	4.5
Lat	3.2	4.0	4.8	5.2	5.4	5.5	5.6	5.5	5.0	4.4	3.2	2.8	4.6
Lat + 15	3.4	4.1	4.6	4.8	4.8	4.8	4.9	5.0	4.8	4.4	3.3	3.0	4.3

ROCHESTER, NY—Latitude 43.12

0	1.6	2.4	3.4	4.6	5.5	6.1	6.0	5.2	4.0	2.7	1.6	1.3	3.7
Lat - 15	2.3	3.1	4.1	5.0	5.6	5.9	6.0	5.5	4.6	3.4	2.1	1.8	4.1
Lat	2.5	3.3	4.2	4.9	5.2	5.4	5.5	5.3	4.6	3.6	2.3	2.0	4.1
Lat + 15	2.6	3.4	4.1	4.5	4.6	4.7	4.8	5.0	4.4	3.6	2.3	2.0	3.8

	Jan	Feb	Mar	Apr	May	Jun	Jul	Aug	Sep	Oct	Nov	Dec	Year
SYRACUSE, NY—Latitude 43.12													
0	1.7	2.5	3.5	4.6	5.5	6.1	6.0	5.2	4.0	2.7	1.6	1.3	3.7
Lat - 15	2.4	3.3	4.2	5.0	5.6	5.9	6.0	5.5	4.7	3.5	2.1	1.8	4.2
Lat	2.7	3.5	4.3	4.9	5.2	5.4	5.6	5.3	4.7	3.7	2.3	2.0	4.1
Lat + 15	2.8	3.6	4.2	4.5	4.6	4.7	4.9	4.8	4.5	3.7	2.3	2.1	3.9
ASHEVILLE, NC—Latitude 35.43													
0	2.5	3.3	4.3	5.4	5.8	6.0	5.8	5.3	4.5	3.8	2.7	2.2	4.3
Lat - 15	3.5	4.2	4.9	5.7	5.8	5.9	5.7	5.5	5.0	4.6	3.6	3.1	4.8
Lat	3.9	4.6	5.1	5.6	5.4	5.4	5.3	5.3	5.0	5.0	4.1	3.6	4.9
Lat + 15	4.2	4.7	5.0	5.2	4.8	4.7	4.7	4.8	4.8	5.0	4.3	3.9	4.7
CAPE HATTERAS, NC—Latitude 35.27													
0	2.4	3.3	4.4	5.6	6.1	6.4	6.2	5.6	4.8	3.7	2.8	2.2	4.5
Lat - 15	3.3	4.1	5.1	6.0	6.1	6.2	6.1	5.8	5.3	4.6	3.7	3.2	5.0
Lat	3.8	4.5	5.2	5.9	5.8	5.7	5.7	5.6	5.4	4.9	4.2	3.6	5.0
Lat + 15	4.0	4.6	5.2	5.5	5.1	5.0	5.0	5.1	5.1	4.9	4.5	3.9	4.8
CHARLOTTE, NC—Latitude 35.22													
0	2.5	3.3	4.4	5.5	6.0	6.3	6.1	5.6	4.7	3.9	2.8	2.3	4.4
Lat - 15	3.4	4.1	5.0	5.9	6.0	6.1	6.0	5.8	5.2	4.7	3.7	3.1	4.9
Lat	3.8	4.5	5.2	5.7	5.7	5.7	5.6	5.6	5.3	5.1	4.2	3.6	5.0
Lat + 15	4.1	4.6	5.1	5.3	5.0	4.9	4.9	5.1	5.0	5.2	4.4	3.9	4.8
GREENSBORO, NC—Latitude 36.08													
0	2.4	3.2	4.3	5.4	6.0	6.3	6.1	5.5	4.6	3.7	2.7	2.2	4.4
Lat - 15	3.3	4.1	5.0	5.8	6.0	6.1	6.0	5.7	5.2	4.6	3.6	3.1	4.9
Lat	3.8	4.5	5.2	5.7	5.6	5.7	5.6	5.5	5.2	5.0	4.1	3.6	5.0
Lat + 15	4.1	4.6	5.1	5.2	5.0	4.9	4.9	5.1	5.0	5.0	4.3	3.8	4.8
RALEIGH, NC—Latitude 35.87													
0	2.4	3.2	4.4	5.5	6.0	6.3	6.1	5.5	4.6	3.8	2.7	2.2	4.4
Lat - 15	3.4	4.1	5.0	5.8	6.0	6.2	6.0	5.7	5.1	4.6	3.7	3.1	4.9
Lat	3.8	4.5	5.2	5.7	5.7	5.7	5.6	5.5	5.2	4.9	4.1	3.6	5.0
Lat + 15	4.1	4.6	5.1	5.3	5.0	4.9	4.9	5.0	5.0	5.0	4.3	3.8	4.8
WILMINGTON, NC—Latitude 34.27													
0	2.6	3.4	4.5	5.7	6.1	6.3	6.0	5.4	4.6	3.9	2.9	2.4	4.5
Lat - 15	3.5	4.2	5.2	6.0	6.1	6.1	5.9	5.6	5.1	4.7	3.9	3.3	5.0
Lat	4.0	4.6	5.4	5.9	5.8	5.6	5.5	5.4	5.2	5.0	4.4	3.8	5.0
Lat + 15	4.2	4.7	5.3	5.5	5.1	4.9	4.8	4.9	5.0	5.1	4.6	4.1	4.9

BISMARCK, ND—Latitude 46.77													
0	1.7	2.6	3.8	4.9	6.0	6.6	6.8	5.8	4.2	2.8	1.7	1.4	4.0
Lat - 15	3.1	4.0	5.0	5.6	6.1	6.5	6.8	6.4	5.3	4.2	2.9	2.6	4.9
Lat	3.5	4.4	5.2	5.5	5.7	5.9	6.3	6.1	5.4	4.5	3.2	3.0	4.9
Lat + 15	3.7	4.5	5.1	5.1	5.1	5.1	5.5	5.5	5.1	4.5	3.4	3.2	4.7
FARGO, ND—Latitude 46.90													
0	1.6	2.5	3.7	4.7	5.7	6.2	6.4	5.5	4.0	2.7	1.6	1.3	3.8
Lat - 15	2.9	3.9	4.8	5.3	5.9	6.1	6.5	6.1	4.9	3.9	2.6	2.3	4.6
Lat	3.4	4.3	5.0	5.2	5.5	5.6	6.0	5.8	5.0	4.1	2.9	2.7	4.6
Lat + 15	3.6	4.5	5.0	4.8	4.9	4.8	5.2	5.3	5.0	4.1	3.0	2.9	4.4
MINOT, ND—Latitude 48.27													
0	1.5	2.4	3.6	4.9	5.8	6.4	6.6	5.6	4.0	2.7	1.6	1.2	3.9
Lat - 15	2.9	3.8	4.9	5.6	6.0	6.3	6.6	6.2	5.0	4.1	2.8	2.4	4.7
Lat	3.3	4.1	5.1	5.5	5.6	5.8	6.1	6.0	5.1	4.4	3.1	2.8	4.7
Lat + 15	3.5	4.3	5.0	5.1	5.0	5.0	5.3	5.4	4.8	4.4	3.3	3.0	4.5
AKRON, OH—Latitude 40.92													
0	1.7	2.4	3.4	4.6	5.5	6.1	6.0	5.2	4.2	2.9	1.8	1.4	3.8
Lat - 15	2.3	3.1	4.0	4.9	5.5	5.9	5.9	5.5	4.8	3.8	2.3	1.8	4.2
Lat	2.5	3.4	4.1	4.8	5.2	5.5	5.5	5.3	4.8	4.0	2.5	2.0	4.1
Lat + 15	2.7	3.4	3.9	4.4	4.6	4.7	4.8	4.8	4.6	4.0	2.6	2.1	3.9
CLEVELAND, OH—Latitude 41.40													
0	1.6	2.4	3.3	4.6	5.6	6.2	6.1	5.3	4.1	2.8	1.7	1.3	3.8
Lat - 15	2.2	3.1	3.9	5.0	5.6	6.0	6.1	5.6	4.7	3.6	2.2	1.7	4.2
Lat	2.4	3.3	4.0	4.9	5.3	5.5	5.6	5.3	4.8	3.8	2.4	1.9	4.1
Lat + 15	2.5	3.3	3.9	4.5	4.7	4.8	4.9	4.9	4.6	3.8	2.4	2.0	3.9
COLUMBUS, OH—Latitude 40.00													
0	1.8	2.5	3.5	4.6	5.5	6.0	5.9	5.3	4.3	3.1	1.9	1.5	3.8
Lat - 15	2.5	3.2	4.0	4.9	5.5	5.9	5.8	5.5	4.9	4.0	2.6	2.0	4.2
Lat	2.7	3.4	4.1	4.8	5.2	5.4	5.4	5.3	4.9	4.3	2.8	2.2	4.2
Lat + 15	2.9	3.5	4.0	4.4	4.6	4.7	4.8	4.8	4.7	4.3	2.9	2.3	4.0
DAYTON, OH—Latitude 39.40													
0	1.9	2.6	3.6	4.7	5.7	6.2	6.0	5.4	4.4	3.2	2.0	1.5	3.9
Lat - 15	2.7	3.4	4.2	5.1	5.7	6.1	6.0	5.7	5.1	4.1	2.7	2.1	4.4
Lat	3.0	3.7	4.3	4.9	5.3	5.6	5.6	5.5	5.1	4.4	2.9	2.4	4.4
Lat + 15	3.1	3.8	4.1	4.6	4.7	4.8	4.9	5.0	4.9	4.4	3.0	2.5	4.2

	Jan	Feb	Mar	Apr	May	Jun	Jul	Aug	Sep	Oct	Nov	Dec	Year
TOLEDO, OH —Latitude 41.60													
0	1.7	2.6	3.5	4.7	5.8	6.3	6.2	5.4	4.3	3.0	1.8	1.4	3.9
Lat - 15	2.5	3.4	4.2	5.1	5.8	6.2	6.2	5.8	5.0	3.9	2.5	2.0	4.4
Lat	2.8	3.7	4.3	5.0	5.5	5.7	5.7	5.6	5.0	4.1	2.7	2.2	4.4
Lat + 15	3.0	3.8	4.2	4.6	4.9	5.0	5.0	5.0	4.8	4.1	2.8	2.3	4.1
YOUNGSTOWN, OH—Latitude 41.27													
0	1.6	2.4	3.3	4.4	5.3	5.9	5.8	5.0	4.0	2.8	1.7	1.3	3.6
Lat - 15	2.2	3.0	3.8	4.7	5.4	5.8	5.7	5.3	4.6	3.6	2.2	1.7	4.0
Lat	2.4	3.2	3.9	4.6	5.0	5.3	5.3	5.0	4.6	3.8	2.4	1.9	3.9
Lat + 15	2.5	3.2	3.8	4.3	4.5	4.6	4.7	4.6	4.3	3.7	2.4	1.9	3.7
OKLAHOMA CITY, OK—Latitude 35.40													
0	2.8	3.5	4.6	5.7	6.2	6.8	6.9	6.2	5.0	4.0	2.9	2.4	4.8
Lat - 15	3.9	4.4	5.3	6.0	6.2	6.6	6.8	6.5	5.6	5.0	3.9	3.5	5.3
Lat	4.4	4.9	5.5	5.9	5.8	6.0	6.3	6.3	5.7	5.4	4.4	4.1	5.4
Lat + 15	4.7	5.0	5.4	5.5	5.2	5.2	5.5	5.7	5.5	5.5	4.7	4.4	5.2
TULSA, OK—Latitude 36.20													
0	2.5	3.3	4.3	5.3	5.9	6.4	6.7	6.0	4.7	3.8	2.7	2.2	4.5
Lat - 15	3.5	4.2	5.0	5.7	5.9	6.3	6.6	6.3	5.3	4.7	3.6	3.2	5.0
Lat	4.0	4.5	5.2	5.6	5.5	5.8	6.1	6.0	5.4	5.1	4.0	3.7	5.1
Lat + 15	4.3	4.7	5.1	5.2	4.9	5.0	5.3	5.5	5.2	5.1	4.2	4.0	4.9
ASTORIA, OR—Latitude 46.15													
0	1.1	1.8	2.8	3.9	4.9	5.3	5.4	4.8	3.8	2.4	1.3	1.0	3.2
Lat - 15	1.7	2.4	3.4	4.3	5.0	5.2	5.4	5.1	4.7	3.3	2.0	1.5	3.7
Lat	1.9	2.6	3.4	4.2	4.7	4.7	5.0	4.9	4.7	3.5	2.1	1.7	3.6
Lat + 15	2.0	2.6	3.3	3.8	4.1	4.1	4.4	4.4	4.5	3.5	2.2	1.8	3.4
BURNS, OR—Latitude 43.58													
0	1.8	2.6	3.8	5.2	6.4	7.1	7.5	6.5	5.1	3.4	1.9	1.5	4.4
Lat - 15	2.8	3.7	4.7	5.8	6.5	6.9	7.5	7.0	6.3	4.8	2.8	2.4	5.1
Lat	3.1	4.0	4.9	5.7	6.1	6.3	6.9	6.8	6.4	5.2	3.1	2.7	5.1
Lat + 15	3.3	4.1	4.8	5.3	5.4	5.4	6.0	6.1	6.1	5.2	3.3	2.9	4.8
EUGENE, OR—Latitude 44.12													
0	1.3	2.0	3.1	4.4	5.5	6.2	6.7	5.8	4.4	2.7	1.4	1.0	3.7
Lat - 15	1.8	2.6	3.8	4.8	5.6	6.0	6.7	6.3	5.4	3.6	1.9	1.4	4.2
Lat	2.0	2.7	3.8	4.7	5.3	5.5	6.2	6.0	5.5	3.8	2.1	1.6	4.1
Lat + 15	2.0	2.8	3.7	4.3	4.6	4.8	5.4	5.5	5.2	3.8	2.1	1.6	3.8

MEDFORD, OR—Latitude 42.37													
0	1.5	2.4	3.7	5.2	6.5	7.3	7.7	6.7	5.2	3.3	1.7	1.2	4.4
Lat - 15	2.1	3.2	4.5	5.7	6.6	7.1	7.7	7.2	6.3	4.5	2.3	1.7	4.9
Lat	2.3	3.5	4.6	5.6	6.2	6.5	7.1	6.9	6.4	4.8	2.4	1.9	4.9
Lat + 15	2.4	3.5	4.5	5.2	5.5	5.6	6.2	6.3	6.2	4.8	2.5	2.0	4.5
NORTH BEND, OR—Latitude 43.42													
0	1.5	2.2	3.4	4.7	5.7	6.2	6.5	5.6	4.5	3.0	1.8	1.3	3.9
Lat - 15	2.4	3.0	4.1	5.1	5.8	6.1	6.5	6.0	5.4	4.0	2.6	2.1	4.4
Lat	2.6	3.2	4.2	5.0	5.5	5.6	6.1	5.8	5.5	4.3	2.8	2.4	4.4
Lat + 15	2.7	3.3	4.1	4.6	4.8	4.8	5.3	5.3	5.2	4.3	2.9	2.5	4.2
PENDLETON, OR—Latitude 45.68													
0	1.4	2.1	3.4	4.9	6.2	6.9	7.4	6.3	4.8	3.0	1.6	1.1	4.1
Lat - 15	2.0	3.0	4.4	5.5	6.3	6.8	7.4	6.9	6.1	4.4	2.3	1.7	4.7
Lat	2.2	3.2	4.5	5.4	5.9	6.2	6.8	6.7	6.2	4.7	2.5	1.9	4.7
Lat + 15	2.2	3.2	4.3	5.0	5.2	5.3	5.9	6.0	5.9	4.8	2.6	2.0	4.4
PORTLAND, OR—Latitude 45.60													
0	1.2	1.9	3.0	4.2	5.3	5.9	6.3	5.4	4.1	2.5	1.4	1.0	3.5
Lat - 15	1.7	2.5	3.6	4.6	5.4	5.8	6.3	5.9	5.1	3.4	1.9	1.4	4.0
Lat	1.9	2.6	3.7	4.5	5.0	5.3	5.8	5.6	5.1	3.6	2.1	1.6	3.9
Lat + 15	1.9	2.7	3.5	4.1	4.4	4.6	5.1	5.1	4.9	3.6	2.1	1.6	3.6
REDMOND, OR—Latitude 44.27													
0	1.7	2.5	3.8	5.3	6.5	7.2	7.6	6.6	5.1	3.3	1.9	1.4	4.4
Lat - 15	2.6	3.5	4.8	5.9	6.6	7.0	7.6	7.1	6.3	4.7	2.9	2.4	5.1
Lat	3.0	3.8	4.9	5.7	6.2	6.4	7.0	6.9	6.4	5.1	3.2	2.7	5.1
Lat + 15	3.1	3.9	4.8	5.3	5.5	5.5	6.1	6.2	6.1	5.1	3.3	2.9	4.8
SALEM, OR—Latitude 44.92													
0	1.3	2.0	3.1	4.4	5.5	6.1	6.6	5.7	4.4	2.7	1.4	1.1	3.7
Lat - 15	1.8	2.7	3.8	4.8	5.6	6.0	6.6	6.2	5.3	3.6	2.0	1.5	4.2
Lat	2.0	2.8	3.9	4.7	5.2	5.5	6.1	6.0	5.4	3.8	2.2	1.6	4.1
Lat + 15	2.1	2.8	3.7	4.3	4.6	4.7	5.3	5.4	5.1	3.8	2.2	1.7	3.8
ALLENTOWN, PA—Latitude 40.65													
0	1.9	2.7	3.7	4.7	5.4	6.0	5.9	5.2	4.2	3.1	2.0	1.6	3.9
Lat - 15	2.8	3.6	4.4	5.1	5.5	5.8	5.8	5.5	4.8	4.0	2.7	2.3	4.4
Lat	3.1	3.9	4.5	5.0	5.2	5.4	5.4	5.3	4.9	4.2	3.0	2.6	4.4
Lat + 15	3.3	4.0	4.4	4.6	4.6	4.7	4.8	4.8	4.6	4.2	3.1	2.8	4.2

	Jan	Feb	Mar	Apr	May	Jun	Jul	Aug	Sep	Oct	Nov	Dec	Year
ERIE, PA—Latitude 42.08													
0	1.6	2.4	3.4	4.6	5.7	6.3	6.2	5.3	4.1	2.7	1.6	1.3	3.8
Lat - 15	2.1	3.1	4.0	5.0	5.7	6.1	6.2	5.6	4.7	3.5	2.1	1.6	4.2
Lat	2.3	3.3	4.1	4.9	5.4	5.6	5.8	5.4	4.7	3.7	2.2	1.8	4.1
Lat + 15	2.4	3.4	4.0	4.5	4.8	4.9	5.0	4.9	4.5	3.7	2.3	1.8	3.9
HARRISBURG, PA—Latitude 40.22													
0	2.0	2.8	3.8	4.8	5.5	6.1	5.9	5.3	4.3	3.2	2.0	1.6	3.9
Lat - 15	2.9	3.7	4.5	5.2	5.6	6.0	5.9	5.5	4.9	4.1	2.9	2.4	4.5
Lat	3.2	4.0	4.6	5.1	5.3	5.5	5.5	5.3	4.9	4.3	3.2	2.7	4.5
Lat + 15	3.5	4.1	4.5	4.7	4.7	4.8	4.8	4.8	4.7	4.4	3.3	2.9	4.3
PHILADELPHIA, PA—Latitude 39.88													
0	2.0	2.8	3.8	4.8	5.5	6.1	6.0	5.4	4.4	3.2	2.1	1.7	4.0
Lat - 15	2.9	3.7	4.5	5.2	5.6	6.0	5.9	5.7	5.0	4.2	3.0	2.6	4.5
Lat	3.3	4.0	4.7	5.1	5.3	5.5	5.5	5.5	5.1	4.4	3.4	2.9	4.6
Lat + 15	3.5	4.1	4.6	4.7	4.7	4.8	4.8	5.0	4.8	4.5	3.5	3.1	4.3
PITTSBURGH, PA—Latitude 40.50													
0	1.7	2.5	3.5	4.6	5.5	6.1	5.9	5.2	4.2	3.0	1.8	1.4	3.8
Lat - 15	2.4	3.2	4.1	4.9	5.5	5.9	5.9	5.5	4.8	3.8	2.4	1.9	4.2
Lat	2.6	3.4	4.2	4.8	5.2	5.4	5.5	5.3	4.8	4.1	2.6	2.1	4.2
Lat + 15	2.7	3.5	4.1	4.4	4.6	4.7	4.8	4.8	4.6	4.1	2.7	2.2	3.9
WILKES-BARRE, PA—Latitude 41.33													
0	1.8	2.5	3.6	4.6	5.4	6.0	5.9	5.2	4.1	2.9	1.8	1.4	3.8
Lat - 15	2.5	3.4	4.2	5.0	5.4	5.8	5.8	5.5	4.7	3.8	2.4	2.0	4.2
Lat	2.8	3.6	4.3	4.8	5.1	5.4	5.4	5.3	4.7	4.0	2.6	2.2	4.2
Lat + 15	3.0	3.7	4.2	4.5	4.6	4.7	4.8	4.8	4.5	4.0	2.7	2.4	4.0
WILLIAMSPORT, PA—Latitude 41.27													
0	1.8	2.6	3.6	4.6	5.4	6.0	5.9	5.1	4.0	2.9	1.8	1.4	3.8
Lat - 15	2.6	3.4	4.3	5.0	5.5	5.9	5.9	5.4	4.6	3.7	2.4	2.1	4.2
Lat	2.9	3.7	4.4	4.9	5.1	5.4	5.5	5.2	4.6	3.9	2.6	2.3	4.2
Lat + 15	3.0	3.8	4.3	4.5	4.6	4.7	4.8	4.8	4.4	3.9	2.7	2.4	4.0
PROVIDENCE, RI—Latitude 41.73													
0	1.9	2.7	3.7	4.7	5.6	6.0	5.9	5.2	4.2	3.1	1.9	1.6	3.9
Lat - 15	3.0	3.7	4.5	5.1	5.6	5.9	5.9	5.5	4.9	4.1	2.9	2.5	4.5
Lat	3.4	4.1	4.7	5.0	5.3	5.4	5.5	5.3	5.0	4.4	3.2	2.9	4.5
Lat + 15	3.6	4.2	4.6	4.6	4.7	4.7	4.8	4.8	4.7	4.4	3.3	3.1	4.3

CHARLESTON, SC—Latitude 32.90													
0	2.7	3.5	4.7	5.9	6.2	6.2	6.1	5.5	4.7	4.1	3.1	2.5	4.6
Lat - 15	3.5	4.3	5.3	6.2	6.2	6.1	6.0	5.6	5.1	4.8	3.9	3.4	5.0
Lat	4.0	4.7	5.5	6.1	5.8	5.6	5.6	5.4	5.2	5.2	4.5	3.9	5.1
Lat + 15	4.3	4.9	5.4	5.7	5.2	4.9	4.9	4.9	5.0	5.2	4.7	4.2	4.9
COLUMBIA, SC—Latitude 33.95													
0	2.6	3.4	4.5	5.7	6.1	6.3	6.1	5.5	4.8	4.0	2.9	2.4	4.5
Lat - 15	3.4	4.2	5.1	6.0	6.1	6.1	6.0	5.7	5.2	4.8	3.8	3.3	5.0
Lat	3.9	4.6	5.3	5.9	5.7	5.7	5.6	5.5	5.3	5.2	4.3	3.8	5.1
Lat + 15	4.1	4.8	5.2	5.5	5.1	4.9	4.9	5.0	5.1	5.2	4.6	4.1	4.9
GREENVILLE, SC—Latitude 34.90													
0	2.6	3.3	4.4	5.6	6.0	6.3	6.0	5.5	4.7	3.9	2.8	2.3	4.5
Lat - 15	3.5	4.2	5.1	5.9	6.0	6.1	5.9	5.7	5.2	4.8	3.8	3.2	5.0
Lat	4.0	4.6	5.3	5.8	5.6	5.6	5.5	5.5	5.2	5.2	4.3	3.7	5.0
Lat + 15	4.2	4.8	5.2	5.4	5.0	4.9	4.8	5.1	5.0	5.2	4.5	4.0	4.8
HURON, SD—Latitude 44.38													
0	1.8	2.6	3.7	4.9	5.8	6.5	6.6	5.8	4.4	3.0	1.9	1.5	4.1
Lat - 15	3.0	3.8	4.7	5.4	6.0	6.4	6.7	6.3	5.3	4.3	3.0	2.6	4.8
Lat	3.4	4.2	4.8	5.3	5.6	5.9	6.2	6.1	5.4	4.6	3.3	3.0	4.8
Lat + 15	3.7	4.3	4.8	4.9	5.0	5.1	5.4	5.5	5.2	4.6	3.5	3.2	4.6
PIERRE, SD—Latitude 44.38													
0	1.8	2.7	3.9	5.0	6.0	6.7	6.8	6.0	4.5	3.1	2.0	1.5	4.2
Lat - 15	3.1	3.9	4.9	5.6	6.1	6.6	6.8	6.5	5.6	4.5	3.2	2.7	4.9
Lat	3.6	4.3	5.1	5.4	5.7	6.0	6.3	6.3	5.6	4.8	3.5	3.1	5.0
Lat + 15	3.8	4.4	5.0	5.0	5.1	5.2	5.5	5.7	5.4	4.8	3.7	3.3	4.7
RAPID CITY, SD—Latitude 44.05													
0	1.9	2.8	4.0	5.1	6.0	6.7	6.8	6.1	4.7	3.3	2.1	1.6	4.3
Lat - 15	3.2	4.1	5.1	5.7	6.1	6.6	6.8	6.6	5.8	4.7	3.4	3.0	5.1
Lat	3.7	4.5	5.3	5.6	5.7	6.0	6.3	6.4	5.9	5.1	3.9	3.4	5.2
Lat + 15	4.0	4.7	5.2	5.2	5.0	5.2	5.5	5.8	5.7	5.2	4.1	3.7	4.9
SIOUX FALLS, SD—Latitude 43.57													
0	1.9	2.7	3.8	4.8	5.8	6.5	6.6	5.7	4.3	3.1	1.9	1.5	4.1
Lat - 15	3.1	3.9	4.7	5.4	5.9	6.4	6.6	6.1	5.2	4.3	3.0	2.6	4.8
Lat	3.6	4.3	4.9	5.2	5.5	5.8	6.1	5.9	5.3	4.6	3.3	3.0	4.8
Lat + 15	3.8	4.5	4.8	4.8	4.9	5.1	5.3	5.3	5.1	4.6	3.5	3.2	4.6

	Jan	Feb	Mar	Apr	May	Jun	Jul	Aug	Sep	Oct	Nov	Dec	Year
CHATTANOOGA, TN—Latitude 35.03													
0	2.4	3.1	4.1	5.3	5.8	6.1	5.9	5.5	4.5	3.8	2.6	2.1	4.3
Lat - 15	3.1	3.8	4.6	5.6	5.8	6.0	5.8	5.7	5.0	4.6	3.4	2.8	4.7
Lat	3.5	4.1	4.8	5.4	5.4	5.5	5.4	5.5	5.0	4.9	3.8	3.2	4.7
Lat + 15	3.7	4.2	4.6	5.1	4.9	4.8	4.7	5.0	4.8	4.9	4.0	3.4	4.5
KNOXVILLE, TN—Latitude 35.82													
0	2.3	3.0	4.0	5.2	5.8	6.2	5.9	5.5	4.5	3.7	2.5	2.0	4.2
Lat - 15	3.0	3.7	4.6	5.5	5.8	6.1	5.8	5.7	5.0	4.6	3.3	2.8	4.7
Lat	3.4	4.0	4.7	5.4	5.5	5.6	5.4	5.5	5.1	4.9	3.7	3.1	4.7
Lat + 15	3.6	4.1	4.6	5.0	4.9	4.9	4.8	5.0	4.8	4.9	3.8	3.3	4.5
MEMPHIS, TN—Latitude 35.05													
0	2.5	3.2	4.2	5.4	6.1	6.6	6.5	6.0	4.8	4.0	2.7	2.2	4.5
Lat - 15	3.3	4.0	4.8	5.7	6.1	6.4	6.4	6.2	5.4	4.9	3.5	3.0	5.0
Lat	3.7	4.4	5.0	5.6	5.8	5.9	6.0	6.0	5.4	5.2	3.9	3.4	5.0
Lat + 15	4.0	4.5	4.9	5.2	5.1	5.1	5.2	5.5	5.2	5.3	4.1	3.6	4.8
NASHVILLE, TN—Latitude 36.12													
0	2.3	3.1	4.1	5.4	6.0	6.5	6.3	5.7	4.7	3.8	2.5	2.0	4.4
Lat - 15	3.1	3.9	4.7	5.7	6.0	6.4	6.2	5.9	5.2	4.6	3.3	2.8	4.8
Lat	3.5	4.2	4.8	5.6	5.7	5.9	5.8	5.7	5.3	4.9	3.6	3.1	4.9
Lat + 15	3.7	4.3	4.7	5.2	5.0	5.1	5.1	5.2	5.0	5.0	3.8	3.4	4.6
ABILENE, TX—Latitude 32.43													
0	3.1	3.9	5.1	6.1	6.5	7.0	7.0	6.3	5.2	4.4	3.3	2.9	5.1
Lat - 15	4.1	4.8	5.8	6.4	6.5	6.8	6.8	6.5	5.7	5.3	4.3	3.9	5.6
Lat	4.7	5.3	6.0	6.3	6.1	6.2	6.3	6.3	5.8	5.7	4.9	4.5	5.7
Lat + 15	5.1	5.5	6.0	5.8	5.4	5.4	5.5	5.7	5.6	5.8	5.2	4.9	5.5
AMARILLO, TX—Latitude 35.23													
0	3.0	3.8	4.9	6.1	6.6	7.1	7.0	6.3	5.2	4.4	3.2	2.7	5.0
Lat - 15	4.2	4.9	5.8	6.5	6.6	6.9	6.9	6.5	5.9	5.5	4.4	3.9	5.7
Lat	4.9	5.4	6.0	6.4	6.2	6.3	6.4	6.3	6.0	5.9	5.0	4.6	5.8
Lat + 15	5.3	5.7	5.9	6.0	5.5	5.5	5.5	5.7	5.8	6.0	5.3	5.0	5.6
AUSTIN, TX—Latitude 30.30													
0	3.0	3.8	4.7	5.4	5.9	6.6	6.8	6.3	5.2	4.4	3.3	2.8	4.9
Lat - 15	3.7	4.4	5.2	5.6	5.8	6.4	6.7	6.5	5.7	5.0	4.1	3.5	5.2
Lat	4.2	4.8	5.4	5.5	5.5	5.9	6.2	6.3	5.8	5.4	4.6	4.0	5.3
Lat + 15	4.4	5.0	5.3	5.1	4.9	5.1	5.4	5.7	5.5	5.5	4.8	4.3	5.1

BROWNSVILLE, TX—Latitude 25.90													
0	2.9	3.7	4.6	5.3	5.8	6.4	6.5	6.0	5.2	4.5	3.4	2.7	4.8
Lat-15	3.2	4.0	4.8	5.4	5.7	6.2	6.4	6.0	5.4	4.9	3.9	3.1	4.9
Lat	3.6	4.3	5.0	5.3	5.4	5.7	5.9	5.8	5.5	5.3	4.4	3.6	5.0
Lat+15	3.8	4.4	4.9	5.0	4.8	5.0	5.2	5.3	5.3	5.4	4.6	3.8	4.8
CORPUS CHRISTI, TX—Latitude 27.77													
0	2.8	3.6	4.4	5.0	5.5	6.1	6.3	5.8	5.0	4.3	3.3	2.7	4.6
Lat-15	3.2	4.0	4.7	5.2	5.4	5.9	6.1	5.9	5.3	4.8	3.9	3.1	4.8
Lat	3.6	4.3	4.9	5.1	5.1	5.5	5.7	5.7	5.4	5.2	4.3	3.6	4.9
Lat+15	3.8	4.5	4.8	4.7	4.6	4.8	5.0	5.2	5.3	5.3	4.6	3.8	4.7
EL PASO, TX—Latitude 31.80													
0	3.5	4.5	5.9	7.1	7.8	8.0	7.4	6.8	5.9	4.9	3.8	3.2	5.7
Lat-15	4.6	5.6	6.6	7.4	7.8	7.7	7.2	6.9	6.4	5.9	4.9	4.4	6.3
Lat	5.3	6.2	7.0	7.3	7.3	7.1	6.7	6.7	6.6	6.4	5.7	5.1	6.5
Lat+15	5.8	6.5	6.9	6.8	6.4	6.0	5.8	6.1	6.3	6.6	6.1	5.6	6.2
FORT WORTH, TX—Latitude 32.83													
0	2.9	3.7	4.7	5.6	6.2	6.9	7.0	6.4	5.2	4.2	3.1	2.7	4.9
Lat-15	3.8	4.5	5.3	5.9	6.2	6.7	6.9	6.5	5.7	5.0	4.0	3.5	5.3
Lat	4.3	4.9	5.5	5.7	5.8	6.2	6.4	6.3	5.8	5.4	4.5	4.1	5.4
Lat+15	4.6	5.1	5.4	5.3	5.2	5.3	5.6	5.7	5.6	5.5	4.8	4.4	5.2
HOUSTON, TX—Latitude 29.98													
0	2.7	3.4	4.2	5.0	5.6	6.0	5.9	5.6	4.9	4.2	3.1	2.5	4.4
Lat-15	3.2	3.9	4.6	5.2	5.6	5.9	5.8	5.7	5.2	4.8	3.7	3.0	4.7
Lat	3.6	4.3	4.7	5.1	5.3	5.4	5.4	5.5	5.3	5.2	4.1	3.5	4.8
Lat+15	3.8	4.4	4.6	4.7	4.7	4.7	4.8	5.0	5.1	5.2	4.3	3.7	4.6
LUBBOCK, TX—Latitude 33.65													
0	3.1	3.9	5.1	6.2	6.7	7.1	7.0	6.4	5.2	4.4	3.3	2.8	5.1
Lat-15	4.2	5.0	5.8	6.6	6.7	6.9	6.8	6.3	5.8	5.4	4.4	3.9	5.7
Lat	4.9	5.5	6.1	6.5	6.3	6.3	6.3	5.9	5.9	5.9	5.1	4.6	5.8
Lat+15	5.3	5.7	6.0	6.0	5.6	5.4	5.5	5.7	5.7	6.0	5.4	5.0	5.6
LUFKIN, TX—Latitude 31.23													
0	2.7	3.5	4.5	5.3	5.9	6.4	6.4	6.0	5.2	4.3	3.1	2.5	4.6
Lat-15	3.4	4.1	4.9	5.5	5.9	6.2	6.3	6.2	5.8	5.0	3.9	3.2	5.0
Lat	3.8	4.5	5.1	5.4	5.6	5.8	5.9	5.9	5.9	5.4	4.3	3.7	5.1
Lat+15	4.0	4.6	5.0	5.0	4.9	5.0	5.2	5.4	6.0	5.4	4.6	4.0	4.9

THE NEW SOLAR ELECTRIC HOME

	Jan	Feb	Mar	Apr	May	Jun	Jul	Aug	Sep	Oct	Nov	Dec	Year
MIDLAND, TX—Latitude 31.93													
0	3.3	4.2	5.5	6.5	7.0	7.3	7.0	6.5	5.4	4.6	3.6	3.0	5.3
Lat-15	4.3	5.1	6.2	6.8	7.0	7.1	6.9	6.6	5.9	5.5	4.6	4.1	5.8
Lat	5.0	5.7	6.5	6.7	6.5	6.5	6.4	6.4	6.0	6.0	5.3	4.8	6.0
Lat+15	5.4	5.9	6.4	6.2	5.8	5.6	5.5	5.8	5.8	6.1	5.6	5.2	5.8
PORT ARTHUR, TX—Latitude 29.95													
0	2.7	3.5	4.3	5.2	5.8	6.3	6.1	5.7	5.0	4.3	3.1	2.6	4.6
Lat-15	3.3	4.1	4.7	5.3	5.8	6.1	6.0	5.8	5.4	4.9	3.8	3.2	4.9
Lat	3.7	4.4	4.9	5.2	5.5	5.7	5.5	5.6	5.5	5.3	4.2	3.6	4.9
Lat+15	3.9	4.5	4.8	4.9	4.9	4.9	4.9	5.2	5.3	5.4	4.4	3.9	4.7
SAN ANGELO, TX—Latitude 31.37													
0	3.2	4.1	5.2	6.1	6.5	7.0	6.9	6.4	5.3	4.5	3.5	3.0	5.1
Lat-15	4.1	4.9	5.8	6.4	6.5	6.7	6.8	6.5	5.7	5.3	4.4	3.9	5.6
Lat	4.7	5.4	6.1	6.3	6.1	6.2	6.3	6.3	5.8	5.7	5.0	4.6	5.7
Lat+15	5.1	5.6	6.0	5.8	5.4	5.3	5.5	5.7	5.6	5.8	5.4	5.0	5.5
SAN ANTONIO, TX—Latitude 29.53													
0	3.1	3.9	4.8	5.5	6.0	6.7	6.9	6.4	5.4	4.5	3.4	2.9	4.9
Lat-15	3.7	4.5	5.2	5.7	5.9	6.5	6.7	6.6	5.8	5.1	4.1	3.5	5.3
Lat	4.3	4.9	5.4	5.6	5.6	6.0	6.3	6.3	5.9	5.5	4.6	4.1	5.4
Lat+15	4.5	5.0	5.3	5.2	5.0	5.2	5.5	5.8	5.7	5.6	4.9	4.4	5.2
VICTORIA, TX—Latitude 28.85													
0	2.8	3.6	4.4	5.1	5.7	6.2	6.2	5.8	5.0	4.3	3.3	2.7	4.6
Lat-15	3.3	4.1	4.8	5.2	5.6	6.1	6.1	5.9	5.4	4.9	3.9	3.2	4.9
Lat	3.7	4.5	4.9	5.1	5.3	5.6	5.7	5.7	5.4	5.3	4.3	3.6	4.9
Lat+15	3.9	4.6	4.8	4.8	4.7	4.9	5.0	5.2	5.2	5.4	4.6	3.8	4.7
WACO, TX—Latitude 31.62													
0	2.9	3.7	4.7	5.5	6.0	6.7	6.9	6.4	5.2	4.3	3.2	2.7	4.9
Lat-15	3.7	4.4	5.3	5.7	6.0	6.5	6.8	6.5	5.7	5.0	4.0	3.6	5.3
Lat	4.2	4.8	5.4	5.6	5.6	6.0	6.3	6.3	5.8	5.4	4.5	4.1	5.4
Lat+15	4.5	5.0	5.3	5.2	5.0	5.2	5.5	5.7	5.6	5.5	4.8	4.4	5.1
WICHITA FALLS, TX—Latitude 33.97													
0	2.9	3.7	4.8	5.8	6.4	6.9	7.0	6.3	5.2	4.2	3.1	2.6	4.9
Lat-15	3.9	4.6	5.4	6.1	6.4	6.7	6.8	6.5	5.7	5.1	4.1	3.6	5.4
Lat	4.5	5.0	5.6	6.0	6.0	6.2	6.3	6.3	5.8	5.5	4.6	4.2	5.5
Lat+15	4.8	5.2	5.5	5.6	5.3	5.3	5.5	5.7	5.6	5.6	4.9	4.5	5.3

CEDAR CITY, UT—Latitude 37.70												
0	2.7	3.5	4.6	6.0	7.0	7.8	7.3	6.5	5.7	4.3	2.9	2.4	5.0
Lat - 15	4.0	4.7	5.5	6.4	7.0	7.5	7.1	6.8	6.6	5.6	4.2	3.7	5.8
Lat	4.6	5.2	5.7	6.3	6.5	6.8	6.6	6.6	6.7	6.1	4.8	4.4	5.9
Lat + 15	5.0	5.4	5.7	5.9	5.7	5.9	5.7	6.0	6.5	6.2	5.1	4.8	5.7

SALT LAKE CITY, UT—Latitude 40.77												
0	1.9	2.9	4.1	5.4	6.5	7.4	7.3	6.5	5.2	3.7	2.2	1.7	4.6
Lat - 15	2.9	4.0	5.0	5.9	6.6	7.2	7.3	7.0	6.3	5.0	3.3	2.5	5.2
Lat	3.2	4.3	5.2	5.8	6.2	6.6	6.7	6.7	6.4	5.4	3.7	2.9	5.3
Lat + 15	3.4	4.4	5.1	5.4	5.5	5.6	5.8	6.1	6.1	5.5	3.9	3.1	5.0

BURLINGTON, VT—Latitude 44.47												
0	1.6	2.6	3.6	4.6	5.5	6.0	6.1	5.2	4.0	2.6	1.6	1.2	3.7
Lat - 15	2.6	3.6	4.5	5.0	5.6	5.9	6.1	5.6	4.7	3.5	2.2	1.9	4.3
Lat	2.9	3.9	4.7	4.9	5.3	5.4	5.6	5.3	4.8	3.7	2.4	2.1	4.3
Lat + 15	3.1	4.1	4.6	4.5	4.6	4.7	4.9	4.8	4.5	3.7	2.4	2.2	4.0

LYNCHBURG, VA—Latitude 37.33												
0	2.4	3.2	4.3	5.4	6.0	6.5	6.2	5.6	4.7	3.7	2.6	2.1	4.4
Lat - 15	3.4	4.2	5.1	5.8	6.1	6.3	6.1	5.9	5.3	4.6	3.6	3.1	5.0
Lat	3.9	4.6	5.3	5.7	5.7	5.8	5.7	5.7	5.3	5.0	4.0	3.6	5.0
Lat + 15	4.2	4.8	5.2	5.3	5.1	5.0	5.0	5.2	5.1	5.0	4.3	3.8	4.8

NORFOLK, VA—Latitude 36.90												
0	2.3	3.0	4.1	5.1	5.8	6.2	5.9	5.4	4.5	3.5	2.5	2.0	4.2
Lat - 15	3.2	3.9	4.8	5.5	6.0	6.0	5.8	5.6	5.0	4.3	3.5	2.9	4.7
Lat	3.6	4.3	4.9	5.4	5.5	5.6	5.4	5.4	5.1	4.6	3.9	3.4	4.8
Lat + 15	3.8	4.4	4.8	5.0	4.8	4.8	4.8	4.9	4.9	4.7	4.1	3.6	4.6

RICHMOND, VA—Latitude 37.32												
0	2.3	3.0	4.1	5.2	5.8	6.3	6.0	5.4	4.5	3.5	2.5	2.0	4.2
Lat - 15	3.2	3.9	4.8	5.5	5.8	6.1	5.9	5.7	5.1	4.4	3.5	2.9	4.7
Lat	3.6	4.3	5.0	5.4	5.5	5.6	5.5	5.5	5.2	4.7	3.9	3.3	4.8
Lat + 15	3.9	4.4	4.9	5.0	4.9	4.9	4.8	5.0	4.9	4.8	4.1	3.6	4.6

ROANOKE, VA—Latitude 37.32												
0	2.3	3.1	4.1	5.2	5.8	6.2	5.9	5.5	4.5	3.6	2.5	2.0	4.2
Lat - 15	3.3	4.0	4.8	5.6	5.8	6.0	5.9	5.7	5.1	4.6	3.5	2.9	4.8
Lat	3.7	4.3	5.0	5.5	5.5	5.6	5.5	5.5	5.1	4.9	3.9	3.4	4.8
Lat + 15	3.9	4.5	4.9	5.1	4.9	4.8	4.8	5.0	4.9	4.9	4.1	3.6	4.6

	Jan	Feb	Mar	Apr	May	Jun	Jul	Aug	Sep	Oct	Nov	Dec	Year
OLYMPIA, WA—Latitude 46.97													
0	1.0	1.7	2.8	4.0	5.0	5.6	5.9	5.1	3.8	2.2	1.2	0.9	3.3
Lat - 15	1.4	2.3	3.4	4.4	5.1	5.5	5.9	5.5	4.6	3.0	1.6	1.2	3.7
Lat	1.5	2.4	3.4	4.2	4.7	5.0	5.5	5.2	4.6	3.1	1.7	1.3	3.6
Lat + 15	1.6	2.4	3.3	3.9	4.2	4.3	4.8	4.7	4.4	3.0	1.8	1.4	3.3
SEATTLE, WA—Latitude 47.45													
0	1.0	1.7	2.8	4.1	5.3	5.8	6.1	5.2	3.8	2.2	1.2	0.8	3.3
Lat - 15	1.5	2.3	3.5	4.6	5.4	5.7	6.1	5.6	4.7	3.0	1.7	1.3	3.8
Lat	1.6	2.5	3.6	4.4	5.1	5.2	5.7	5.4	4.7	3.2	1.8	1.4	3.7
Lat + 15	1.7	2.5	3.5	4.1	4.5	4.5	4.9	4.9	4.5	3.2	1.8	1.4	3.5
SPOKANE, WA—Latitude 47.63													
0	1.3	2.0	3.2	4.6	5.8	6.5	7.0	5.9	4.4	2.7	1.4	1.1	3.8
Lat - 15	2.1	3.0	4.2	5.3	6.0	6.4	7.0	6.6	5.7	4.0	2.1	1.7	4.5
Lat	2.3	3.2	4.4	5.2	5.6	5.9	6.5	6.3	5.7	4.3	2.3	1.9	4.5
Lat + 15	2.4	3.3	4.3	4.8	4.9	5.1	5.6	5.7	5.5	4.4	2.4	2.0	4.2
YAKIMA, WA—Latitude 46.57													
0	1.4	2.2	3.6	5.0	6.2	6.9	7.2	6.2	4.7	3.0	1.6	1.1	4.1
Lat - 15	2.2	3.3	4.7	5.7	6.4	6.8	7.3	6.8	6.0	4.4	2.5	1.9	4.8
Lat	2.5	3.5	4.8	5.5	6.0	6.2	6.7	6.6	6.1	4.7	2.7	2.2	4.8
Lat + 15	2.6	3.6	4.7	5.1	5.3	5.3	5.8	5.9	5.9	4.8	2.8	2.3	4.5
CHARLESTON, WV—Latitude 38.37													
0	2.0	2.7	3.7	4.8	5.6	6.0	5.8	5.3	4.3	3.3	2.1	1.7	3.9
Lat - 15	2.7	3.3	4.3	5.1	5.6	5.9	5.7	5.5	4.9	4.1	2.9	2.3	4.4
Lat	2.9	3.6	4.4	5.0	5.3	5.4	5.3	5.3	4.9	4.4	3.1	2.6	4.4
Lat + 15	3.1	3.7	4.3	4.6	4.7	4.7	4.7	4.9	4.7	4.4	3.3	2.7	4.1
ELKINS, WV—Latitude 38.88													
0	1.9	2.6	3.6	4.5	5.3	5.7	5.5	5.0	4.1	3.1	2.0	1.6	3.8
Lat - 15	2.6	3.3	4.1	4.8	5.3	5.6	5.5	5.2	4.6	3.9	2.7	2.2	4.2
Lat	2.9	3.5	4.2	4.7	5.0	5.2	5.1	5.1	4.7	4.1	2.9	2.4	4.2
Lat + 15	3.0	3.6	4.1	4.4	4.4	4.5	4.5	4.6	4.5	4.1	3.0	2.6	3.9
HUNTINGTON, WV—Latitude 38.37													
0	2.0	2.7	3.7	4.8	5.6	6.0	5.8	5.2	4.3	3.3	2.1	1.7	3.9
Lat - 15	2.6	3.4	4.3	5.1	5.6	5.9	5.7	5.4	4.9	4.1	2.8	2.3	4.4
Lat	2.9	3.6	4.4	4.9	5.3	5.4	5.3	5.2	4.9	4.4	3.1	2.5	4.3
Lat + 15	3.1	3.7	4.3	4.6	4.7	4.7	4.7	4.8	4.7	4.4	3.2	2.7	4.1

EAU CLAIRE, WI—Latitude 44.87

	Jan	Feb	Mar	Apr	May	Jun	Jul	Aug	Sep	Oct	Nov	Dec	Avg
0	1.7	2.7	3.7	4.6	5.6	6.1	6.1	5.2	3.9	2.7	1.6	1.4	3.8
Lat - 15	2.9	4.0	4.7	5.1	5.7	6.0	6.1	5.6	4.7	3.7	2.4	2.3	4.4
Lat	3.3	4.4	4.9	5.0	5.3	5.5	5.6	5.4	4.7	3.9	2.7	2.6	4.4
Lat + 15	3.6	4.6	4.8	4.6	4.7	4.7	4.9	4.8	4.5	3.9	2.8	2.8	4.2

GREEN BAY, WI—Latitude 44.48

	Jan	Feb	Mar	Apr	May	Jun	Jul	Aug	Sep	Oct	Nov	Dec	Avg
0	1.7	2.6	3.7	4.7	5.7	6.3	6.1	5.2	3.9	2.7	1.6	1.4	3.8
Lat - 15	2.9	3.8	4.6	5.2	5.8	6.1	6.1	5.6	4.7	3.6	2.4	2.3	4.4
Lat	3.3	4.2	4.8	5.0	5.4	5.6	5.7	5.4	4.7	3.8	2.6	2.6	4.4
Lat + 15	3.5	4.3	4.7	4.7	4.8	4.9	4.9	4.9	4.5	3.8	2.7	2.8	4.2

LA CROSSE, WI—Latitude 43.87

	Jan	Feb	Mar	Apr	May	Jun	Jul	Aug	Sep	Oct	Nov	Dec	Avg
0	1.8	2.7	3.7	4.7	5.7	6.3	6.2	5.4	4.0	2.8	1.7	1.4	3.9
Lat - 15	2.9	3.9	4.6	5.2	5.8	6.1	6.2	5.8	4.8	3.8	2.6	2.3	4.5
Lat	3.3	4.3	4.8	5.0	5.4	5.6	5.8	5.5	4.8	4.1	2.8	2.6	4.5
Lat + 15	3.6	4.4	4.7	4.7	4.8	4.9	5.0	5.0	4.6	4.1	2.9	2.8	4.3

MADISON, WI—Latitude 43.13

	Jan	Feb	Mar	Apr	May	Jun	Jul	Aug	Sep	Oct	Nov	Dec	Avg
0	1.9	2.8	3.7	4.7	5.8	6.4	6.2	5.4	4.1	2.8	1.7	1.5	3.9
Lat - 15	3.0	3.9	4.5	5.1	5.8	6.2	6.2	5.7	4.8	3.8	2.5	2.3	4.5
Lat	3.4	4.3	4.7	5.0	5.5	5.7	5.8	5.5	4.8	4.0	2.8	2.6	4.5
Lat + 15	3.6	4.4	4.6	4.6	4.8	4.9	5.0	5.0	4.6	4.0	2.9	2.8	4.3

MILWAUKEE, WI—Latitude 42.95

	Jan	Feb	Mar	Apr	May	Jun	Jul	Aug	Sep	Oct	Nov	Dec	Avg
0	1.8	2.6	3.5	4.6	5.8	6.4	6.3	5.4	4.1	2.9	1.8	1.4	3.9
Lat - 15	2.8	3.6	4.3	5.1	5.9	6.2	6.3	5.8	4.9	3.8	2.5	2.2	4.5
Lat	3.2	3.9	4.4	4.9	5.5	5.7	5.8	5.6	4.9	4.0	2.8	2.5	4.5
Lat + 15	3.4	4.1	4.4	4.6	4.9	5.0	5.1	5.0	4.7	4.0	2.9	2.7	4.2

CASPER, WY—Latitude 42.92

	Jan	Feb	Mar	Apr	May	Jun	Jul	Aug	Sep	Oct	Nov	Dec	Avg
0	2.0	2.9	4.1	5.2	6.1	7.0	7.0	6.3	4.9	3.4	2.2	1.7	4.4
Lat - 15	3.4	4.3	5.2	5.8	6.2	6.8	7.0	6.8	6.0	4.8	3.6	3.1	5.2
Lat	3.9	4.7	5.4	5.7	5.8	6.2	6.5	6.5	6.1	5.2	4.1	3.6	5.3
Lat + 15	4.3	4.9	5.3	5.3	5.1	5.3	5.6	5.9	5.9	5.3	4.3	3.9	5.1

CHEYENNE, WY—Latitude 41.15

	Jan	Feb	Mar	Apr	May	Jun	Jul	Aug	Sep	Oct	Nov	Dec	Avg
0	2.2	3.1	4.2	5.3	6.0	6.7	6.7	5.9	4.9	3.6	2.4	1.9	4.4
Lat - 15	3.6	4.4	5.3	5.9	6.0	6.5	6.6	6.3	5.8	5.0	3.8	3.3	5.2
Lat	4.1	4.9	5.5	5.8	5.6	6.0	6.1	6.1	6.0	5.4	4.3	3.9	5.3
Lat + 15	4.5	5.1	5.5	5.4	5.0	5.1	5.3	5.5	5.7	5.5	4.6	4.2	5.1

	Jan	Feb	Mar	Apr	May	Jun	Jul	Aug	Sep	Oct	Nov	Dec	Year
ROCK SPRINGS, WY—Latitude 41.60													
0	2.1	3.0	4.2	5.4	6.4	7.2	7.2	6.4	5.2	3.7	2.3	1.9	4.6
Lat - 15	3.5	4.4	5.3	6.0	6.5	7.0	7.1	6.9	6.3	5.2	3.7	3.2	5.4
Lat	4.0	4.8	5.5	5.9	6.1	6.4	6.6	6.6	6.4	5.6	4.1	3.7	5.5
Lat + 15	4.3	5.1	5.5	5.5	5.4	5.5	5.7	6.0	6.1	5.7	4.4	4.1	5.3
SHERIDAN, WY—Latitude 44.77													
0	1.8	2.7	3.9	5.0	5.8	6.7	6.9	6.0	4.6	3.1	2.0	1.6	4.2
Lat - 15	3.1	4.0	5.0	5.6	5.9	6.5	6.9	6.6	5.7	4.5	3.2	2.8	5.0
Lat	3.5	4.4	5.2	5.5	5.6	6.0	6.4	6.3	5.8	4.8	3.6	3.2	5.0
Lat + 15	3.7	4.6	5.1	5.1	4.9	5.1	5.5	5.7	5.5	4.9	3.8	3.5	4.8

Appendix B
CONVERSION FACTORS

LENGTH	meters	inches	feet
1 meter =	1	39.37	3.281
1 inch =	0.0254	1	0.0833
1 foot =	0.3048	12	1

AREA	meters^2	cm^2	feet^2	inches^2	circular mil
1 square meter=	1	10^4	10.76	1,550	
1 square centimeter =	10^{-4}	1	1.076×10^{-3}	0.1550	
1 square foot =	9.290×10^{-2}	929.0	1	144	
1 square inch =	6.452×10^{-4}	6.452	6.944×10^{-3}	1	1.274×10^5
1 circular mil =				7.854×10^{-7}	1

VOLUME	meters^3	cm^3 (cc)	feet^3	inches^3
1 cubic meter=	1	10^6	35.31	6.102×10^4
1 cubic centimeter =	10^{-6}	1	3.531×10^{-5}	0.06102
1 cubic foot =	2.832×10^{-2}	28,320	1	1,728
1 cubic inch =	1.639×10^{-5}	16.39	5.787×10^{-4}	1

ENERGY	BTU	fp	J	kcal	kWh
1 British thermal unit =	1	777.9	1,055	0.2520	2.930×10^{-4}
1 foot-pound =	1.285×10^{-3}	1	1.356	3.240×10^{-4}	3.766×10^{-7}
1 joule =	9.481×10^{-4}	0.7376	1	2.390×10^{-4}	2.778×10^{-7}
1 kilocalorie =	3.968	3,086	4,184	1	1.163×10^{-3}
1 kilowatt-hour =	3,413	2.655×10^{6}	3.6×10^{6}	860.2	1

POWER	BTU/h	fp/s	hp	kcal/s	kW	W
1 BTU/h =	1	0.2161	3.929×10^{-4}	7.000×10^{-5}	2.930×10^{-4}	0.2930
1 fp/s =	4.628	1	1.8180×10^{-3}	3.239×10^{-4}	1.356×10^{-3}	1.356
1 horsepower =	2,545	550	1	0.1782	0.7457	745.7
1 kcal/s =	1.429×10^{4}	3,087	5.163	1	4.184	4,184
1 kilowatt =	3,413	737.6	1.341	0.2390	1	1,000
1 watt	3.413	0.7376	1.341×10^{-3}	2.390×10^{-4}	0.001	1

Decimal Prefixes

10^{1}	deca (da)	10^{-1}	deci (d)
10^{2}	hecto (h)	10^{-2}	centi (c)
10^{3}	kilo (k)	10^{-3}	milli (m)
10^{6}	mega (M)	10^{-6}	micro (μ)
10^{9}	giga (G)	10^{-9}	nano (n)
10^{12}	tera (T)	10^{-12}	pico (p)
10^{15}	peta(P)	10^{-15}	femto (f)
10^{18}	exa (E)	10^{-18}	atto (a)

Temperature—Farhenheit & Celsius Conversions

$$°F = (°C \times 9/5) + 32$$
$$°C = (°F - 32) \times 5/9$$

Appendix C
WEB SITES

There are many excellent sources of information on PV and the renewable energies, conservation and the environment. Following is a listing of agencies, organizations, businesses, periodicals, books, and individuals mentioned in this book.

Agencies, Organizations, Businesses & Periodicals

American Council for an Energy-Efficient Economy (ACEEE)—www.aceee.org
American Society of Civil Engineers (ASCE)—www.asce.org
American Solar Energy Society—www.ases.org
Anderson General Store, Guemes Island, WA—www.guemesislandstore.com

California Energy Commission—www.energy.ca.gov
  Guide to Photovoltaic System Design & Installation—
    www.energy.ca.gov/reports/2001-09-04_500-01-020.PDF

Database of State Incentives for Renewable Energy (DSIRE)—www.dsireusa.org
Direct Power & Water—www.directpower.com

Electron Connection—www.electronconnection.com
Energy Outfitters—www.energyoutfitters.com

Gaiam—www.gaiam.com
groSolar—http://grosolar.com

Institute of Electrical and Electronics Engineers (IEEE)—http://standards.ieee.org
International Energy Association (IEA)—www.iea.org
International Mechanical Code (IMC)—www.iccsafe.org
International Residential Code (IRC)—www.iccsafe.org

Kiss + Cathcart, Architects—www.kisscathcart.com

Light Energy Systems—www.lightenergysystems.com

National Electrical Code (NEC)—www.nfpa.org
National Institute for Occupational Safety and Health—www.cdc.gov/NIOSH

Occupational Safety and Health Administration (OSHA)—www.osha.gov
Offline Independent Energy Systems—www.psnw.com/~ofln
Outback Power Systems—www.outbackpower.com

Positive Energy—www.positiveenergysolar.com
Prometheus Institute for Sustainable Development—www.prometheus.org

Real Goods—www.gaiam.com/realgoods
Rising Sun Enterprises—www.rselight.com
RV Solar Electric—www.rvsolarelectric.com

Sagrillo Power & Light—www.awea.org/smallwind/sagrillo
Sierra Club—www.sierraclub.org
Sandia National Laboratories—http://photovoltaics.sandia.gov
Sanyo—www.sanyo.co.jp/clean/solar
Solar Depot—www.solardepot.com
Solar Design Associates—www.solardesign.com
Solar Electrical Systems—www.solarelectricalsystems.com
Solar Energy Industries Association—www.seia.org
Solar Energy International—www.solarenergy.org
SOLutions in Solar Electricity—www.solarsolar.com
Specialty Concepts—www.specialtyconcepts.com
Sunnyside Solar—www.sunnysidesolar.com
SunWatt—sunwatt@juno.com

Talmage Solar Engineering—www.solarmarket.com/solarmarket/talmagesolar.com

Underwriters Laboratories Standards (UL)—www.ul.com
Uniform Building Code (UBC)—www.iccsafe.org
Union of Concerned Scientists—www.ucsusa.org
U.S. Department of Agriculture —www.usda.gov
U.S. Department of Energy—www.doe.gov
    Energy Information Agency (EAI)—www.eia.doe.gov
    Energy Star (U.S. EPA/U.S. DOE)—www.energystar.gov
    National Renewable Energy Laboratories—www.nrel.gov/pv
        PVWATTS, Calculator for Grid-Connected PV Systems—
            http://rredc.nrel.gov/solar/codes_algs/PVWATTS

Office of Energy Efficiency and Renewable Energy—www.eere.energy.gov
Energy Guide Label—www1.eere.energy.gov/consumer/tips/energyguide.html
Million Solar Roofs (MSR)—www1.eere.energy.gov/solar/deployment.html
Solar America Initiative—www1.eere.energy.gov/solar/solar_america
Tribal Energy Grants—www.eere.energy.gov/tribalenergy/government_grants.cfm
Public Utility Regulatory Policies Act (PURPA)—www.oe.energy.gov/purpa.htm
Zomeworks—http://zomeworks.com

Individuals

Joel Davidson—www.solarsolar.com
Stephen Heckeroth—www.renewables.com
Greg Johanson—www.solarelectricalsystems.com
Andy Kerr—www.andykerr.net
Richard Komp—Sunwatt@juno.com
Don Loweburg—www.psnw.com/~ofln
Paul Maycock—www.prometheus.org
Fran Orner—www.solarsolar.com
John Perlin—http://johnperlin.com
Mike Sagrillo—www.awea.org/smallwind/sagrillo
Robert Sardinsky—www.rselight.com/rselight
Allan Sindelar—www.positiveenergysolar.com
Peter Talmage—www.solarmarket.com/solarmarket/talmagesolar.com
John Wiles—www.nmsu.edu/~tdi

Books & Periodicals

Home Energy magazine—www.homeenergy.org
Home Power magazine—www.homepower.com
Remodeling magazine—http://remodeling.hw.net
Cost vs. Value reports—www.costvsvalue.com

Jennifer Thorne Amann, Alex Wilson & Katie Ackerly, *Consumer Guide to Home Energy Savings*, 9th ed. (Gabriola Island, BC, Canada: New Society Publishers, 2007). www. newsociety.com

aatec publications—www.Solar Electric Books.com
Practical Photovoltaics: Electricity from Solar Cells—Richard Komp, Ph.D.
RVers Guide to Solar Battery Charging—Noel & Barbara Kirkby
From Space to Earth: The Story of Solar Electricity—John Perlin

APPENDIX D

The National Electrical Code®
Excerpts from
Article 690. Solar Photovoltaic Systems

For more than 100 years, *The National Electrical Code (NEC®)* has set the standards for electrical construction and operation.

Before its official release, a draft of the first *NEC* Article 690 for Solar Photovoltaic Systems was published in the 1983 edition of *The Solar Electric Home*. This edition of *The New Solar Electric Home* references Article 690 and other *NEC* articles relevant to PV systems.

The following excerpts from the 2008 edition of *NEC* 690 explain circuit requirements and identifying (marking) PV systems.

Note: This appendix does not supplant or replace the *NEC*. Get your own copy of the *NEC* and the *NEC Handbook* from The National Fire Prevention Association, One Batterymarch Park, Quincy, MA 02269; www.nfpa.org/catalog/.

The National Electrical Code
Article 690. Solar Photovoltaic Systems
II. Circuit Requirements

690.7 Maximum Voltage

(A) **Maximum Photovoltaic System Voltage.** In a dc photovoltaic source circuit or output circuit, the maximum photovoltaic system voltage for that circuit shall be calculated as the sum of the rated open-circuit voltage of the series-connected photovoltaic modules corrected for the lowest expected ambient temperature. For crystalline and multicrystalline silicon modules, the rated open-circuit voltage shall be multiplied by the correction factor provided in Table 690.7. This voltage shall be used to determine the voltage rating of cables, disconnects, overcurrent devices, and other equipment. Where the lowest expected ambient temperature is below -40°C (-40°F), or where other than crystalline or multicrystalline silicon photovoltaic modules are used, the system voltage adjustment shall be made in accordance with the manufacturer's instructions.

When open-circuit voltage temperature coefficients are supplied in the instructions for listed PV modules, they shall be used to calculate the maximum photovoltaic sytem voltage as required by 110.3(B) instead of using Table 690.7.

Table 690.7
Voltage Correction Factors for Crystalline and Multicrystalline Silicon Modules

Correction Factors for Ambient Temperatures Below 25°C (77°F) (Multiply the rated open-circuit voltage by the appropriate correction factor shown below.)		
Ambient Temperature (°C)	Factor	Ambient Temperature (°F)
24 to 20	1.02	76 to 68
19 to 15	1.04	67 to 59
14 to 10	1.06	58 to 50
9 to 5	1.08	49 to 41
4 to 0	1.10	40 to 32
−1 to −5	1.12	31 to 23
−6 to −10	1.14	22 to 14
−11 to −15	1.16	13 to 5
−16 to −20	1.18	4 to −4
−21 to −25	1.20	−5 to −13
−26 to −30	1.21	−14 to −22
−31 to −35	1.23	−23 to −31
−36 to −40	1.25	−32 to −40

(B) Direct-Current Utilization Circuits. The voltage of dc utilization circuits shall conform to 210.6.

(C) Photovoltaic Source and Output Circuits. In one- and two-family dwellings, photovoltaic source circuits and photovoltaic output circuits that do not include lampholders, fixtures, or receptacles shall be permitted to have a maximum photovoltaic system voltage up to 600 volts. Other installations with a maximum photovoltaic system voltage over 600 volts shall comply with Article 690, Part I.

(D) Circuits Over 150 Volts to Ground. In one- and two-family dwellings, live parts in photovoltaic source circuits and photovoltaic output circuits over 150 volts to ground shall not be accessible to other than qualified persons while energized.

> FPN: See 110.27 for guarding of live parts, and 210.6 for voltage to ground and between conductors.

(E) Bipolar Source and Output Circuits. For 2-wire circuits connected to bipolar systems, the maximum system voltage shall be the highest voltage between the conductors of the 2-wire circuit if all of the following conditions apply:

(1) One conductor of each circuit is solidly grounded.

(2) Each circuit is connected to a separate subarray.

(3) The equipment is clearly marked with a label as follows:

WARNING
BIPOLAR PHOTOVOLTAIC ARRAY.
DISCONNECTION OF NEUTRAL OR
GROUNDED CONDUCTORS MAY RESULT IN
OVERVOLTAGE ON ARRAY OR INVERTER.

690.8 Circuit Sizing and Current.

(A) Calculation of Maximum Circuit Current. The maximum current for the specific circuit shall be calculated in accordance with 690.8(A)(1) through (A)(4).

> FPN: Where the requirements of 690.8(A)(1) and (B)(1) are both applied, the resulting multiplication factor is 156 percent.

(1) Photovoltaic Source Circuit Currents. The maximum current shall be the sum of parallel module rated short-circuit currents multiplied by 125 percent.

(2) Photovoltaic Output Circuit Currents. The maximum current shall be the sum of parallel source circuit maximum currents as calculated in 690.8(A)(1).

(3) Inverter Output Circuit Current. The maximum current shall be the inverter continuous output current rating.

(4) Stand-Alone Inverter Input Circuit Current. The maximum current shall be the stand-alone continuous inverter input current rating when the inverter is producing rated power at the lowest input voltage.

(B) Ampacity and Overcurrent Device Ratings. Photovoltaic system currents shall be considered to be continuous.

(1) Sizing of Conductors and Overcurrent Devices. The circuit conductors and overcurrent devices shall be sized to carry not less than 125 percent of the maximum currents as calculated in 690.8(A). The rating or setting of overcurrent devices shall be permitted in accordance with 240.4(B) and (C).

Exception: Circuits containing an assembly, together with its overcurrent device(s), that is listed for continuous operation at 100 percent of its rating shall be permitted to be utilized at 100 percent of its rating.

(2) Internal Current Limitation. Overcurrent protection for photovoltaic output circuits with devices that internally limit the current from the photovoltaic output circuit shall be permitted to be rated at less than the value calculated in 690.8(B)(1). This reduced rating shall be at least 125 percent of the limited current value. Photovoltaic output circuit conductors shall be sized in accordance with 690.8(B)(1).

Exception: An overcurrent device in an assembly listed for continuous operation at 100 percent of its rating shall be permitted to be utilized at 100 percent of its rating.

(C) Systems with Multiple Direct-Current Voltages. For a photovoltaic power source that has multiple output circuit voltages and employs a common-return conductor, the ampacity of the common-return conductor shall not be less than the sum of the ampere ratings of the overcurrent devices of the individual output circuits.

(D) Sizing of Module Interconnection Conductors. Where a single overcurrent device is used to protect a set of two or more parallel-connected module circuits, the ampacity of each of the module interconnection conductors shall not be less than the sum of the rating of the single fuse plus 125 percent of the short-circuit current from the other parallel-connected modules.

690.9 Overcurrent Protection

(A) Circuits and Equipment. Photovoltaic source circuit, photovoltaic output circuit, inverter output circuit, and storage battery circuit conductors and equipment shall be protected in accordance with the requirements of Article 240. Circuits connected to more than one electrical source shall have overcurrent devices located so as to provide overcurrent protection from all sources.

Exception: An overcurrent device shall not be required for circuit conductors sized in accordance with 690.8(B) and located where one of the following apply:

(a) There are no external sources such as parallel-connected source circuits, batteries, or backfeed from inverters.

(b) The short-circuit currents from all sources do not exceed the ampacity of the conductors.

> FPN: Possible backfeed of current from any source of supply, including a supply through an inverter into the photovoltaic output circuit and photovoltaic source circuits, is a consideration in determining whether adequate overcurrent protection from all sources is provided for conductors and modules.

(B) Power Transformers. Overcurrent protection for a transformer with a source(s) on each side shall be provided in accordance with 450.3 by considering first one side of the transformer, then the other side of the transformer, as the primary.

Exception: A power transformer with a current rating on the side connected toward the photovoltaic power source, not less than the short-circuit output current rating of the inverter, shall be permitted without overcurrent protection from that source.

(C) Photovoltaic Source Circuits. Branch-circuit or supplementary-type overcurrent devices shall be permitted to provide overcurrent protection in photovoltaic source circuits. The overcurrent devices shall be accessible but shall not be required to be readily accessible.

Standard values of supplementary overcurrent devices allowed by this section shall be in one ampere size increments, starting at one ampere up to and including 15 amperes. Higher standard values above 15 amperes for supplementary overcurrent devices shall be based on the standard sizes provided in 240.6(A).

(D) Direct-Current Rating. Overcurrent devices, either fuses or circuit breakers, used in any dc portion of a photovoltaic power system shall be listed for use in dc circuits and shall have the appropriate voltage, current, and interrupt ratings.

(E) Series Overcurrent Protection. In series-connected strings of two or more modules, a single overcurrent protection device shall be permitted.

690.10 Stand-Alone Systems.

The premises wiring system shall be adequate to meet the requirements of this *Code* for a similar installation connected to a service. The wiring on the supply side of the building or structure disconnecting means shall comply with this *Code* except as modified by 690.10(A) through (D).

(A) Inverter Output. The ac inverter output from a stand-alone system shall be permitted to supply ac power to the building or structure disconnecting means at current levels less than the calculated load connected to that disconnect. The inverter output rating or the rating of an alternate energy source shall be equal to or greater than the load posed by the largest single utilization equipment connected to the system. Calculated general lighting loads shall not be considered as a single load.

• 690.10(A) was changed in 2008 to clarify that the ac output of a stand-alone inverter may be less than the calculated load, provided the output is adequate to supply the largest single utilization equipment.

(B) Sizing and Protection. The circuit conductors between the inverter output and the building or structure disconnecting means shall be sized based on the output rating of the inverter. These conductors shall be protected from overcurrents in accordance with Article 240. The overcurrent protection shall be located at the output of the inverter.

(C) Single 120-Volt Supply. The inverter output of a stand-alone solar photovoltaic system shall be permitted to supply 120 volts to single-phase, 3-wire, 120/240-volt service equipment or distribution panels where there are no 240-volt outlets and where there are no multiwire branch circuits. In all installations, the rating of the overcurrent device connected to the output of the inverter shall be less than the rating of the neutral bus in the service equipment. This equipment shall be marked with the following words or equivalent:

WARNING
SINGLE 120-VOLT SUPPLY. DO NOT CONNECT
MULTIWIRE BRANCH CIRCUITS!

(D) Energy Storage or Backup Power System Requirements. Energy storage or backup power supplies are not required.

VI. Marking

690.53 Direct-Current Photovoltaic Power Source.

A permanent label for the direct-current photovoltaic power source indicating items (1) through (5) shall be provided by the installer at the photovoltaic disconnecting means:

(1) Rated maximum power-point current
(2) Rated maximum power-point voltage
(3) Maximum system voltage
      FPN to (3): See 690.7(A) for maximum photovoltaic system voltage.
(4) Short-circuit current
      FPN to (4): See 690.8(A) for calculation of maximum circuit current.
(5) Maximum rated output current of the charge controller (if installed)
      FPN: Reflecting systems used for irradiance enhancement may result in increased levels of output current and power.

690.54 Interactive System Point of Interconnection.

All interactive system(s) points of interconnection with other sources shall be marked at an accessible location at the disconnecting means as a power source and with the rated ac output current and the nominal operating ac voltage.

690.55 Photovoltaic Power Systems Employing Energy Storage.

Photovoltaic power systems employing energy storage shall also be marked with the maximum operating voltage, including any equalization voltage and the polarity of the grounded circuit conductor.

690.56 Identification of Power Sources.

(A) Facilities with Stand-Alone Systems. Any structure or building with a photovoltaic power system that is not connected to a utility service source and is a stand-alone system shall have a permanent plaque or directory installed on the exterior of the building or structure at a readily visible location acceptable to the authority having jurisdiction. The plaque or directory shall indicate the location of system disconnecting means and that the structure contains a stand-alone electrical power system.

(B) Facilities with Utility Services and PV Systems. Buildings or structures with both utility service and a photovoltaic system shall have a permanent plaque or directory providing the location of the service disconnecting means and the photovoltaic system disconnecting means, if not located at the same location.

GLOSSARY

AC (alternating current)—Electric current that reverses its direction of flow. The standard current used by utilities in the U.S. is 60 cycles per second, also expressed as 60 hertz or 60 Hz.

Ah—*See* **ampere-hour**

AHJ. *See* **authority having jurisdiction**

Air Mass—The relative distance that sunlight must travel through the atmosphere to a given location. **Air Mass 1 (AM1)** corresponds to the spectrum of sunlight that has been filtered by passing through one thickness of the earth's atmosphere. **Air Mass 1.5 (AM1.5)** is the spectrum of sunlight (400–1,100 nanometers wavelength) passing through 1.5 thicknesses of the earth's atmosphere and corresponds to noon on a clear sunny day with the sun about 60 degrees above the horizontal and the PV module directly facing the sun. This standard is used to rate solar cells and modules. *Also see* STC

alternating current—*See* **AC**

altitude—The angular elevation of an array from the horizontal.

ammeter—An analog or digital current flow indicator.

amorphous solar cell—Silicon cell with no crystalline atomic matrix.

ampacity—The current that a conductor in use can carry continuously without exceeding its temperature rating.

ampere (amp, A)—A unit of electrical current or the rate of flow of electrons. One volt across one ohm of resistance causes a current flow of ampere. One ampere equals 6.25×10^{18} electrons per second passing a given point in a circuit.

ampere-hour (amp-hour, Ah)—A current of one ampere running for one hour, typically used to describe battery capacity; e.g., 200 Ah.

ampere-hour (coulombic) efficiency—The ratio of the ampere-hours removed from a cell or battery during a discharge to the ampere-hours required to restore the initial capacity.

angle of incidence—Angle at which light strikes a solar cell or array.

array—A set of modules or panels assembled for a specific application; may consist of modules in series for increased voltage, or in parallel for increased current, or a combination of both.

as-built drawing—A final drawing of wiring or construction work made after installation and modifications have been made.

authority having jurisdiction (AHJ)—The individual, designated by local government, with legal powers to administer, interpret, and enforce building codes.

autonomous power system—A self-contained power system not connected to the utility grid, also known as a stand-alone power system.

azimuth—The horizontal direction of the angular distance from a fixed point expressed in degrees.

balance of system (BOS)—The parts of a photovoltaic system other than the solar array.

battery—A DC voltage source consisting of two or more cells that convert chemical energy into electrical energy.

battery, marine—A rechargeable battery for use in non-motive power systems.

battery, motive—A battery intended to power electrically operated mobile equipment (e.g., forklift or automobile) and designed to be operated in a daily deep-cycle regime at moderate discharge rates.

battery, primary—A non-rechargeable battery.

battery, secondary—A rechargeable battery.

battery autonomy—The amount of battery storage required to provide power during cloudy days and at night for a specific load at a specific location.

battery capacity—The total amount of electricity that can be drawn from a fully charged battery until it is discharged to a specific voltage; expressed in ampere-hours.

battery capacity, energy—Total amount of electricity that can be drawn from a fully charged battery; expressed in watt-hours (kilowatt-hours).

battery cell—The simplest operating unit in a storage battery consisting of one or more positive electrodes, an electrolyte that permits ionic conduction, one or more negative electrodes, and separators enclosed in a single container.

battery cycle life—The number of cycles to a specified depth of discharge a battery can undergo before efficiency is affected.

battery life—Period when a battery is operating above specific efficiency levels. Measured in cycles or years, depending on intended use.

blackout—An electric power outage.

block diagram—A pictorial representation, usually boxes and lines, showing the interconnection between system components.

blocking diode—A device that prevents current from reversed flow, often used to prevent current from flowing backward through an array and draining the storage battery.

brownout—An electric utility voltage reduction usually caused by excess load on the electric grid.

BTU (British Thermal Unit)—The unit of heat energy sufficient to raise the temperature of one pound of water 1°F.

building code—A set of regulations that specifies materials, standards, and methods used in construction, maintenance, and repair of structures.

bypass diode—A device placed in parallel that allows current to flow around a device that is not conducting electricity, such as a shaded solar cell or module.

candela—The unit of luminous intensity defined as 1/60 of the luminous intensity per square centimeter of a blackbody radiator operating at the temperature of freezing platinum.

charge controller—A device that regulates or controls DC voltage and/or current used to charge a battery.

charge rate—The current applied to a cell or battery to restore its available capacity. This rate is commonly normalized with respect to the rated capacity of the cell or battery.

charging—The conversion of electrical energy into chemical potential energy within a cell by the passage of a direct current in the direction opposite to that of discharge.

combiner box—A junction box used as a connection point for two or more circuits.

conservation—Careful preservation and protection of a natural resource to prevent exploitation, destruction, or neglect.

contractor—A person or business hired to perform work and often supply materials to perform the work.

converter—A device that changes AC to DC or reduces a DC voltage to a lower DC voltage.

coulombic efficiency—*See* **ampere-hour (coulombic) efficiency; efficiency**

cycle life—The number of cycles, to a specified depth of discharge, that a cell or battery can undergo before failing to meet its specified capacity or efficiency performance criteria.

DC (direct current)—Electric current that always flows in the same direction. Photovoltaic cells and batteries are DC devices.

dead load—Static structural weight of permanent building members.

deep-discharge cycles—Cycles in which a battery is nearly completely discharged.

depth of discharge (DOD)—The ampere-hours removed from a fully charged cell or battery expressed as a percentage of rated capacity. For example, the removal of 25 ampere-hours from a fully charged 100 ampere-hour-rated cell results in a a 25% depth of discharge.

diode—A two-terminal device that conducts electricity much more readily in one direction than in the other.

direct current—*See* **DC**

discharge rate—Current removed from a cell or battery. Can be expressed in amperes, but more commonly is normalized to rated capacity (C), expressed as C/X. For example, drawing 20 amperes from a cell with a rated capacity of 100 ampere-hours is called the C/5 discharge rate (100 Ah/20 amps). Similarly, discharge currents of 5, 10, and 33.3 amperes would be designated as the C/20, C/10, and C/33.3 rates, respectively.

DOE—U.S. Department of Energy.

do-it-yourself (DIY)—Performing work on your own home by yourself or with the help of others.

efficiency—The ratio of the useful output to the input.

> **ampere-hour (coulombic) efficiency**—The ratio of the ampere-hours removed from a chemical cell or battery during a discharge to the ampere-hours required to restore the initial capacity.

> **energy (watt-hour) efficiency**—The ratio of the energy delivered by a chemical cell or battery during a discharge to the total energy required to restore the initial state of charge. The watt-hour efficiency is approximately equal to the product of the voltage and ampere-hour efficiencies. This is sometimes referred to as **round-trip efficiency**. Round-trip energy efficiencies usually do not include energy losses resulting from self-discharge, auxiliary equipment (parasitic losses), or battery equalization.

> **solar cell efficiency**—The ratio of solar power striking a solar cell to the electrical power produced by the cell.

> **solar power system efficiency**—The ratio of solar power input to the electrical power output of the total power system including all losses (reflectance, dust and dirt, cell temperature, module mismatch, wire resistance loss, battery efficiency, inverter efficiency, transformer efficiency).

> **voltage efficiency**—The ratio of the average discharge voltage of a chemical cell or battery to the average charge voltage during the subsequent restoration of an equivalent capacity.

electric current—The rate at which electricity flows through an electrical conductor; expressed in amps.

electricity—A fundamental entity of nature consisting of negative and positive electrons and protons or possibly electrons and positrons.

electrolyte—The medium that provides the ion transport mechanism between the positive and negative electrodes of a chemical cell.

EMF—Electromagnetic field.

EMI—Electromagnetic interference.

EMT—Electro-metallic tube conduit.

energy—Capacity to perform work over a period of time. *See* watt-hour

EPA—U.S. Environmental Protection Agency.

equalization—The process of restoring all cells in a battery to an equal state of charge. For lead-acid batteries, this is a charging process designed to bring all cells to 100% state of charge. Some battery types may require a complete discharge as a part of the equalization process.

equalizing charge—A continuation of normal battery charging at a voltage level slightly higher than the normal end-of-charge voltage in order to provide cell equalization within a battery.

feed-in tariff—A special rate paid by an electric utility or government for electricity fed into the utility grid.

fill factor (FF)—The ratio of maximum power to the product of the open-circuit voltage and short-circuit current.

flat-plate collector—A PV array in which the incident solar radiation strikes a flat surface with no concentration of sunlight.

footcandle—A unit of illumination equal to illumination of a surface one square foot in area and approximately 10.7639 lux.

fossil fuel—Coal, oil, and natural gas that were formed in the earth from plant or animal remains.

full sun—The amount of sunlight falling on the earth at sea level when the sun is shining straight down through a dry clean atmosphere. A close approximation is desert sun at high noon. The sunlight intensity of one kilowatt per square meter ($1 \, kW/m^2$) is the standard used to rate solar cells and modules. *Also see* STC

fuse—A protective device, usually a short piece of wire, that melts and breaks the circuit when the current exceeds its rated value.

fuse block—An insulating base on which clips for holding fuses are mounted.

galvanic corrosion—An electrochemical process that causes electrical current to flow between dissimilar metals eventually corroding one of the materials; the anode.

gassing—The evolution of gas from one or more electrodes in a cell. Gassing commonly results from local (self-discharge) or from the electrolysis of water in the electrolyte during charging.

generator (genset)—A machine that produces electricity, usually referring to a fossil fuel internal combustion engine driven generator.

grid—Utility network of transmission lines that distribute electricity.

grid-connected—An electrical system connected in parallel to the utility grid; also grid-interconnected, grid-tied.

ground—An electrical current connection to the earth.

ground fault—An unintentional electrical path between a part and some potential to earth ground.

ground fault circuit interrupter (GFCI)—A device that opens the ungrounded and grounded conductor when a ground fault exceeds a certain current, typically 4 mA to 6 mA.

ground rod—A grounding electrode; a steel or copper rod driven into the earth to make an electrical connection with the earth. *Also see* **grounding electrode**

ground wire—A conductor leading to an electric connection with the earth. *Also see* **grounding conductor**

grounding conductor—A wire or cable that, under normal conditions, carries no current but serves to connect equipment to the earth ground. *Also see* **ground wire**

grounding electrode—A conductor embedded in the earth used for maintaining a ground potential. *Also see* **ground rod**

hybrid system—A system composed of multiple power sources.

hydroelectricity—Electricity generated by the flow of water typically from a reservoir and dam.

hydrometer—An instrument used to measure the specific gravity of a liquid such as battery electrolyte.

incentive—A monetary inducement to invest in an improvement such as a PV system or energy efficiency equipment.

insolation—Solar input (irradiance) on a given area expressed in watts per square meter (W/m^2) over a period of time, for example, five kilowatt-hours per square meter per day $(5\ kWh/m^2/day)$, also known as five sun hours. *Also see* sun hours.

$\mathbf{I_{max}}$—Maximum output current of a solar cell, module, or system.

$\mathbf{I_{op}}$—Operating current of a solar cell, module, or system under load.

$\mathbf{I_{sc}}$—Short-circuit current of a solar cell, module, or system under no load.

insulation—Material used to prevent heat transfer.

inverter—A device that converts DC to AC.

inverter, synchronous—A device that converts DC to AC in synchronization with the power line.

irradiance—Solar input on a given surface expressed in watts per square meter (W/m^2).

islanding—The condition that occurs when a grid-connected inverter continues to transfer power to the utility grid during a utility outage.

IV curve—A graphic representation of current and voltage.

junction box—A protective enclosure into which wires or cables are fed or connected.

kilowatt (kW)—One thousand watts.

kilowatt-hour—The amount of electrical energy that performs work equal to 1,000 watts for one hour.

kilowatt-hour meter—An accumulator for measuring the flow of electric energy.

life-cycle cost—Estimated cost of owning and operating a device for the period of its useful life.

live load—Dynamic structural load due to wind (downward, lateral, or lifting forces), snow accumulation, people, and any temporary weights.

load—A device or combination of devices that consume electrical power.

lumen—The unit of luminous flux equal to the luminous flux emitted within a unit solid angle from a source of one candela.

lux—A unit of illumination equal to the illumination on a surface one square meter in area on which there is luminous flux of one lumen.

magnetic declination—The angle between the direction a compass needle points toward magnetic north and true geographic north.

maximum power current (I_{max})—The maximum current on the current-voltage (IV) curve.

maximum power point (MPP)—The operating point on the current-voltage (IV) curve where the product of current and voltage are maximized.

maximum power voltage (V_{max})—The maximum voltage on the current-voltage (IV) curve.

megawatt (MW)—One million watts.

megawatt-hour (MWh)—The amount of electrical energy needed to perform work equal to 1,000,000 watts for one hour.

module or **solar module**—A complete solar photovoltaic electric power unit.

mount or **mounting structure**—A frame or rack on which to assemble and fasten a solar module or solar array.

multimeter—An analog or digital indicator used to measure voltage, currrent, resistance, and other electrical properties.

National Electrical Code (NEC)—A comprehensive set of safety standards for electrical installations. NEC Article 690 specifies photovoltaic standards.

net metering—A method of connecting a generator in parallel to the utility grid through one meter in which current can flow both from the grid to the load and from the generator to the grid resulting in the meter indicating the net flow.

nominal operating cell temperature (NOCT)—Cell temperature at 1 kWh per square meter irradiance, 20°C ambient temperature, 1 meter per second wind, and electrically open circuit.

non-renewable power—Power from a source that is not replaced or restored once it is consumed, such as fossil fuels.

off-grid—Not connected to the utility electrical grid, autonomous, stand-alone.

ohm—A unit of resistance. One ohm equals the resistance of one volt at one ampere current.

one-line diagram—A simplified schematic wiring diagram of an electrical system.

on-grid—Connected to the utility electrical grid.

open-circuit voltage (V_{oc})—The voltage output of a photovoltaic device when no current is flowing through the circuit.

orientation—The direction or azimuth in which a solar array faces.

overcharge—Charging of a battery cell continued after 100% state of charge has been reached. Overcharging does not increase the energy stored in a cell and usually results in gassing and/or excessive heat generation, both of which reduce battery life.

overcurrent device—A protective fuse or circuit breaker.

panel—A collection of solar modules. *Also see* **service panel**

parallel connected—A method of connection in which the positive terminals are connected together and the negative terminals are connected together. Current output adds and voltage remains the same.

payback—A return on investment equal to the original capital outlay.

payback period—The amount of time required to receive a return on investment equal to the outlay.

peak sun hours—The equivalent number of hours of peak sun conditions of one kilowatt per meter square (1 kW/m²) that produces the same total insolation as actual sun conditions.

permit—Permission from the AHJ that authorizes construction work to begin and establishes the inspection requirements.

photon—A unit of electromagnetic radiation.

photovoltaic cell—Semiconductor device that converts light into electricity. *Also see* **polycrystalline, ribbon & thin-film solar cells**

photovoltaic effect—The movement of electrons within a material when it absorbs photos with energy above a certain level.

photovoltaic system—The total combination of components and subsystems that convert solar energy into electrical energy.

plane of array (POA)—The angle of a solar array from the horizontal; the altitude or pitch angle of a solar array.

polycrystalline solar cell—A cell sliced from a cast ingot consisting of many small crystals of silicon in no orderly arrangement.

power (watts)—Energy dissipated in an electrical circuit to perform work: voltage × current = watts.

primary cell/battery—Cell or battery whose initial capacity cannot be significantly restored by charging and therefore is limited to a single discharge.

PVC conduit—Polyvinyl chloride plastic tube electrical conduit.

PVUSA Test Conditions (PTC)—A standard reference developed at the Davis, California, PVUSA test site: one kilowatt per square meter (1 kW/m^2), 20°C ambient temperature, one meter per second (1 m/sec) wind speed at 10 meters above grade, and Air Mass 1.5 (AM1.5).

pyranometer—An instrument for measuring solar irradiance.

radioactivity—The emission of alpha, beta, and gamma rays by the disintegration of the nuclei of atoms; atomic energy.

rebate—A one-time refund for a portion of the purchase price.

regulator—A device that controls current to prevent overcharging batteries. *Also see* **charge controller**

renewable energy certificate (REC)—A tradeable commodity that represents a certain amount of electricity generated from renewable resources.

renewable power—Power from sources that are part of a natural cycle and virtually never ending—wind, water, sunlight, and plant matter.

resistance—The property of a conductor expressed in ohms that impedes current and results in the dissipation of power in the form of heat. One ohm equals the difference in potential of one volt at one ampere.

resistor—A component that opposes the flow of current.

rreturn on investment (ROI)—Net profit after taxes divided by assets.

ribbon solar cell—Polycrystalline cells formed by pulling a wide ribbon from molten silicon.

roundtrip efficiency—*See* **efficiency; energy (watt-hour) efficiency**

safety—Taking precautions to prevent injury or loss. **BE SAFE.**

secondary cell or battery—A cell or battery that is capable of being charged repeatedly.

self-discharge rate—The rate at which a battery will discharge on standing; affected by temperature and battery design.

series connected—A method of connection in which the positive terminal of one device is connected to the negative terminal of another. The voltages add and the current is limited to the least of any device in the string.

service panel or **electric service panel**—An electric distribution panel connected to the electrical grid.

short-circuit current (I_{sc})—The current flowing freely from a solar cell, module, or array with no load.

silicon—The second most abundant element in the earth's crust.

single-crystal solar cell—Cell sliced from a silicon single-crystal ingot.

SLI battery—A shallow-cycle battery designed for automotive starting, lighting, and ignition loads.

solar cell—*See* **photovoltaic cell**

stand-alone system—A solar photovoltaic system that supplies power independently of an electrical distribution network.

Standard Test Conditions (STC)—A standard reference of one kilowatt per square meter (1 kW/m2), 25°C cell temperature, and Air Mass 1.5 (AM1.5).

state of charge—A battery's available capacity, stated as a percentage of rated capacity.

sulfation—A condition that affects unused and discharged batteries; large crystals of lead sulfate grow on the plate, instead of the usual tiny crystals, making the battery extremely difficult to recharge.

sun hours—Also **peak sun hours.** *See* **insolation**

synchronous inverter—*See* **inverter, synchronous**

tax incentive—A measure to reduce the amount of taxes owed as an inducement for investment.

temperature coefficient—The rate of change in voltage, current, or power output from a PV device due to change in temperature.

thin-film solar cell—Solar cells made by the gaseous deposition of silicon or other semiconductor materials.

transformer—A device that transforms electric energy by electromagnetic induction from one or more circuits to one or more other circuits at the same frequency, but usually at a different voltage and current.

troubleshooting—A systematic method of investigating the cause of a problem and determining the best solution.

Underwriters Laboratories (UL)—An electrical safety testing organization. Other testing organizations include **Factory Mutual (FM)** and **ETL Testing Laboratories.**

V_{max}—The maximum voltage of a solar cell, module, or system.

V_{oc}—*See* **open-circuit voltage**

V_{op}—The operating voltage of a solar cell, module, or system under load.

volt—A unit of electromotive force required to produce a current of one ampere through a resistance of one ohm.

voltage efficiency—The ratio of the average discharge voltage of a cell or battery to the average charge voltage during the subsequent restoration of an equivalent capacity.

watt (W)—A unit of power at which electrical energy is used to perform work. *Also see* **power (watts)**

INDEX

ABOUT THE AUTHORS

JOEL DAVIDSON has worked in the photovoltaics industry for 30 years and the building industry for more than 40 years. He is a PV consultant to homeowners, businesses, manufacturers, governments, and utility companies. A PV industry founder, he was named "one of this country's most experienced, hands-on pioneers" by the Rocky Mountain Institute. He had one of the first PV homes in the U.S.

FRAN ORNER has over 20 years of PV system design and installation experience. She illustrated the 1987 edition of *The New Solar Electric Home* and co-authored and illustrated this latest editon. Fran is an accomplished construction do-it-yourselfer and designer. Her home is one of the first net-metered PV systems in California to receive the solar rebate.

The authors own **Solutions in Solar Electricity**—www.solarsolar.com. Together they have helped thousands of people go solar.